Interpolation and Approximation
with Splines and Fractals

Interpolation and Approximation with Splines and Fractals

Peter Massopust

Helmholtz Zentrum München
Institute of Biomathematics and Biometry

and

Centre of Mathematics
Technische Universität München, Germany

OXFORD

UNIVERSITY PRESS

2010

OXFORD
UNIVERSITY PRESS

Oxford University Press, Inc., publishes works that further
Oxford University's objective of excellence
in research, scholarship, and education.

Oxford   New York
Auckland   Cape Town   Dar es Salaam   Hong Kong   Karachi
Kuala Lumpur   Madrid   Melbourne   Mexico City   Nairobi
New Delhi   Shanghai   Taipei   Toronto

With offices in
Argentina   Austria   Brazil   Chile   Czech Republic   France   Greece
Guatemala   Hungary   Italy   Japan   Poland   Portugal   Singapore
South Korea   Switzerland   Thailand   Turkey   Ukraine   Vietnam

Published by Oxford University Press, Inc.
198 Madison Avenue, New York, New York 10016

www.oup.com

Oxford is a registered trademark of Oxford University Press

Library of Congress Cataloging-in-Publication Data
Massopust, Peter Robert, 1958-
Interpolation and approximation with splines and fractals / Peter Massopust.
     p. cm.
Includes bibliographical references and index.
ISBN 978-0-19-533654-2
1. Approximation theory. 2. Interpolation. 3. Spline theory.
4. Fractals. I. Title.
QA221.M35 2010
511$'$.42–dc22   2009007637

9 8 7 6 5 4 3 2 1
Printed in the United States of America
on acid-free paper

Dedicated to

Maritza,
Jonathan,
and
Sarah

Contents

Preface

This textbook is based on a script for 27 two-hour lectures that I gave during the summer semester 2007 at the Centre of Mathematics at the Technische Universität München. These lectures were intended for students who had successfully completed their first two years of studies in mathematics or physics, as well as for students in the engineering sciences and biomathematics with an interest in mathematical modeling. The students who attended the lectures had a solid background in real analysis and linear algebra, knew basic numerical analysis, and were familiar with the rudimentaries of functional analysis including some metric topology. Based upon the same prerequisites, this book is intended for a one or two year course on the subject of interpolation and approximation theory with splines and fractals.

There are numerous books on splines and their applications and also a large number of books dealing with fractals and their properties. This textbook is unique: It presents both theories as different aspects of the same basic idea, namely the approximation and interpolation of mathematical or natural objects by means of fundamental building blocks or, equivalently, by classes of functions with certain desirable properties reflecting the nature of the object that is to be approximated or interpolated, or more generally, modeled.

It is the intend of this book to supplement the classical theory of splines and their approximation and interpolation properties with those of fractals and fractal functions. This synthesis will complement currently required courses dealing with these topics and expose the prospective reader to some new and deep relationships. In addition to providing a classical introduction to the main issues involving approximation and interpolation with uni- and multivariate splines, which comprises the first half of the book, the second half will describe fractal sets, fractal functions, and their properties, including the newly discovered theory of superfractals. Throughout the book, connections between these two apparently different areas will be exposed and presented. In this way, more options are given to the prospective reader who will encounter complex approximation and interpolation problems in areas such as geometric modeling, image analysis, and biomathematics.

The material covered in this textbook represents the applied but also the pure branch of mathematics. There are several instances where one may follow a more abstract approach and then there are instances when applications are at the core of considerations.

At the end of each chapter, the reader finds a list of exercises ranging in level from straight-forward to more challenging, some with hints. On occasion, there will be problems that extend or supplement a concept introduced in the chapter and the reader is particularly encouraged to work through these exercises.

Many examples accompany the material discussed in this book. They are intended to give the reader a quick glance of the ideas and methods developed in the text. For a full understanding, the reader is sometimes asked to fill in the details. When needed and deemed necessary, figures and illustrations are also included.

In some instances, there is a need to introduce material and concepts that lie beyond the scope of the prerequisites mentioned above. Whenever this occurs, a brief self-contained exposition of the topics with vital hints to the literature is provided and the reader is particularly encouraged to work through the designated exercises.

The text is divided into eight chapters. The first three present polynomial and spline interpolation and approximation, and the remaining five deal primarily with fractal sets, fractal functions, and fractal surfaces. As with any textbook, the presentation and the material become slightly more challenging as the reader progresses. The last chapters on superfractals and superfractal functions contain material that is at the forefront of current research into the topic. There are many open questions in the field of superfractals on which the reader can work once the fundamental theory has been mastered.

Chapter 1 is intended as an introduction to the main themes of this textbook, namely interpolation and approximation theory with certain classes of functions. A short section outlining the notation and terminology used in this text, together with a brief introduction to semi-normed and normed spaces, metric spaces, and pre-Hilbert and Hilbert spaces, set the stage for the main themes. Some readers may already be familiar with this material, but it is provided for the sake of coherence and completeness of presentation. The abstract interpolation and approximation problem is presented and subjects such as existence and uniqueness of univariate polynomial interpolation, divided differences, and error estimates for univariate polynomial interpolation are discussed and the relevant theorems proved. An example of Runge's is included that shows the shortcomings of polynomial interpolation. The existence and uniqueness of a best approximation is also presented, as is the Weierstrass Approximation Theorem. In addition, the concept of approximation space, the approximation order, the approximation by step functions, and the modulus of continuity and the modulus of smoothness are introduced and their properties presented.

Univariate spline functions are then defined in Chapter 2. To obtain a numerically stable and analytically satisfactory basis for spline space, B-splines are introduced and their properties discussed. The case where knots are repeated is presented and the relevant piecewise polynomial spaces are defined. Cardinal spline and Hermite spline interpolation is also considered in this context. Error estimates for polynomial spline interpolation are derived and the minimality property of spline approximation is shown. In addition, we define exponential splines, relate them to distributional derivates and construct exponential B-splines. A short section generalizes the concept of exponential spline to that of $\mathcal{L}$-spline. The chapter ends with a discussion of splines and wavelets. It is shown that a certain collection of cardinal splines forms a multresolution analysis of $L^2(\mathbb{R})$.

Interpolation in the s-dimensional setting is presented in Chapter 3. First, polynomial interpolation in $\mathbb{R}^s$ is introduced and theorems regarding the uniqueness and existence of multidimensional polynomial interpolation are proven. Formulae for the error of multidimensional polynomial interpolation over certain types of domains are also derived. A multidimensional analogue of the univariate spline, called a box spline, is considered next. Some properties of box splines are presented. Next, tensor product B-splines are constructed and the shortcomings of this tensor product approach are discussed. Extending Newton's approach of univariate polynomial interpolation using divided differences to higher dimensions gives rise to Kergin interpolation, which is considered in the final section of this chapter.

Fractal sets generated by iterated function systems are presented in Chapter 4. First, the concept of dimension is discussed and then iterated function systems are introduced. The approximation of geometric objects or the modeling of naturally occuring objects by fractals is considered next. In this context, Barnsley's Collage Theorem is proven. Each fractal set can also be described using an underlying algebraic structure, called the code space. This mathematical object is defined and its properties derived. This then gives rise to fractal transformation between fractal sets. After a brief introduction to the basic elements of measure theory, the concept of an iterated function system with probabilities and that of a fractal measure is presented. The underlying algorithm for the construction of fractal sets that are the supports of fractal measures is known as the "chaos game."

Chapter 5 is about fractal functions and their properties. Fractal functions provide examples for nowhere differentiable functions, objects that had already been investigated earlier at the end of the 19th and beginning of the

20th century. Fractal functions are first defined by an interpolation process and then as fixed points of a certain class of operators, called Read-Bajraktarević operators. It is shown that there exist natural bases for certain classes of fractal functions, and that to every continuous function one can in a natural way associate an entire family of fractal functions. Moreover, it is also possible to construct fractal B-splines and even fractal functions of class C^k whose k-th derivatives are the fractal functions defined earlier. A formula for the fractal or box dimension of affine fractal functions is proven. Some approximation-theoretic results for fractal functions are derived. In addition, the indefinite integral and the Fourier transform of fractal functions is explicitly computed. Another section deals with an interesting and deep relationship between fractal functions and wavelets. The final section gives conditions for a class of fractal functions to be elements of Besov and Triebel-Lizorkin spaces.

Fractal surfaces are defined and constructed in Chapter 6. The emphasis will be primarily on affine fractal surfaces and their properties. The fractal or box dimension of affine fractal surfaces and their Hölder exponent is computed. The second half of this chapter deals with bilinear fractal surfaces, fractal surfaces arising from quadratic forms, and the construction of smooth fractal surfaces by means of indefinite integrals.

In Chapter 7, the new theory of superfractal sets is presented. It is shown that there exists a parameter V, called the variability parameter, that allows the characterization of fractal sets in terms of their "randomness." Up to that point, only fractals with $V = 1$ were considered. Superfractals can be thought of as attractors of superIFSs, i.e., iterated function systems whose mappings are themselves iterated function systems. We construct such superfractals and V-variable fractals and also their associated measures. A graph-theoretic construction of such fractals using code trees is also given.

Superfractal functions are constructed in Chapter 8. These objects are relatively new and the discussion of their properties is restricted to V-variable interpolation.

This book was written while I was the senior research scientist on the Marie Curie Excellence in Research Team MAMEBIA (Mathematical Methods in Biological Image Analysis) at the Institute of Biomathematics and Biometry at the Helmholtz Zentrum München – German Research Center for Environmental Health, and a guest professor at the Centre of Mathematics, Research Unit 6, Mathematical Modeling, at the Technische Universität München.

At this point, I would also like to acknowledge that this work was partially supported by the grant MEXT-CT-2004-013477, Acronym MAMEBIA, of the European Commission.

I am very grateful to Michael Barnsley for arranging a research visit to the Mathematical Sciences Institute at The Australian National University in Canberra, where I learned more about superfractals. My thanks also go to John Hutchinson for discussions regarding superfractals and their applications.

I would also like to thank Dipl. Math. Stefan Held for proof-reading the original script for my lectures, Ms. Phyllis Cohen at Oxford University Press and Ms. Kavitha Ashok at Glyph International for assistance during the final preparation of this text.

Last, but not least, I wish to thank my wife Maritza, my son Jonathan, and my daughter Sarah for their continuous support and encouragement, and for their many sacrifices during the preparation of this textbook.

Peter Massopust
Munich, Germany
September 2008

Interpolation and Approximation with Splines and Fractals

1

The General Interpolation and Approximation Problem

This chapter introduces the main concepts of the interpolation and approximation theory for real functions and presents some of the main results. In particular, the abstract interpolation problem is defined and the abstract approximation problem considered. As an example for both concepts, we consider polynomial interpolation and approximation and discuss their properties. To this end, Lagrange[1] interpolation, divided differences, and Newton[2] interpolation are presented. The material in this chapter forms the basis for the later developments and serves as a reference. For this purpose, we also need to introduce some classes of spaces such as semi-normed and normed spaces, metric spaces, and pre-Hilbert[3] and Hilbert spaces, as they are indispensible in the modern theory of interpolation and approximation. In addition, we present the moduli of continuity and smoothness and relate them to the order of approximation. Several function spaces that play an important role in the theoretical and also practical considerations of interpolation and

1 JOSEPH LOUIS LAGRANGE, 25 January 1736–10 April 1813. Italian mathematician who made major contributions to analytical mechanics, the calculus of variations, partial differential equations, complex function theory, and algebra.

2 SIR ISAAC NEWTON, 4 January 1643–31 March 1727. English mathematician, physicist, natural philosopher, and alchemist who is considered to be one of the greatest scientists of all times. He was one of the founders of differential and integral calculus and the discoverer of the theory of gravitation. Newton also made major contributions to mechanics, optics, and astronomy.

3 DAVID HILBERT, 23 January 1862–14 February 1943. A German mathematician who is considered to be one of the most influential and universal mathematicians of the 19th and 20th century. He made fundamental contributions to the axiomatization of geometry, to invariant theory, mathematical physics, and functional analysis.

approximation theory are defined and their properties discussed. One of the major results in approximation theory, the Weierstrass[4] Approximation Theorem, is stated and proved. Finally, the approximation by step functions is considered as it serves as the basis for later developments. However, only concepts that are necessary for this textbook are presented, and more information about interpolation and approximation can be found in the texts [26, 79, 148], and [159], which were consulted as references. The reader who is interested in additional details or a more advanced treatment of interpolation and approximation theory is invited to consult the books by Jackson [89], Cheney [36], Lorentz [103], Rivlin [129], Bergh and Løfstrøm [24], Butzer [31, 32], and Butzer and Scherer [33].

1.1 Notation and Terminology

The set of natural numbers is denoted by $\mathbb{N} := \{1, 2, 3, \ldots\}$ and the ring of integers by $\mathbb{Z} := \{0, \pm 1, \pm 2, \pm 3, \ldots\}$. The fields of rational, real, and complex numbers are written as $\mathbb{Q}$, $\mathbb{R}$, and $\mathbb{C}$, respectively. The set of nonnegative real numbers and positive real numbers is denoted by $\mathbb{R}_0^+$ and $\mathbb{R}^+$, respectively. The symbol $\mathbb{K}$ denotes either the field $\mathbb{R}$ or the field $\mathbb{C}$. The letter I represents the unit interval $[0, 1]$ on the real line $\mathbb{R}$. The cardinality of a set S is denoted by card S. The n-fold Cartesian product of a collection of sets $\{S_i \mid i = 1, \ldots, n\}$ is written as $\displaystyle\mathop{\mathsf{X}}_{i=1}^{n} S_i$.

The letter K always denotes a nonempty compact subset of $\mathbb{R}^n$, $n \in \mathbb{N}$, and the Greek letter Ω always a nonempty, open, and connected subset of $\mathbb{R}^n$. The latter sets are also called *regions* or *domains*. The closure of a set S, i.e., the intersection of all closed sets C containing S, is written as $\overline{S}$: $\displaystyle\overline{S} := \bigcap_{S \subseteq C} C$. The interior of a set S, written S°, is the union of all open sets O contained in S: $\displaystyle S^\circ = \bigcup_{O \subseteq S} O$. The boundary ∂S of a set S is defined as $\partial S := \overline{S} \setminus S^\circ$. The complement of a set A in a set B is written as $B \setminus A$ or, if B is fixed, as $\complement A$. The power set of a set S, i.e., the collection of all subsets of S, is denoted by $\mathscr{P}(S)$.

For sets A and B, we refer to the association which assigns to each $a \in A$ exactly one $b \in B$ as a *map* or *mapping*. In case $B \subseteq \mathbb{K}^n$, $n \in \mathbb{N}$, we call a mapping a *function*. The collection of all mappings $f : A \to B$ is denoted by $\mathrm{Map}(A, B)$ or B^A. The set A is called the *domain* of a mapping f and sometimes written as dom f, and the set B is called the *codomain* of f. The set $\{f(a) \mid a \in A\}$ is termed the *range of* f and written as rg f. The mapping (function) $\mathrm{id}_A : A \to A$ with $A \ni a \mapsto a$ is called the identity mapping (function) for A.

---

4 KARL THEODOR WILHELM WEIERSTRASS, 31 October 1815–19 February 1897. German mathematician who made major contributions to the foundations of analysis, complex function theory, the theory of elliptic functions, differential geometry, and the calculus of variation.

Mappings between linear spaces, syn. vector spaces, are called *operators*. An operator from a vector space to a field $\mathbb{K}$ is termed a *functional*. The *characteristic function* or *indicator function* of a set A is defined by

$$\chi_A(x) := \begin{cases} 1, & x \in A; \\ 0, & x \notin A. \end{cases}$$

Vectors are always written as column vectors and $^\top$ denotes the transpose of a vector or matrix. Fractal sets and collections of fractals are denoted by upper-case fraktur letters such as $\mathfrak{A}, \mathfrak{B}, \ldots$, fractal functions by lower-case fraktur letters such as $\mathfrak{f}, \mathfrak{g}, \mathfrak{h}, \ldots$ and fractal measures by lower-case fraktur letters such as $\mathfrak{m}$ and $\mathfrak{m}'$.

1.2 Semi-Normed and Normed Spaces

In this section we introduce semi-normed and normed linear spaces, as they play an important role in mathematics and are an indispensable tool in this textbook.

Definition 1.1 (Semi-Normed Linear Space)   Let V be a linear space over a field $\mathbb{K}$. A functional $p : \mathsf{V} \to \mathbb{K}$ is called a *semi-norm* iff the following axioms are fulfilled.

1. Positive Homogeneity: $\forall v \in \mathsf{V}$ and $\forall \lambda \in \mathbb{K}$: $p(\lambda v) = |\lambda| p(v)$;

2. Triangle Inequality: $\forall v, w \in \mathsf{V}$: $p(v + w) \le p(v) + p(w)$.

Both axioms imply that $p(0) = 0$ and thus $p(v) \ge 0$, for all $v \in \mathsf{V}$. (The reader is encouraged to verify this statement.) A semi-norm is called a *norm* provided $p(v) = 0$ iff $v = 0$.

Definition 1.2 (Normed Linear Space)   A linear space V (over a field $\mathbb{K}$) with a semi-norm (norm) is called a *semi-normed (normed) linear space* and denoted by (V, p). A complete normed linear space is called a *Banach[5] space.*

If a normed linear space (V, p) is not complete, one can construct its *completion* $(\widehat{\mathsf{V}}, d)$ by adding to V the limits of all Cauchy sequences in V. More precisely, the points of $\widehat{\mathsf{V}}$ are equivalence classes of Cauchy sequences, where two Cauchy sequences $x := \{x_\nu \mid \nu \in \mathbb{N}\}$ and $y := \{y_\nu \mid \nu \in \mathbb{N}\}$ are equivalent, in symbols, $x \sim y$, if $\lim\limits_{\nu \to \infty} p(x_\nu - y_\nu) = 0$. The set V is then *dense* in $\widehat{\mathsf{V}}$, i.e., for every $\widehat{x} \in \widehat{\mathsf{V}}$ and for all $\varepsilon > 0$, there exists an $x \in \mathsf{V}$ so that $p(\widehat{x} - x) < \varepsilon$.

---

5 STEFAN BANACH, 30 March 1892–31 August 1945. Polish mathematician who worked in the areas of measure theory and harmonic analysis. He was one of the founders of functional analysis.

Example 1.3   An important class of normed linear spaces is given by $\mathsf{V} := \mathbb{R}^n$, $n \in \mathbb{N}$, endowed with various norms. To this end, let $1 \le p \in \mathbb{R}$ and $x := (x_1, \ldots, x_n)^\top \in \mathbb{R}^n$. Define the *p-norms* $\| \; \|_p : \mathbb{R}^n \to \mathbb{R}$ by

$$\|x\|_p := \left(\sum_{i=1}^{n} |x_i|^p \right)^{1/p},$$

and the ∞-norm $\| \; \|_\infty : \mathbb{R}^n \to \mathbb{R}$ by

$$\|x\|_\infty := \max\{|x_i| \mid i = 1, \ldots, n\}.$$

It is not difficult to show that the following relationships exist between these norms.

$$\|x\|_p \le \|x\|_1 \le \sqrt[p]{n^{p-1}} \, \|x\|_p \quad \text{and} \quad \|x\|_\infty \le \|x\|_p \le \sqrt[p]{n} \, \|x\|_\infty. \qquad (1.1)$$

The second pair of inequalities allows the interpretation of the ∞-norm as a limit of p-norms:

$$\|x\|_\infty = \lim_{p \to \infty} \|x\|_p. \qquad (1.2)$$

Example 1.4   The generalization of the p-norms $\| \; \|_p$, $p \in [1, \infty]$, to infinite-dimensional linear spaces, such as $\mathsf{V} := \mathbb{R}^\mathbb{N} = \mathrm{Map}(\mathbb{N}, \mathbb{R})$ or $\mathsf{V} := \mathbb{R}^\mathbb{Z} = \mathrm{Map}(\mathbb{Z}, \mathbb{R})$, produces the *sequence spaces* $\ell^p(\mathbb{N})$, respectively, $\ell^p(\mathbb{Z})$. For $1 \le p \in \mathbb{R}$ and for the symbol ∞, define

$$\ell^p(\mathbb{N}) := \left\{ x := (x_i \mid i \in \mathbb{N}) \in \mathbb{R}^\mathbb{N} \,\middle|\, \left(\sum_{i=1}^{\infty} |x_i|^p \right)^{1/p} < \infty \right\}, \quad p \ge 1$$

$$\ell^\infty(\mathbb{N}) := \left\{ x := (x_i \mid i \in \mathbb{N}) \in \mathbb{R}^\mathbb{N} \,\middle|\, \sup\{|x_i| \mid i \in \mathbb{N}\} < \infty \right\}.$$

In a similar fashion, one defines the *bi-infinite* sequence spaces $\ell^p(\mathbb{Z})$ and $\ell^\infty(\mathbb{Z})$.

Function spaces provide another important class of semi-normed and normed linear spaces. To this end, let K be a nonempty compact subset of $\mathbb{R}^n$, $n \in \mathbb{N}$.

Example 1.5   Consider the set $C(K) := C(K, \mathbb{R})$ of all continuous functions $f : K \to \mathbb{R}$. Under the usual operations of function addition, $(f + g)(x) := f(x) + g(x)$, for all $x \in K$, and scalar multiplication of functions, $(cf)(x) := cf(x)$, for all $c \in \mathbb{R}$ and all $x \in K$, the set $C(K)$ becomes a linear space over $\mathbb{R}$. On the linear space $C(K)$ we introduce the so-called *Chebyshev*[6]

---

6 PAFNUTY CHEBYSHEV, 26 May 1821–8 December 1894. Russian mathematician who worked primarily in the fields of probability theory, mathematical statistics, and number theory. He is considered as one of the founding fathers of Russian mathematics.

norm $\| \; \|_{\infty,K} : C(K) \to \mathbb{R}$, defined by

$$\|f\|_{\infty,K} := \sup\{|f(\boldsymbol{x})| \,|\, \boldsymbol{x} \in K\}.$$

Endowed with this norm, the linear space $(C(K), \| \; \|_{\infty,K})$ becomes a Banach space. (The reader is encouraged to verify this statement.)

Example 1.6   A function space related to $C(K)$ is the linear space $C_b(\Omega) := C_b(\Omega, \mathbb{R})$ of *bounded real-valued continuous functions* defined by

$$C_b(\Omega) := \{f \in C(\Omega) \,|\, |f(\boldsymbol{x})| < \infty, \forall \boldsymbol{x} \in \Omega\}.$$

$C_b(\Omega)$ becomes a Banach space when endowed with the Chebychev norm $\| \; \|_{\infty,\Omega} : C_b(\Omega) \to \mathbb{R}$,

$$\|f\|_{\infty,\Omega} := \sup\{|f(\boldsymbol{x})| \,|\, \boldsymbol{x} \in \Omega\}.$$

If $\Omega = \mathbb{R}^n$, we write $\| \; \|_\infty$ instead of $\| \; \|_{\infty,\mathbb{R}^n}$.

Example 1.7   Now, let $\Omega := (a, b)$, $a, b \in \mathbb{R}$, $a < b$, and denote by $C^k(a, b) := C^k((a, b), \mathbb{R})$, $k \in \mathbb{N}$, the linear space of all *k-times continuously differentiable functions* $f : (a, b) \to \mathbb{R}$. Note that functions in $C^k(a, b)$ need not be bounded on (a, b). If, however, f and all its derivatives up to order k are bounded and uniformly continuous on (a, b), then we can extend f to the closure of (a, b), i.e., the closed interval $[a, b]$. Thus, we define the space $C^k[a, b] := C^k([a, b], \mathbb{R})$ to consist of all those functions f in $C^k(a, b)$, for which f and all its derivatives up to order k are bounded and uniformly continuous on (a, b).

The functional $p_k : C^k[a, b] \to \mathbb{R}$ given by

$$p_k(f) := \sup\{|D^k f| \,|\, x \in [a, b]\} \tag{1.3}$$

defines a semi-norm on $C^k[a, b]$. Here, D denotes the *derivative* or *differential operator* of a function:

$$Df := \frac{df}{dx} \quad \text{and} \quad D^k := DD^{k-1}, \; k \in \mathbb{N}, \quad D^0 f := f.$$

Note that p_k is only a semi-norm since the vanishing of the k-th derivative of a function $f \in C^k[a, b]$ implies that f is a polynomial of degree $k - 1$.

The last class of normed linear spaces we introduce here are the *Lebesgues*[7] *spaces* $L^p(\Omega)$, $p \in [1, \infty]$, where Ω is a subset of $\mathbb{R}^n$. For a given $p \in [1, \infty)$, we introduce on the space $C(\Omega)$ the norms

$$\|f\|_{L^p,\Omega} := \left(\int_\Omega |f(\boldsymbol{x})|^p \, d\boldsymbol{x} \right)^{1/p}.$$

(As before, when $\Omega = \mathbb{R}^n$ we write $\| \; \|_{L^p}$ instead of $\| \; \|_{L^p,\mathbb{R}^n}$.) The resulting spaces $(C(\Omega), \| \; \|_{L_p,\Omega})$ are normed linear spaces (See Exercise 6!) but they

---

7 HENRI LÉON LEBESGUE, 28 June 1875–26 July 1941. French mathematician who is the founder of modern measure theory. He also worked in the theory of Fourier series and in potential theory.

are not complete with respect to these norms, i.e., sequences $\{f_\nu \mid \nu \in \mathbb{N}\}$ of functions in $C(\Omega)$ need not converge in the norm $\|\ \|_{L_p,\Omega}$, $p \in [1, \infty)$, to a limit function f that is again in $C(\Omega)$. Completing the space $C(\Omega)$ with respect to the norms $\|\ \|_{L_p,\Omega}$ produces the Lebesgue spaces $(L^p(\Omega), \|\ \|_{L_p,\Omega})$.

Definition 1.8 (Lebesgue Spaces) The *Lebesgue spaces* $L^p(\Omega)$ are the completion of the space $C(\Omega)$ with respect to the norms $\|\ \|_{L_p,\Omega}$, $p \in [1, \infty)$.

The norms on the spaces $L^p(\Omega)$ resemble those defined on the linear spaces $\mathbb{R}^n$ and the sequence spaces $\ell^p(\mathbb{N})$, respectively, $\ell^p(\mathbb{Z})$. To define a corresponding norm for the symbol ∞, we need to introduce the concept of *dual space* of a normed linear space.

Definition 1.9 (Topological Dual Space) Let $(\mathsf{X}, \|\ \|)$ be a normed linear space. The set X^*, defined as

$$\mathsf{X}^* := \{\varphi : \mathsf{X} \to \mathbb{R} \mid \varphi \text{ is linear and continuous}\}$$

and endowed with the norm $\|\ \|_* : \mathsf{X}^* \to \mathbb{R}$ given by

$$\|\varphi\|_* := \sup\{|\varphi(x)| \mid \|x\| \leq 1\}$$

is called the *(topological) dual space* to $(\mathsf{X}, \|\ \|)$.

Remark 1.10 Note that for a given linear space X there are two dual spaces: the algebraic dual X' and the topological dual X^*. The algebraic dual consists of all *linear functionals* from X to $\mathbb{R}$, whereas for the topological dual one requires in addition that the linear functionals are continuous. If X is finite-dimensional then $\mathsf{X}' = \mathsf{X}^*$, but when X is infinite-dimensional $\mathsf{X}' \neq \mathsf{X}^*$ in general. (See also Exercise 8 at the end of this chapter!)

The following theorem, whose proof is left to the reader, gives more information about the topological dual space of a normed linear space.

Theorem 1.11 Let $(\mathsf{X}, \|\ \|)$ be a normed linear space. The topological dual $(\mathsf{X}^*, \|\ \|_*)$ is a complete normed linear space, i.e., a Banach space.

Proof Exercise! *(Hint: Use that $(\mathbb{R}, |\ |)$ is a complete normed linear space.)* □

Example 1.12 Let $1 < p < \infty$ and let q be such that $1/p + 1/q = 1$. Let $\Omega \subseteq \mathbb{R}^n$. Then the topological dual space to $L^p(\Omega)$ is $L^q(\Omega)$. This statement is the content of the *Riesz*[8] *representation theorem*. The identification of

---

8 FRIGYES RIESZ, 22 January 1880–28 February 1956. Hungarian mathematician who made fundamental contributions to functional analysis. He also worked in ergodic theory. He was the older brother of the Hungarian mathematician Marcel Riesz.

$\varphi \in (L^p(\Omega))^*$ with an element of $L^q(\Omega)$ is done via

$$\varphi(f) = \int_\Omega f(\boldsymbol{x})g(\boldsymbol{x})dx =: \langle f, g \rangle,$$

for some $g \in L^q(\Omega)$. In other words, every continuous linear functional $\varphi : L^p(\Omega) \to \mathbb{R}$ can be represented in the form $\varphi = \langle \bullet, g \rangle$, for some $g \in L^q(\Omega)$.

Now, we are in a position to define the Lebesgue space $L^\infty(\Omega)$, $\Omega \subseteq \mathbb{R}^n$, $n \in \mathbb{N}$.

Definition 1.13   The Lebesgue space $L^\infty(\Omega)$ is defined as the topological dual space of $L^1(\Omega)$.

To motivate the definition of a norm for $L^\infty(\Omega)$, suppose that $f \in L^1(\Omega)$ and $g \in C_b(\Omega)$. Then the mapping $\varphi : L^1(\Omega) \to \mathbb{R}$ given by

$$\varphi(f) := \int_\Omega f(\boldsymbol{x})g(\boldsymbol{x})dx$$

is a linear functional on $L^1(\Omega)$, i.e., an element of $L^\infty(\Omega)$, satisfying

$$|\varphi(f)| \leq \int_\Omega |f(\boldsymbol{x})||g(\boldsymbol{x})|dx \leq \int_\Omega |f(\boldsymbol{x})|dx \cdot \sup\{|g(\boldsymbol{x})| \,|\, \boldsymbol{x} \in \Omega\}$$

$$= \|f\|_{L^1,\Omega} \|g\|_{\infty,\Omega}.$$

Thus, $g \in C_b(\Omega)$ represents the continuous linear functional φ and $\|g\|_{\infty,\Omega}$ is its norm. However, not all continuous linear functionals in $L^1(\Omega)$ can be represented by functions g from $C_b(\Omega)$; g may be discontinuous across m-dimensional surfaces in $\mathbb{R}^m$, $m = 1, \ldots, n-1$. To define the correct norm on $L^\infty(\Omega)$, which takes into account these discontinuities, we need to introduce the concept of a *null set* or a *set of measure zero* in $\mathbb{R}^n$.

Definition 1.14 (Null Set)   Suppose that $S \subseteq \mathbb{R}^n$, $n \in \mathbb{N}$. The set S is called a *null set* or a *set of measure zero* if for any given $\varepsilon > 0$, there exists a covering of S by at most countably many cubes of the form $W := (a_1, b_1) \times \cdots \times (a_n, b_n)$, $a_i, b_i \in \mathbb{R}$, $a_i < b_i$, $i = 1, \ldots, n$, such that the sum of all their n-dimensional volumes is less than ε.

It is not difficult to realize that every finite set has measure zero. As a more interesting example, we consider the set $S := \mathbb{Z} \subseteq \mathbb{R}$ and show that it has measure zero. For, given any $\varepsilon > 0$, every $z \in \mathbb{Z}$ is contained in the interval $I_z := (z - 2^{-(|z|+2)}\varepsilon, z + 2^{-(|z|+2)}\varepsilon)$ and $\mathbb{Z} \subseteq \bigcup_{z \in \mathbb{Z}} I_z$. Furthermore,

$$\sum_{z=-\infty}^{+\infty} 2^{-(|z|+2)}\varepsilon = \sum_{z=-\infty}^{-1} 2^{-(|z|+2)}\varepsilon + 2^{-2}\varepsilon + \sum_{z=1}^{\infty} 2^{-(|z|+2)}\varepsilon = \frac{3}{4}\varepsilon < \varepsilon.$$

Analogously, one can show that $\mathbb{Q}$ has measure zero. However, there are other larger sets, which are not countably finite, but possess measure zero. One such example is the ternary *Cantor*[9] *set.*

A function f is said to have property P *almost everywhere*, written f has P a.e., if f has property P except for a set of measure zero. For instance, suppose $f : \mathbb{R} \to \mathbb{R}$ is defined by

$$f(x) := \begin{cases} x^2, & x \in \mathbb{R} \setminus \mathbb{Z}; \\ 0, & x \in \mathbb{Z}. \end{cases}$$

Then f is continuous a.e., i.e., continuous except for $x \in \mathbb{Z}$, a set of measure zero.

It is shown in measure theory that the norm on $L^\infty(\Omega)$, $\Omega \subseteq \mathbb{R}^n$, $n \in \mathbb{N}$ is given by

$$\|f\|_{L^\infty,\Omega} := \inf\{c \in \mathbb{R}^+ \,|\, f(\boldsymbol{x}) \leq c, \text{ for a.e. } \boldsymbol{x}\}.$$

In other words, $(L^\infty(\Omega), \| \; \|_{L^\infty,\Omega})$ consists of all functions $f \in L^\infty(\Omega)$ that are bounded almost everywhere on Ω.

Suppose that Ω has finite n-dimensional volume and $p, q \in [1, \infty)$ with $p \leq q$. Then

$$L^\infty(\Omega) \subseteq \cdots \subseteq L^q(\Omega) \subseteq L^p(\Omega) \subseteq \cdots \subseteq L^1(\Omega). \tag{1.4}$$

These inclusions can be easily verified using the *Hölder*[10] *Inequality*, Theorem 1.16, below. (The reader is encouraged to verify this statement.) Note that (1.4) only holds when Ω has finite n-dimensional volume.

Example 1.15 Let Ω have finite n-dimensional volume. The space $C(\Omega)$ is not contained in any of the spaces $L^p(\Omega), p \in [1, \infty]$. A simple example is given by $\Omega := (0, 1)$ and the function $f : (0, 1) \to \mathbb{R}$, $x \mapsto x^{-1}$. Then, $f \in C(0, 1)$ but clearly not in $L^\infty(0, 1)$, and thus also not in any of the $L^p(0, 1)$-spaces. However, the space $C_b(\Omega)$ is properly contained in $L^\infty(\Omega)$.

The spaces $(L^p(\Omega), \| \; \|_{L^p,\Omega})$, $1 \leq p < \infty$, are, by definition, complete and hence they are Banach spaces. Since $(L^\infty(\Omega), \| \; \|_{L^\infty,\Omega})$ is the dual space of the normed linear space $(L^1(\Omega), \| \; \|_{L^1,\Omega})$, it is by Theorem 1.11 also a Banach space.

The following two inequalities are often encountered when working with the Lebesgue spaces $(L^p(\Omega), \| \; \|_{L^p,\Omega})$, $1 \leq p \leq \infty$.

---

9 GEORG FERDINAND LUDWIG PHILIPP CANTOR, 3 March 1845–6 January 1918. German mathematician who is considered the founder of modern set theory. He also contributed to number theory and the theory of functions.
10 OTTO LUDWIG HÖLDER, 22 December 1859–29 August 1937. German mathematician whose research interests were the theory of groups and the convergence of Fourier series.

Theorem 1.16 (Hölder Inequality) Let $\Omega \subseteq \mathbb{R}^n$ and let $1 \leq p, q \leq \infty$ with $\frac{1}{p} + \frac{1}{q} = 1$. Suppose that $f \in L^p(\Omega)$ and $g \in L^q(\Omega)$. Then $fg \in L^1(\Omega)$ and

$$\|fg\|_{L^1,\Omega} \leq \|f\|_{L^p,\Omega} \, \|g\|_{L^q,\Omega}.$$

Proof Exercise! □

Remark 1.17 Hölder's inequality is also valid for finite sums of real or complex numbers and for the sequence spaces $\ell^p(\mathbb{N})$ and $\ell^p(\mathbb{Z})$. In the former case, it is given by

$$\sum_{k=1}^{n} |x_k y_k| \leq \left(\sum_{k=1}^{n} |x_k|^p \right)^{\frac{1}{p}} \left(\sum_{k=1}^{n} |y_k|^q \right)^{\frac{1}{q}},$$

with $x_k, y_k \in \mathbb{R}$ or $\mathbb{C}$, $k = 1, \ldots, n$, and in the latter case by

$$\|a \cdot b\|_{\ell^1} \leq \|a\|_{\ell^p} \, \|b\|_{\ell^q},$$

where we set $a \cdot b := \{a_k b_k \mid k \in \mathbb{I}\}$, for sequences $a := \{a_k \mid k \in \mathbb{I}\} \in \ell^p(\mathbb{I})$ and $b := \{b_k \mid k \in \mathbb{I}\} \in \ell^q(\mathbb{I})$, with $\mathbb{I} = \mathbb{N}$ or $\mathbb{Z}$. Note that for $p = q = 2$, we obtain the Cauchy[11]-Schwarz[12] inequality.

Theorem 1.18 (Minkowski[13] Inequality) Let $\Omega \subseteq \mathbb{R}^n$ and let $1 \leq p \leq \infty$. Suppose that $f, g \in L^p(\Omega)$. Then $f + g \in L^p(\Omega)$ and

$$\|f + g\|_{L^p,\Omega} \leq \|f\|_{L^p,\Omega} + \|g\|_{L^p,\Omega}.$$

Proof Exercise! □

Remark 1.19 Minkowski's inequality is also valid for finite sums of real or complex numbers and for the sequence spaces $\ell^p(\mathbb{N})$ and $\ell^p(\mathbb{Z})$. In the former case, it is given by

$$\left(\sum_{k=1}^{n} |x_k + y_k|^p \right)^{\frac{1}{p}} \leq \left(\sum_{k=1}^{n} |x_k|^p \right)^{\frac{1}{p}} + \left(\sum_{k=1}^{n} |y_k|^p \right)^{\frac{1}{p}},$$

with $x_k, y_k \in \mathbb{R}$ or $\mathbb{C}$, $k = 1, \ldots, n$, and in the latter case by

$$\|a + b\|_{\ell^p} \leq \|a\|_{\ell^p} + \|b\|_{\ell^p},$$

---

11 Augustin Louis Cauchy, 21 August 1789–23 May 1857. French mathematician who made major contributions to real and complex analysis, differential equations, mathematical physics, and elasticity theory.

12 Hermann Amandus Schwarz, 25 January 1843–30 November 1921. German mathematician who worked in complex analysis and the calculus of variations.

13 Hermann Minkowski, 22 June 1864–12 January 1909. German mathematician and physicist who contributed enormously to quadratic forms and convex bodies. He introduced the concept of the four-dimensional space–time continuum, known as Minkowski space–time, into the theory of special relativity.

for sequences $a := \{a_k \mid k \in \mathbb{I}\}$ and $b := \{b_k \mid k \in \mathbb{I}\}$ in $\ell^p(\mathbb{I})$, with $\mathbb{I} = \mathbb{N}$ or $\mathbb{Z}$. Here, we defined $a + b := \{a_k + b_k \mid k \in \mathbb{I}\}$. Note that Minkowski's inequality states that the triangle inequality is valid in the L^p- and ℓ^p-spaces.

1.3 The Abstract Interpolation Problem

Let $m \in \mathbb{N}$ and $n \in \mathbb{N}_0 := \mathbb{N} \cup \{0\}$. Suppose that

$$F : \mathbb{R}^m \times \mathbb{C}^{n+1} \to \mathbb{C},$$

$$(\boldsymbol{x}; a_0, a_1, \ldots, a_n) \mapsto F(\boldsymbol{x}; a_0, a_1, \ldots, a_n)$$

is a function. The collection $(a_0, a_1, \ldots, a_n) \in \mathbb{C}^{n+1}$ is called a *parameter set*. The *interpolation problem for F* is defined as follows:

Given the set $\{(\boldsymbol{x}_\nu, y_\nu) \in \mathbb{R}^m \times \mathbb{C} \mid \nu = 0, 1, \ldots, n\}$ where $\boldsymbol{x}_\mu \neq \boldsymbol{x}_\nu$ for $\mu \neq \nu$, find values for $a_0, a_1, \ldots, a_n \in \mathbb{C}$ such that

$$\forall \nu = 0, 1, \ldots, n : \quad F(\boldsymbol{x}_\nu; a_0, a_1, \ldots, a_n) = y_\nu. \tag{1.5}$$

The set $X := \{\boldsymbol{x}_\nu \in \mathbb{R}^m \mid \nu = 0, 1, \ldots, n\}$ is called the set of *support points* or *knots*, and the set $Y := \{y_i \in \mathbb{C} \mid \nu = 0, 1, \ldots, n\}$ the set of *support* or *interpolation values*. The set product $X \times Y$ is referred to as the *interpolation set*. We will abbreviate (1.5) by writing $F(\boldsymbol{x}; a_0, , a_1, \ldots, a_n)|_X = Y$.

The interpolation problem for F is called *linear* if there exist functions $F_i : \mathbb{R}^m \to \mathbb{C}$ such that

$$\forall \boldsymbol{x} \in \operatorname{conv} X : \quad F(\boldsymbol{x}; a_0, a_1, \ldots, a_n) = \sum_{i=0}^{n} a_i F_i(\boldsymbol{x}), \tag{1.6}$$

where $\operatorname{conv} X$ denotes the *convex hull of X*, i.e., the smallest convex set containing X:

$$\operatorname{conv} X := \bigcap_{\substack{X \subseteq K \\ K \text{ convex}}} K.$$

Next, we present some examples of interpolation problems.

Example 1.20 Let $m := 1$. Then the following are linear interpolation problems.

1. POLYNOMIAL INTERPOLATION PROBLEM:

$$F(x; a_0, a_1, \ldots, a_n) := \sum_{\nu=0}^{n} a_\nu x^\nu.$$

2. TRIGONOMETRIC INTERPOLATION PROBLEM:

$$F(x; a_0, a_1, \ldots, a_n) := \sum_{\nu=0}^{n} a_\nu e^{i\nu x}.$$

Example 1.21   Let $m := 1$. The following are examples of non-linear interpolation problems.

1. RATIONAL INTERPOLATION PROBLEM: Let $p, q \in \mathbb{N}$ with $p + q = n$.

$$F(x; a_0, a_1, \ldots, a_p, b_0, b_1 \ldots b_q) := \frac{\displaystyle\sum_{\nu=0}^{p} a_\nu x^\nu}{\displaystyle\sum_{\mu=0}^{q} b_\mu x^\mu}.$$

2. EXPONENTIAL INTERPOLATION PROBLEM: Let $n := 2p$.

$$F(x; a_0, a_1, \ldots, a_p, \lambda_0, \lambda_1, \ldots, \lambda_p) := \sum_{\nu=0}^{p} a_\nu e^{\lambda_\nu x}.$$

In the following, we will primarily deal with real-valued interpolation and approximation, but we emphasize that many results also hold in the complex setting.

1.3.1 Interpolation with Polynomials and Real Parameters

In order to proceed, we need to introduce the linear space of real polynomials. To this end, let $k \in \mathbb{N}$ and let $\deg p$ denote the degree of a real polynomial p. Set

$$\Pi^k := \{p : \mathbb{R} \to \mathbb{R} \mid p \text{ is a real polynomial with } \deg p < k\}$$

and

$$\Pi^k[a, b] := \{p \in \Pi^k : \operatorname{dom} p = [a, b]\}.$$

The properties below are straight-forward to show:

- Π^k is a linear subspace of $C^\varkappa(\mathbb{R})$, for any $\varkappa \in \mathbb{N}_0$;

- The set $\{1, x, \ldots, x^{k-1}\}$ is a basis for Π^k and thus $\dim \Pi^k = k$.

We will mostly work with the *order of a polynomial* rather then its degree. The order of a polynomial $p \in \Pi^k$, written as $\operatorname{ord} p$, is defined as the dimension of Π^k. Hence $\operatorname{ord} p = \deg p + 1 = k$.

In the following, let $n \in \mathbb{N}$ and $X := \{x_\nu \mid \nu = 0, 1 \ldots, n\} \subseteq I$, where $I \subseteq \mathbb{R}$ is a nonempty compact interval, be a given knot set. The polynomial interpolation problem F on the real line with real parameters $a_0, a_1, \ldots, a_n$ for a function $f : I \to \mathbb{R}$ is defined by

$$F(x; a_0, a_1, \ldots, a_n) := \sum_{\mu=0}^{n} a_\mu x^\mu \tag{1.7}$$

and

$$\forall \nu = 0, 1, \ldots, n: \quad y_\nu := f(x_\nu). \tag{1.8}$$

More precisely, given $n + 1$ values $Y := \{y_\nu \in \mathbb{R} \mid \nu = 0, 1, \ldots, n\}$ of a function f on a knot set X, determine the coefficients a_ν, $\nu = 0, 1, \ldots, n$, in (1.7) so that

$$\forall \nu = 0, 1, \ldots, n: \quad \sum_{\mu=0}^{n} a_\mu x_\nu^\mu = y_\nu = f(x_\nu). \tag{1.9}$$

In connection with the above interpolation problem the following questions arise naturally.

Question 1. Given $n \in \mathbb{N}$ and an interpolation set $X \times Y$, does there always exist a polynomial $p \in \Pi^{n+1}$ so that $p(x_\nu) = y_\nu$, $\forall \nu = 0, 1, \ldots, n$?

Question 2. If $p \in \Pi^{n+1}$ exists, is it unique?

Question 3. Is it possible to determine p numerically and computationally efficiently?

Question 4. Let $I := \operatorname{conv} X$. Under what conditions on $f \in C^m(I)$, some $m \in \mathbb{N}_0$, can the error $f(x) - p(x)$, $x \in I$, be estimated?

We will see shortly that all these questions can be answered affirmatively. The first result provides the answer to Questions 1 and 2.

Theorem 1.22 Let $n \in \mathbb{N}$. For any given interpolation set $\{(x_\nu, y_\nu) \mid \nu = 0, 1, \ldots, n\} \subseteq \mathbb{R}^2$, with $x_\mu \neq x_\nu$ for $\mu \neq \nu$, there exists exactly one polynomial $p \in \Pi^{n+1}$ so that $p(x_\nu) = y_\nu$, $\nu = 0, 1, \ldots, n$.

Proof Exercise! $\square$

The simple example below illustrates the concept of polynomial interpolation.

Example 1.23 Suppose that $X \times Y := \{(1, 3), (2, 5), (4, -1), (6, 0), (7, 2)\}$ is an interpolation set. To compute a polynomial of order $n + 1 = 5$, i.e., degree 4, we express (1.9) in matrix form

$$\begin{pmatrix} 1 & x_0 & \cdots & x_0^n \\ 1 & x_1 & \cdots & x_1^n \\ \vdots & \vdots & \ddots & \vdots \\ 1 & x_n & \cdots & x_n^n \end{pmatrix} \begin{pmatrix} a_0 \\ a_1 \\ \vdots \\ a_n \end{pmatrix} = \begin{pmatrix} y_0 \\ y_1 \\ \vdots \\ y_n \end{pmatrix}$$

and substitute for $\{x_\nu \mid \nu = 0, 1, \ldots, n\}$ and $\{y_\nu \mid \nu = 0, 1, \ldots, n\}$ the above values. Thus, we arrive at the linear system

$$
\begin{pmatrix}
1 & 1 & 1 & 1 & 1 \\
1 & 2 & 4 & 8 & 16 \\
1 & 4 & 16 & 64 & 256 \\
1 & 6 & 36 & 216 & 1296 \\
1 & 7 & 49 & 343 & 2401
\end{pmatrix}
\begin{pmatrix}
a_0 \\ a_1 \\ a_2 \\ a_3 \\ a_4
\end{pmatrix}
=
\begin{pmatrix}
3 \\ 5 \\ -1 \\ 0 \\ 2
\end{pmatrix}.
$$

Solving the above linear system for the unknown coefficients $\{a_m \mid m = 0, 1, 2, 3, 4\}$ gives

$$
a_0 = -\frac{166}{15}, \; a_1 = \frac{4151}{180}, \; a_2 = -\frac{3841}{360}, \; a_3 = \frac{319}{180}, \; a_4 = -\frac{7}{12},
$$

and, therefore, as the interpolating polynomial

$$
p(x) = -\frac{166}{15} + \frac{4151}{180}\, x - \frac{3841}{360}\, x^2 + \frac{319}{180}\, x^3 - \frac{7}{12}\, x^4. \qquad (1.10)
$$

The interpolation set and the interpolating polynomial are shown in Figure 1.1.

A special case of polynomial interpolation is *cardinal* or *Lagrange interpolation*. Given an interpolation set of cardinality $n + 1$, $n \in \mathbb{N}$, we look for $n + 1$ polynomials l_μ, $\mu = 0, 1, \ldots, n$, so that

$$
\forall \mu, \nu \in \{0, 1, \ldots n\} : l_\mu(x_\nu) = \delta_{\mu\nu}.
$$

The quantity $\delta_{\mu\nu}$ is the *Kronecker*[14] *Delta* defined by

$$
\delta_{\mu\nu} = \begin{cases} 1, & \text{if } \mu = \nu; \\ 0, & \text{if } \mu \neq \nu. \end{cases}
$$

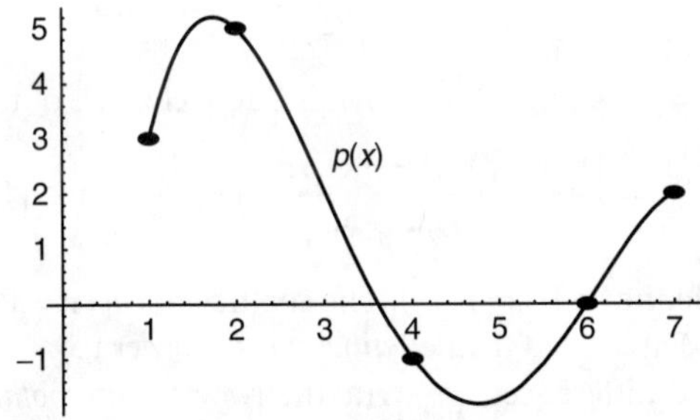

Figure 1.1   An interpolating polynomial.

14 LEOPOLD KRONECKER, 7 December 1823–29 December 1891. German mathematician who contributed to algebra, number theory, and real and complex analysis.

It is easy to verify that the *Lagrange polynomials* l_μ are given by

$$l_\mu(x) = \prod_{\substack{\nu=0 \\ \mu \neq \nu}}^{n} \frac{x - x_\nu}{x_\mu - x_\nu}.$$

The interpolating polynomial p for the interpolation set $\{(x_\nu, y_\nu) \in \mathbb{R}^2 \mid \nu = 0, 1, \ldots, n\}$ can then be written in the form

$$p(x) = \sum_{\nu=0}^{n} y_\nu\, l_\nu(x). \tag{1.11}$$

1.3.2 Divided Differences

Example 1.18 has shown that even in the case of an interpolation set of relatively small size, the computation of the inverse matrix is numerically not efficient and for large interpolation sets even prohibitive. An alternative approach to solving this problem is via Lagrange interpolation. In the current subsection, we present a second approach that is due to Newton. For this purpose, we make the following ansatz for $p \in \Pi^{n+1}$:

$$p(x) = a_0 + a_1(x - x_0) + a_2(x - x_0)(x - x_1)$$
$$+ \cdots + a_n(x - x_0)(x - x_1) \cdots (x - x_{n-1}), \tag{1.12}$$

for real numbers $a_0, a_1, \ldots, a_n$. To determine the coefficients, we substitute $x := x_0$ into (1.12), thus obtaining

$$p(x_0) = y_0 = a_0.$$

Setting $x := x_1$ in (1.12), yields

$$p(x_1) = y_1 = y_0 + a_1(x_1 - x_0)$$

and thus

$$a_1 = \frac{y_1 - y_0}{x_1 - x_0} = \frac{f(x_1) - f(x_0)}{x_1 - x_0} =: [x_0, x_1]f,$$

recalling that $y_\nu = f(x_\nu)$, $\nu = 0, 1, \ldots, n$. Inductively, one obtains

$$a_\nu = [x_0, \ldots, x_\nu]f := \frac{[x_1, \ldots, x_\nu]f - [x_0, \ldots, x_{\nu-1}]f}{x_\nu - x_0}, \quad \nu \in \{1, \ldots, n\}. \tag{1.13}$$

In addition, we define $[x_0]f := f(x_0)$. The coefficients a_ν, $\nu = 0, 1, \ldots, n$, given by formula (1.13) are called *divided differences of order* ν.

Using divided differences, we write the *Newton interpolating polynomial* (1.12) in the form

$$\mathbf{N}(X; f)(x) :- \mathbf{N}(x_0, \ldots, x_n; f)(x) := \sum_{\nu=0}^{n} [x_0, \ldots, x_\nu]f \cdot \prod_{\mu=0}^{\nu-1}(x - x_\mu). \tag{1.14}$$

(The empty product $\prod\limits_{\emptyset}$ is set equal to one.) We will refer to $\mathbf{N}(X; f)$ also as the *Newton interpolant of f on the knot set X.*

Note that the above form of the interpolating polynomial serves a double purpose: One can either interpret the interpolating polynomial $\mathbf{N}(X; f)$ as a mapping acting on functions f for a fixed set of interpolation points $X :=\{x_0, x_1, \ldots, x_n\}$,

$$\mathbf{N}(X; \bullet) : C[x_0, x_n] \to \Pi^n[x_0, x_n]$$

$$f \mapsto \sum_{\nu=0}^{n} [x_0, \ldots, x_n]f \cdot \prod_{\mu=0}^{\nu-1}(x - x_\mu),$$

or as a mapping acting on sets of interpolation points X for a fixed f,

$$\mathbf{N}(\bullet; f) : \mathscr{P}_{n+1}(\mathbb{R}) \to \Pi^n[x_0, x_n]$$

$$\{x_0, x_1, \ldots, x_n\} \mapsto \sum_{\nu=0}^{n} [x_0, \ldots, x_\nu]f \cdot \prod_{\mu=0}^{\nu-1}(x - x_\mu),$$

where $\mathscr{P}_{n+1}(\mathbb{R})$ denotes the collection of all finite subsets of $\mathbb{R}$ containing exactly $n + 1$ elements.

Divided differences provide the answer to Question 3 as their numerical computation can be performed highly efficiently. (For more details, we refer to any text on numerical mathematics such as [148].) To illustrate this property, we revisit Example 1.23 and now compute the polynomial coefficients via divided differences.

Example 1.24 Suppose that $X \times Y := \{(1, 3), (2, 5), (4, -1), (6, 0), (7, 2)\}$ is an interpolation set where the y-values are the ordinates of some continuous function $f : [1, 7] \to \mathbb{R}$. To obtain the coefficients, i.e., the divided differences, in the Newton form of the interpolating polynomial $p \in \Pi^5$ we construct the following table.

$$
\begin{array}{llllll}
x_0 & y_0 = [x_0]f \\[4pt]
 & & [x_0, x_1]f \\[4pt]
x_1 & y_1 = [x_1]f & & [x_0, x_1, x_2]f \\[4pt]
 & & [x_1, x_2]f & & [x_0, x_1, x_2.x_3]f \\[4pt]
x_2 & y_2 = [x_2]f & & [x_1, x_2, x_3]f & & [x_0, x_1, x_2, x_3, x_4]f \\[4pt]
 & & [x_2, x_3]f & & [x_1, x_2, x_3, x_4]f \\[4pt]
x_3 & y_3 = [x_3]f & & [x_2, x_3, x_4]f \\[4pt]
 & & [x_3, x_4]f \\[4pt]
x_4 & y_4 = [x_4]
\end{array}
$$

To compute the divided differences in this table, we proceed from the left to the right using the definition of divided differences. This produces

$$1 \quad 3 = [x_0]f$$

$$\frac{5-3}{2-1} = 2$$

$$2 \quad 5 = [x_1]f \qquad \frac{-2-2}{4-1} = -\frac{4}{3}$$

$$\frac{-1-5}{4-2} = -2 \qquad \frac{-\frac{4}{3} - \frac{3}{8}}{6-1} = -\frac{1435}{24}$$

$$4 \quad -1 = [x_2]f \qquad \frac{\frac{1}{2}-(-2)}{6-2} = \frac{3}{8} \qquad \frac{-\frac{1435}{24} - \frac{1}{40}}{7-1} = -\frac{3589}{360}$$

$$\frac{0-(-1)}{6-4} = \frac{1}{2} \qquad \frac{\frac{1}{2}-\frac{3}{8}}{7-2} = \frac{1}{40}$$

$$6 \quad 0 = [x_3]f \qquad \frac{2-\frac{1}{2}}{7-4} = \frac{1}{2}$$

$$\frac{2-0}{7-6} = 2$$

$$7 \quad 2 = [x_4]$$

The numbers in bold-face are the coefficients in the Newton form of the polynomial. Thus,

$$p(x) = 3 + 2(x-1) - \frac{4}{3}(x-1)(x-2) - \frac{1435}{24}(x-1)(x-2)(x-4)$$
$$- \frac{3589}{360}(x-1)(x-2)(x-4)(x-6).$$

After some algebra, we obtain the polynomial in the form (1.10).

Divided differences may be interpreted as linear functionals $[x_0, \ldots, x_n](\bullet) : C(\mathbb{R}) \to \mathbb{R}$ in the sense that for all $c \in \mathbb{R}$ and all $f, g \in C(\mathbb{R})$

$$[x_0, \ldots, x_n](cf + g) = c[x_0, \ldots, x_n]f + [x_0, \ldots, x_n]g.$$

Moreover, divided differences possess certain symmetry properties that are summarized in the next result.

Proposition 1.25   Let $n \in \mathbb{N}$ and $\sigma : \{0, 1, \ldots, n\} \to \{0, 1, \ldots, n\}$ be a permutation. Then

$$[x_{\sigma(0)}, \ldots, x_{\sigma(n)}]f = [x_0, \ldots, x_n]f.$$

In other words, $[x_0, \ldots, x_n]f$ is a symmetric function of the x_ν, $\nu = 0, 1, \ldots, n$.

Proof   The result follows immediately from the uniqueness of the solution of the interpolation problem. $\square$

Formula (1.13) can also be restated as follows. Let $X := \{x_0, x_1, \ldots, x_n\}$, where $x_\mu \neq x_\nu$ for $\mu \neq \nu, \mu, \nu = 0, 1, \ldots, n$. Then,

$$[x_0, \ldots, x_n]f = \frac{[x_1, \ldots, x_n]f - [x_0, \ldots, x_{n-1}]f}{x_n - x_0},$$

or, written more succinctly,

$$[X]f = \frac{[X \setminus \{x_0\}]f - [X \setminus \{x_n\}]f}{x_n - x_0} = \frac{[X \setminus \{x_\mu\}]f - [X \setminus \{x_\nu\}]f}{x_\nu - x_\mu}. \tag{1.15}$$

The last equality is a consequence of Proposition 1.25.

1.3.3 Error Estimates for Polynomial Interpolation

Next, we state a theorem that gives a formula for the error for polynomial interpolation.

Theorem 1.26   Let $X = \{x_\nu \,|\, \nu = 0, 1 \ldots, n\}$ be a given knot set with pairwise distinct knots and $Y := \{y_\nu \,|\, \nu = 0, 1 \ldots, n\}$ the corresponding set of interpolation values. Furthermore, let $J := \mathrm{conv}\, X$ and $y_\nu = f(x_\nu)$, $\forall \nu = 0, 1, \ldots, n$, where $f \in C^{n+1}(J)$. Let $p \in \Pi^{n+1}$ be the unique polynomial solving the interpolation problem (1.9). Then, for each $x \in J$, there exists $\xi = \xi(x) \in J$ so that

$$f(x) - p(x) = \frac{f^{(n+1)}(\xi)}{(n+1)!} \prod_{\nu=0}^{n} (x - x_\nu). \tag{1.16}$$

Proof   Exercise! $\square$

The above theorem gives a point-wise estimate of the error of interpolation. To obtain the error in the Chebyshev norm $\| \ \|_{\infty, J}$ of $C(J)$, we need the following lemma.

Lemma 1.27   Let $\{x_\nu \,|\, \nu = 0, 1, \ldots, n\}$ be a set of pairwise distinct knots with $a := x_0$ and $b := x_n$. Set $\|x\| := \max\{|x_{\nu+1} - x_\nu| \,|\, \nu = 0, 1, \ldots, n-1\}$. Then

$$\max \left\{ \prod_{\nu=0}^{n} (x - x_\nu) \,|\, x \in [a, b] \right\} \leq \frac{n!}{4} \|x\|^{n+1}.$$

Proof   The proof is by induction on n and left to the reader. $\square$

Theorem 1.26 together with Lemma 1.27 immediately give the next result.

Corollary 1.28 The interpolation error of a function $f \in C^{n+1}[a, b]$ by a polynomial $p \in \Pi^{n+1}$ in the Chebychev norm is given by

$$\|f - p\|_{\infty,[a,b]} \leq \frac{M \|\mathbf{x}\|^{n+1}}{4(n+1)},$$

where $M := \max\{|f^{n+1}(x)| \,|\, x \in [a, b]\}$.

Theorem 1.26 can be used to obtain a second remainder formula for the error.

Theorem 1.29 Under the hypotheses of Theorem 1.26, one also has that

$$f(x) - \mathbf{N}(x_0, \ldots, x_n; f)(x) = [x_0, \ldots, x_n, x]f \cdot \prod_{v=0}^{n}(x - x_v). \qquad (1.17)$$

Proof Set $X^* := \{x_0, \ldots, x_n, x\}$ and apply the Newton interpolation formula (1.14) to X^*. This yields

$$\mathbf{N}(X^*; f)(t) = \mathbf{N}(X; f)(t) + [x_0, \ldots, x_n, x]f \cdot \prod_{v=0}^{n}(t - x_v).$$

The statement follows by taking $t = x$ for $\mathbf{N}(X^*; f)(x) = f(x)$. $\square$

Comparing (1.17) to (1.16) produces the following corollary.

Corollary 1.30 Suppose J is any interval containing the knot set X and $f \in C^{n+1}(J)$. Then there exists $\xi \in J$ so that

$$[X]f = \frac{f^{(n+1)}(\xi)}{(n+1)!}.$$

1.3.4 Runge's Example

The following example is due to C. Runge[15] [133] and shows the shortcomings of polynomial interpolation.

For $n \in \mathbb{N}$, define $f : [-5, 5] \to \mathbb{R}$ by

$$f(x) := \frac{1}{1 + x^2}.$$

Choose knots according to $x_v := -5 + \frac{10v}{n}$ and as interpolation values $y_v := f(x_v)$, $v = 0, 1, \ldots, n$. Now let $p \in \Pi^{n+1}$ be the polynomial uniquely solving the interpolation problem. Figure 1.2 shows the graph of f (solid curve) and the interpolating polynomial p_n, for $n = 5, 10$ (dashed curves).

---

15 CARL RUNGE, 30 August 1856–3 January 1927. German mathematician who made contributions to numerical mathematics and spectroscopy.

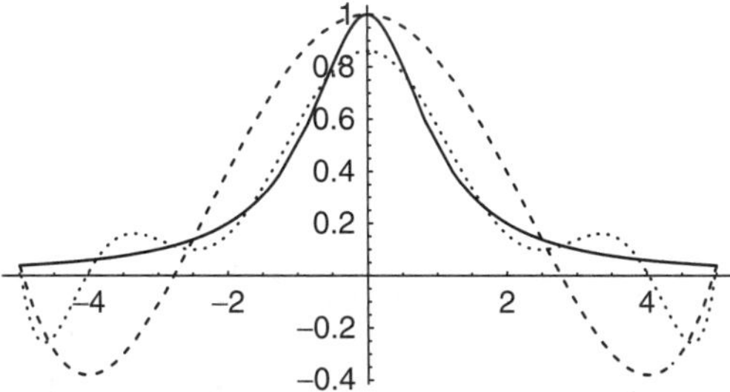

Figure 1.2   The function $f(x) = \frac{1}{1+x^2}$ and its interpolating polynomial for $n = 5, 10$.

Runge showed that

$$\|f(x) - p_n(x)\|_{\infty,[-5,5]} \to \infty \quad \text{as } n \to \infty.$$

Therefore, it is quite possible that an increase in the number of knots or equivalently in the degree of the interpolating polynomial p does not necessarily decrease the interpolation error $\|f - p_n\|_{\infty,[-5,5]}$.

1.4 The Abstract Approximation Problem

In this section, we describe the general approximation problem and investigate the existence and uniqueness of a best approximation. We do this in a general framework that later can be easily applied to spline and fractal approximation. For this purpose, we also consider metric spaces and pre-Hilbert spaces as these spaces provide the correct settings for approximation-theoretic questions.

1.4.1 Metric Spaces

First, we need to introduce the concept of a metric space.

Definition 1.31 (Metric Space)   A *metric space* (M, d) is a pair consisting of a nonempty set M and a function $d : \mathsf{M} \times \mathsf{M} \to \mathbb{R}$, called a *metric*, satisfying the following axioms.

1. Non-negativity: $\forall x, y \in \mathsf{M}: d(x, y) \geq 0$;

2. Identity: $\forall x, y \in \mathsf{M}: d(x, y) = 0$ iff $x = y$;

3. Symmetry: $\forall x, y \in \mathsf{M}: d(x, y) = d(y, x)$;

4. Triangle Inequality: $\forall x, y, z \in \mathsf{M}: d(x, z) \leq d(x, y) + d(y, z)$.

Remark 1.32   Condition 1 in Definition 1.31 is superfluous; it follows from the other axioms. The reader is encouraged to verify this!

Examples of metric spaces are given by the sets $\mathbb{R}^n$, $n \in \mathbb{N}$, endowed with the Euclidean metric $d : \mathbb{R}^n \times \mathbb{R}^n \to \mathbb{R}$,

$$d(\boldsymbol{x}, \boldsymbol{y}) := \sqrt{(x_1 - y_1)^2 + \cdots + (x_n - y_n)^2},$$

where $\boldsymbol{x} := (x_1, \ldots, x_n)$ and $\boldsymbol{y} := (y_1, \ldots, y_n)$.

A metric space (M, d) is called *complete* if every Cauchy sequence of points of M converges to an element that is again in M. If a metric space (M, d) is not complete, one can construct its *completion* $(\widehat{\mathsf{M}}, d)$ analogously to the completion of normed spaces by adding to M the limits of all Cauchy sequences. The points of $\widehat{\mathsf{M}}$ are again equivalence classes of Cauchy sequences, where two Cauchy sequences $x := \{x_\nu \,|\, \nu \in \mathbb{N}\}$ and $y := \{y_\nu \,|\, \nu \in \mathbb{N}\}$ are equivalent, in symbols, $x \sim y$, if $\lim_{\nu \to \infty} d(x_\nu, y_\nu) = 0$. The set M is dense in $\widehat{\mathsf{M}}$, i.e., for every $\widehat{x} \in \widehat{\mathsf{M}}$ and for all $\varepsilon > 0$, there exists an $x \in \mathsf{M}$ so that $d(\widehat{x}, x) < \varepsilon$.

One important observation is that every normed linear space (V, p) is also a metric space, where the metric $d : \mathsf{V} \times \mathsf{V} \to \mathbb{R}$ is defined by $d(u, v) := p(u - v)$, $u, v \in \mathsf{V}$. In particular, all the normed linear spaces introduced in Section 1.1 are metric spaces and those that are Banach spaces are also complete metric spaces. However, not every metric needs to arise from a norm. A simple example of a metric not definable by a norm is the *discrete metric* $\delta : \mathsf{M} \times \mathsf{M} \to \mathbb{R}$ given on *any* nonempty set M by

$$\delta(x, y) := \begin{cases} 1, & x \neq y; \\ 0, & x = y. \end{cases}$$

In the context of metric spaces, we introduce the following geometric and topological concepts. Suppose that (M, d) is a metric space and $\emptyset \neq A \subseteq \mathsf{M}$. The *diameter* diam (A) of A is defined by

$$\operatorname{diam}(A) := \sup\{d(x, y) \,|\, x, y \in A\}.$$

The set $B(x, r) := \{y \in \mathsf{M} \,|\, d(x, y) < r\}$ is called the *open ball around x of radius $r > 0$*. A collection $\mathcal{U}$ of open balls is called a *cover* of the metric space (M, d) provided $\mathsf{M} = \bigcup_{B \in \mathcal{U}} B$. The cover $\mathcal{U}$ is called *finite*, if the number of open balls in $\mathcal{U}$ is finite. A collection $\mathcal{V}$ of open balls is called a *subcover of* $\mathcal{U}$ if $\mathcal{V}$ is a cover of M and $V \in \mathcal{V}$ implies $V \in \mathcal{U}$. A metric space (M, d) is called *compact* if every cover of M by open balls contains a finite subcover. An equivalent characterization of compactness in metric spaces is: (M, d) is compact iff every sequence $\{x_\nu \,|\, \nu \in \mathbb{N}\} \subset \mathsf{M}$ has a convergent subsequence $\{x_{\nu_k} \,|\, k \in \mathbb{N}\}$.

The following results concerning continuous mappings between metric spaces are used later in this text.

Proposition 1.33 Let (M, d) be a compact metric space and (M', d') an arbitrary metric space. Suppose that the mapping $f : (\mathsf{M}, d) \to (\mathsf{M}', d')$ is continuous. Then the set $A \subseteq \mathsf{M}'$ is open iff $f^{-1}(A) \subseteq \mathsf{M}$ is open.

Proof Sufficiency follows immediately from the fact that f is continuous, thus mapping open sets to open sets. For the necessity, we observe that if $f^{-1}(A)$ is open, its complement $\complement f^{-1}(A)$ in M is closed and, being a closed subset of a compact set, compact. The continuity of f now implies that $f(\complement f^{-1}(A))$ is compact, hence closed in M'. As $f(\complement f^{-1}(A)) = \mathsf{M}' \setminus A$, A is open in M'. $\quad\square$

Proposition 1.34 Let (M_1, d_1) be a compact metric space, (M_2, d_2) and (M_3, d_3) arbitrary metric spaces. Suppose that the mapping $f : (\mathsf{M}_1, d_1) \to (\mathsf{M}_2, d_2)$ is continuous. Furthermore, suppose that $g : (\mathsf{M}_2, d_2) \to (\mathsf{M}_3, d_3)$ is a mapping so that $g \circ f : (\mathsf{M}_1, d_1) \to (\mathsf{M}_3, d_3)$ is continuous. Then $g : (\mathsf{M}_2, d_2) \to (\mathsf{M}_3, d_3)$ is also continuous.

Proof Exercise! (*Hint: Use Proposition 1.33!*) $\quad\square$

The next theorem is a fundamental result in functional analysis with far reaching implications in the theory of differential equations and dynamical systems, and is one of the cornerstones in the construction of fractal sets and fractal functions.

Theorem 1.35 (Banach Fixed Point Theorem) Let (M, d) be a nonempty complete metric space, and let $f : (\mathsf{M}, d) \to (\mathsf{M}, d)$ be a mapping with the property that there exists a $\lambda \in [0, 1)$ so that for all $x, x' \in \mathsf{M}$

$$d(f(x), f(x')) \leq \lambda\, d(x, x').$$

Then f has exactly one fixed point $x^* \in \mathsf{M}$, i.e., $f(x^*) = x^*$. Moreover, if $x_0 \in \mathsf{M}$ is arbitrary, then the sequence $\{x_k \mid k \in \mathbb{N}\}$ of iterates defined by $x_k := f(x_{k-1})$, $k \in \mathbb{N}$, converges to x^*.

Proof Exercise! $\quad\square$

1.4.2 Approximation Methods

We first require a few basic definitions.

Definition 1.36 (Approximation Method) Suppose that (X, d) is a metric space and $\mathsf{Y} \subseteq \mathsf{X}$ a nonempty subset. Moreover, suppose that $T : \mathsf{X} \to \mathsf{Y}$ is a mapping associating with every $x \in \mathsf{X}$ a $y := y_T := Tx \in \mathsf{Y}$. The mapping T is called an *approximation method*. In case that Y is a linear subspace of X and T is a linear mapping, T is called a *linear approximation method*. The element

$y = y_T \in \mathsf{Y}$ is referred to as an approximation to x or as an approximant of x (under T).

Among all possible approximants y to a given $x \in \mathsf{X}$, provided there are any, we like to single out those that are "closest" to x (in the sense specified below).

Definition 1.37 (Best Approximation)   Assume that T is an approximation method. An element $y^* \in \mathsf{Y}$ is called a *best approximation* or a *best approximant* to $x \in \mathsf{X}$ if

$$d(x, y^*) = \inf\{d(x, y) \mid y \in \mathsf{Y}\} =: \mathrm{dist}(x, \mathsf{Y}).$$

Definition 1.38 (Projection Method)   Suppose that T is an approximation method. In case that $Ty = y, \forall y \in \mathsf{Y}$, T is called *projection method*.

Let us consider an example of a linear projection method.

Example 1.39   For this purpose, assume that $X := \{a := x_0 < x_1 < \cdots x_{n-1} < x_n =: b\} \subseteq \mathbb{R}$ is a knot set. Define $\mathsf{X} := C[a, b]$ and $\mathsf{Y} := \Pi^n$, for a fixed $n \in \mathbb{N}$. Let $T : C[a, b] \to \Pi^n$ be given by

$$(Tf)(x) := [x_0]f + [x_0, x_1]f \cdot (x - x_0)$$
$$+ \cdots + [x_0, x_1, \ldots, x_n]f \cdot (x - x_0) \cdots (x - x_{n-1}),$$

the interpolating polynomial to f with respect to X in Newton form. Then, T is a linear projection method and the approximant $\Pi^n \ni p_n := Tf$ is unique by Theorem 1.22.

The case where the linear space X are normed is of particular interest.

Theorem 1.40   Suppose $(\mathsf{X}, \|\ \|)$ is a normed linear space, $\mathsf{Y} \subseteq \mathsf{X}$ a linear subspace, and $T : \mathsf{X} \to \mathsf{Y}$ a linear projection method. Then the following estimate holds for all $x \in \mathsf{X}$:

$$\|x - Tx\| \leq (1 + \|T\|)\,\mathrm{dist}(x, \mathsf{Y}). \tag{1.18}$$

Proof   Let $x \in \mathsf{X}$ and $y \in \mathsf{Y}$ be its approximant under T. Then, we have under consideration that $Ty = y$,

$$\|x - Tx\| = \|(x - y) - T(x - y)\| = \|(I - T)(x - y)\| \leq \|I - T\| \cdot \|x - y\|.$$

Taking the supremum over all $y \in \mathsf{Y}$ produces (1.18). (Here and in the following, $I : \mathsf{X} \to \mathsf{X}$ denotes the identity operator $x \mapsto x$.) $\square$

Remark 1.41   The norm $\|T\|$ of the linear projection method T is also called the *Lebesgue constant* of T.

1.4.3 Existence and Uniqueness of a Best Approximation

The next lemma is instrumental in establishing the existence of a best approximation.

Lemma 1.42 Suppose that Y is a compact subset of a metric space (X, d). Then there exists to each $x \in \mathsf{X}$ a best linear approximation $y^* := y_x^* \in \mathsf{Y}$.

Proof Fix $x \in \mathsf{X}$ and set $\delta := \inf\{d(x, y) \mid y \in \mathsf{Y}\}$. We can always choose a sequence $\{y_n \mid n \in \mathbb{N}\} \subseteq \mathsf{Y}$ so that $d(x, y_n) \to \delta$ as $n \to \infty$. The compactness of Y implies that the sequence $\{y_n \mid n \in \mathbb{N}\}$ has at least one limit point $y^* \in \mathsf{Y}$. Since a metric is translation-invariant, we may assume without loss of generality that $d(y_n, y^*) \to 0$ as $n \to \infty$. Thus, for all $\varepsilon > 0$,

$$d(x, y^*) \leq d(x, y_n) + d(y_n, y^*) \leq \delta + \varepsilon,$$

which proves the claim. $\square$

Theorem 1.43 (Existence of a Best Linear Approximation) Suppose that Y is a finite-dimensional subspace of a metric space (X, d). Then there exists to each $x \in \mathsf{X}$ a best linear approximation $y^* := y_x^* \in \mathsf{Y}$.

Proof Let $x \in \mathsf{X}$ and choose $y_0 \in \mathsf{Y}$ arbitrary. The set $\mathsf{Y}_0 := \{y \in \mathsf{Y} \mid \|x - y\| \leq \|x - y_0\|\} \subseteq \mathsf{Y}$ is closed and bounded in X, and since Y is finite dimensional, also compact by the Theorem of Heine[16]-Borel[17]. Lemma 1.42 now guarantees the existence of a best linear approximation $y^* \in \mathsf{Y}$ to $x \in \mathsf{X}$. (Note that a best linear approximation must lie in Y_0.) $\square$

Although Theorem 1.43 guarantees the existence of a best linear approximation, this best linear approximation however need *not* be unique. We next present an example of this scenario. (See also [159].)

Example 1.44 (Non-Uniqueness of a Best Linear Approximation) Let $\mathsf{X} := \mathbb{R}^2$ with norm $\|(u, v)\| := \max\{|u|, |v| \mid u, v \in \mathbb{R}\}$ and let $\mathsf{Y} := \{(u, 0) \mid u \in \mathbb{R}\}$. Let $x := (0, 1) \in \mathsf{X}$. Then every $y := (u, 0) \in \mathsf{Y}$ with $|u| \leq 1$ has distance 1 from x, hence is a best linear approximation. (The approximation method $T : \mathsf{X} \to \mathsf{Y}$ is the projection onto the first coordinate and thus linear.) The situation is graphically depicted in Figure 1.3.

The reason for the approximant not being unique lies in the fact that the norm used in this example is not strictly convex.

The above example justifies the next definition.

---

16 Eduard Heine, 15 March 1821–21 October 1881. German mathematician who worked primarily in potential theory, complex analysis, and the theory of partial differential equations.
17 Félix Édouard Justin Émile Borel, 7 January 1871–3 February 1956. French mathematician who made fundamental contributions to measure theory, probability theory, and game theory.

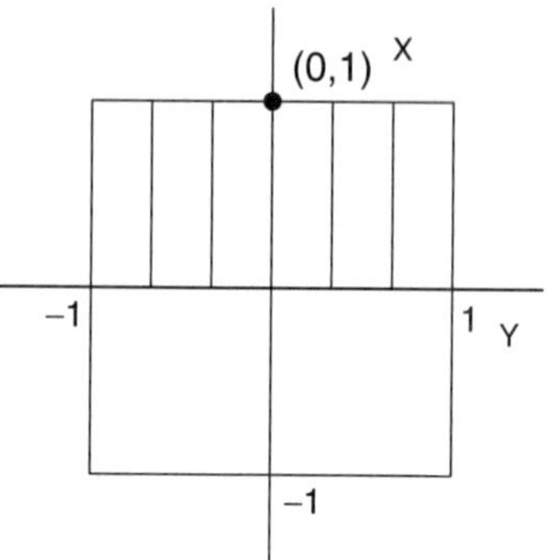

Figure 1.3   The best linear approximations to $(0, 1) \in \mathsf{X}$ from Y.

Definition 1.45 (Strict Convexity)   A normed linear space $(\mathsf{X}, \| \; \|)$ is called *strictly convex* if $\forall x_1, x_2 \in \mathsf{X}$ with $x_1 \neq x_2$ and $\|x_1\| = \|x_2\| = 1$

$$\|x_1 + x_2\| < 2.$$

Before proceeding, let us look at a few examples of normed linear spaces that are strictly convex and some that are not.

Example 1.46   The normed linear spaces $(L^p[a, b], \| \; \|_{L^p})$ are strictly convex for $p \in (1, \infty)$.

Example 1.47   The Lebesgue spaces $(L^1[a, b], \| \; \|_{L^1})$, $(L^\infty[a, b], \| \; \|_{L^\infty})$, and the normed linear space $(C[a, b], \| \; \|_{\infty,[a,b]})$ are *not* strictly convex.

Now we are ready to prove the uniqueness of a best linear approximation under the assumption of strict convexity.

Theorem 1.48 (Uniqueness of a Best Linear Approximation)   Suppose that Y is a finite-dimensional subspace of a strictly convex normed linear space $(\mathsf{X}, \| \; \|)$. Then there exists to each $x \in \mathsf{X}$ a unique best linear approximation $y^* := y_x^* \in \mathsf{Y}$.

Proof   The existence of a best linear approximation follows from Theorem 1.43. To show uniqueness, we assume to the contrary that for some $x \in \mathsf{X}$, $y^* := y_x^*$ and $y^{**} := y_x^{**}$ are two best approximations from Y with $\|x - y^*\| = \|x - y^{**}\| =: \delta$. Then,

$$\left\| x - \frac{y^* + y^{**}}{2} \right\| = \left\| \frac{(x - y^*)}{2} + \frac{(x - y^{**})}{2} \right\|$$

$$< \frac{\delta}{2} \left(\left\| \frac{x - y^*}{\delta} \right\| + \left\| \frac{x - y^{**}}{\delta} \right\| \right) = \delta,$$

in contradiction to y^* and y^{**} being best approximations.   □

Next, we consider the issue of a unique best linear approximation in the context of inner product spaces or pre-Hilbert spaces, as they offer additional geometric insights.

1.4.4 Pre-Hilbert Spaces

Before we define pre-Hilbert spaces, we need to introduce *bilinear forms.* These play an important role in approximation theory.

Definition 1.49 (Bilinear Form)   Assume that V and W are linear spaces over the field of reals $\mathbb{R}$. A functional $B : V \times W \to \mathbb{R}$ is called a *bilinear form* iff it satisfies the following axioms.

1. $B(v_1 + v_2, w) = B(v_1, w) + B(v_2, w), \forall v_1, v_2 \in V$ and $\forall w \in W$;

2. $B(v, w_1 + w_2) = B(v, w_1) + B(v, w_2), \forall v \in V$ and $\forall w_1, w_2 \in W$;

3. $B(\lambda v, w) = B(v, \lambda w) = \lambda B(v, w), \forall \lambda \in \mathbb{R}, \forall v \in V$, and $\forall w \in W$.

A bilinear form B is called *symmetric* if $B(v, w) = B(w, v), \forall v \in V$ and $\forall w \in W$.

If $W := V$, then the bilinear form $B : V \times V \to \mathbb{R}$ is called *positive definite* if $B(v, v) \geq 0, \forall v \in V$, and $B(v, v) = 0$ iff $v = 0$. Finally, a symmetric positive bilinear form $B(\bullet, \bullet)$ is called an *inner product on* V and is usually denoted by $\langle \, , \, \rangle$.

Example 1.50   As an example, we consider $V := W := \mathbb{R}^n, n \in \mathbb{N}$, and define for $x := (x_1, \ldots, x_n)^\top, y := (y_1, \ldots, y_n)^\top \in \mathbb{R}^n, B(x, y) := \sum_{i=1}^{n} x_i y_i$. Then, B is a symmetric positive bilinear form, thus an inner product, on $\mathbb{R}^n$.

Example 1.51   Another example of a bilinear form is given by setting $V := W := C^1[0, 1]$ and by defining $B : C^1[0, 1] \times C^1[0, 1] \to \mathbb{R}$ by

$$B(f, g) := \int_0^1 [f(x)g(x) + (Df)(x)(Dg)(x)]dx.$$

Bilinear forms are used to define *pre-Hilbert spaces* or, as they are sometimes called, *inner product spaces.*

Definition 1.52 (Pre-Hilbert Space and Hilbert Space)   A linear space H over the field of reals $\mathbb{R}$ endowed with a symmetric positive definite bilinear form $\langle \, , \, \rangle : H \times H \to \mathbb{R}$ is called a *(real) pre-Hilbert space* $(H, \langle \, , \, \rangle)$. A pre-Hilbert space is complete iff every Cauchy sequence in H converges to an element of H. Complete pre-Hilbert spaces are called *Hilbert spaces.*

Note that every pre-Hilbert space $(\mathsf{H}, \langle\,,\,\rangle)$ is a normed space and thus also a metric space, since we can define a norm $\|\,\| : \mathsf{H} \to \mathbb{R}$ on V by setting

$$\|v\| := \sqrt{\langle v, v\rangle}.$$

Hence, every Hilbert space is a Banach space and a complete metric space. The inner product also allows the definition of geometric concepts such as orthogonality and orthonormal basis. To this end, let $\mathsf{E} \subseteq \mathsf{H}$ and denote by

$$\mathsf{E}^{\perp} := \{h \in \mathsf{H} \mid \langle e, h\rangle = 0, \forall e \in \mathsf{E}\}$$

the orthogonal complement of E in H. It is not difficult to show that if $(\mathsf{H}, \langle\,,\,\rangle)$ is a Hilbert space, then the following decomposition holds: $\mathsf{H} = \mathsf{E} \oplus \mathsf{E}^{\perp}$ since $\mathsf{E}^{\perp}$ is a closed subspace of H.

Two elements f, g in a pre-Hilbert space $(\mathsf{H}, \langle\,,\,\rangle)$ are called *orthogonal*, in symbols, $f \perp g$, iff $\langle f, g\rangle = 0$, and a sequence $\{e_i \mid i \in \mathbb{N}\} \subseteq \mathsf{H}$ an *orthonormal system* iff $\langle e_i, e_j\rangle = \delta_{ij}$, $i, j \in \mathbb{N}$. If $(\mathsf{H}, \langle\,,\,\rangle)$ is in addition a Hilbert space, then every element $h \in \mathsf{H}$ has a unique representation of the form

$$h = \sum_{i=1}^{\infty} \langle h, e_i\rangle\, e_i.$$

In this case, $\{e_i \mid i \in \mathbb{N}\}$ is called an *orthonormal basis of* H.

Example 1.53   The linear spaces $\mathbb{R}^n$, $n \in \mathbb{N}$, are pre-Hilbert spaces under the inner product $\langle\,,\,\rangle : \mathbb{R}^n \times \mathbb{R}^n \to \mathbb{R}$, $\langle \boldsymbol{x}, \boldsymbol{y}\rangle := \sum_{i=1}^{n} x_i y_i$, for $\boldsymbol{x} := (x_1, \ldots, x_n)^{\top}$, $\boldsymbol{y} := (y_1, \ldots, y_n)^{\top} \in \mathbb{R}^n$. It is not difficult to show that these spaces are also complete, hence Hilbert spaces.

Example 1.54   The sequence spaces $\ell^2(\mathbb{N})$ and $\ell^2(\mathbb{Z})$ are examples of infinite-dimensional Hilbert spaces. An inner product $\langle\,,\,\rangle_{\ell^2} : \ell^2(\mathbb{N}) \times \ell^2(\mathbb{N}) \to \mathbb{R}$ for $\ell^2(\mathbb{N})$ is defined by

$$\langle \boldsymbol{x}, \boldsymbol{y}\rangle_{\ell^2} := \sum_{i=1}^{\infty} x_i y_i.$$

Similarly, one defines an inner product for the spaces $\ell^2(\mathbb{Z})$. A straight-forward computation shows that both spaces are complete, hence Hilbert spaces.

Example 1.55   The Lebesgue space $L^2(\Omega)$, where $\Omega \subseteq \mathbb{R}^n$, $n \in \mathbb{N}$, is a Hilbert space. The inner product $\langle\,,\,\rangle_{L^2} : L^2(\Omega) \times L^2(\Omega) \to \mathbb{R}$ is defined by

$$\langle f, g\rangle_{L^2} := \int_{\Omega} f(\boldsymbol{x}) g(\boldsymbol{x}) d\boldsymbol{x}.$$

Definition 1.56 (Adjoint of Linear Operator) Let $(H, \langle \, , \, \rangle)$ be a Hilbert space and $T : H \to H$ a bounded linear operator. The bounded linear operator $T^* : H \to H$ which satisfies

$$\langle Tx, y \rangle = \langle x, T^*y \rangle, \quad \forall x, y \in H,$$

is called the *adjoint (operator)* of T.

Remark 1.57 The Riesz representation theorem guarantees that to each bounded linear operator T from a Hilbert space into itself there exists a unique adjoint operator T^*.

Definition 1.58 (Unitary Operator) Let $(H, \langle \, , \, \rangle)$ be a Hilbert space and $T : H \to H$ a bounded linear operator. Then T is called *unitary* iff $TT^* = T^*T = I$, where T^* is the adjoint of T and I the identity operator on H.

Remark 1.59 If T is a unitary operator on a Hilbert space $(H, \langle \, , \, \rangle)$, then T leaves the inner product $\langle \, , \, \rangle$ invariant, in the sense that,

$$\langle Tx, Ty \rangle = \langle x, y \rangle, \quad \forall x, y \in H.$$

This then implies that T is an isometry, i.e., a linear operator with the property that

$$\| Tx \| = \| x \|, \quad \forall x \in H.$$

Here $\| \ \|$ denotes the norm that is induced by the inner product $\langle \, , \, \rangle$.

1.4.5 Approximation in Pre-Hilbert Spaces

Lemma 1.60 Assume that $(X, \langle \, , \, \rangle)$ is a pre-Hilbert space. Then the induced norm $\| x \| := \langle x, x \rangle$, $x \in X$, is strictly convex.

Proof Let $x_1, x_2 \in X$ with $x_1 \neq x_2$ and $\| x_1 \| = \| x_2 \| = 1$. Then

$$\left\langle \frac{x_1 + x_2}{2}, \frac{x_1 - x_2}{2} \right\rangle = \frac{\| x_1 \|^2}{4} - \frac{\| x_2 \|^2}{4} = 0,$$

and, thus, $\frac{1}{2}(x_1 + x_2) \perp \frac{1}{2}(x_1 - x_2)$. Since $\frac{1}{2}(x_1 + x_2) = x_1 - \frac{1}{2}(x_1 - x_2)$, we obtain

$$\left\| \frac{x_1 + x_2}{2} \right\|^2 = \| x_1 \|^2 - \left\| \frac{x_1 - x_2}{2} \right\|^2 < \| x_1 \| = 1,$$

proving the lemma. $\square$

The following theorem gives a geometric interpretation of best linear approximation in the context of pre-Hilbert spaces. (See also Figure 1.4.)

Theorem 1.61 (Best Linear Approximation in Hilbert Space) Assume that Y is a (not necessarily finite dimensional) proper subspace of a pre-Hilbert space

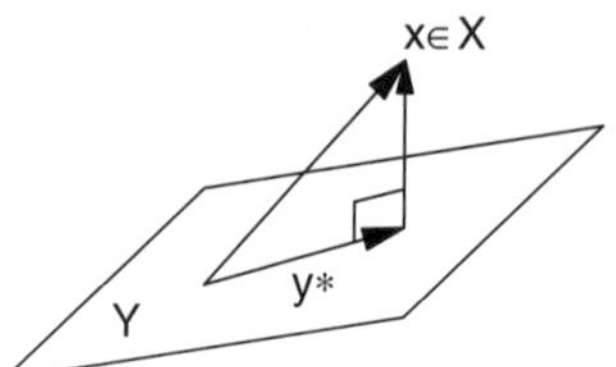

Figure 1.4 Geometric interpretation of a best approximation in a pre-Hilbert space.

$(\mathbf{X}, \langle \, , \, \rangle)$. For each $x \in \mathbf{X}$, an element $y^* := y_x^* \in \mathbf{X}$ is a best linear approximation to x iff

$$\forall y \in \mathbf{Y} : \langle x - y^*, y \rangle = 0 \quad \Longleftrightarrow \quad x - y^* \perp \mathbf{Y}.$$

Proof Let $x \in \mathbf{X}$ and $y^* \in \mathbf{Y}$ be such that $\langle x - y^*, y \rangle = 0$, for all $y \in \mathbf{Y}$. Choose a $y_0 \in \mathbf{Y}$ and take $y := y^* - y_0$. Then, under the above assumption,

$$\|x - y_0\|^2 = \|(x - y^*) + (y^* - y_0)\|^2 = \|x - y^*\|^2 + \|y^* - y_0\|^2 > \|x - y^*\|^2,$$

implying that y^* is a best linear approximation.

Now, suppose that $\langle x - y^*, y \rangle \neq 0$, for some $y \in \mathbf{Y}$. Consider the following expression, with $c \in \mathbb{R}$ to be chosen later:

$$\|(x - y^*) + cy\|^2 = \|x - y^*\|^2 + 2c\langle x - y^*, y \rangle + c^2 \|y\|^2. \tag{1.19}$$

Now choose $c := \dfrac{\langle x - y^*, y \rangle}{\|y\|^2}$. Then the right-hand side of (1.19) becomes

$$\|x - y^*\|^2 - c^2 \|y\|^2 < \|x - y^*\|^2.$$

Hence, y^* is not best linear approximation to $x \in \mathbf{X}$. $\quad\square$

An immediate consequence of Theorem 1.61 is the next result.

Corollary 1.62 Let $\mathbf{Y}$ be a proper subspace of a pre-Hilbert space $(\mathbf{X}, \langle \, , \, \rangle)$ and let $y^* := y_x^* \in \mathbf{Y}$ be a best approximation to $x \in \mathbf{X}$. Then,

$$\|x - y^*\|^2 + \|y^*\|^2 = \|x\|^2.$$

If $x \in \mathbf{X} \setminus \mathbf{Y}$, then $\|y^*\|^2 < \|x\|^2$.

This corollary shows that a best approximation $y^* \in \mathbf{Y} \subsetneq \mathbf{X}$ can never equal, in norm, the element $x \in \mathbf{X}$ that is being approximated. The question now arises whether one can find a sequence of increasing subspaces $\{\mathbf{Y}_n \mid n \in \mathbb{N}\} \subseteq \mathbf{X}$ so that the associated sequence of best approximations $\{y_n^* \in \mathbf{Y}_n \mid n \in \mathbb{N}\}$ approaches x in norm. To this end, we need to consider the concept of *approximation order*.

1.4.6 Order of Approximation

Let (X, d) be a metric space and $\{Y_n \mid n \in \mathbb{N}\} \subseteq X$ a sequence of closed subspaces of X with the property that $Y_n \subseteq Y_{n+1}$, $\forall n \in \mathbb{N}$. Such a sequence of closed subspaces is called a *filtration of* X or a *ladder of subspaces of* X.

The error of approximation for a fixed $x \in X$ by an element of best approximation $y_n^* \in Y_n$ is defined by

$$E_n(x) := E_n(x; Y_n) := \mathrm{dist}(x, Y_n) = \inf\{d(x, y_n) \mid y_n \in Y_n\}.$$

In the case of a pre-Hilbert space $(X, \langle\,,\,\rangle)$, Theorem 1.61 shows that the error $x - y_n^*$ of representing the element $x \in X$ by an element $y_n^* \in Y_n$ of best approximation lies in the orthogonal complement $Y_n^\perp$ of Y_n.

The following questions may now be asked.

Question 1.   How does the error of approximation E_n depend on n and does $E_n \to 0$ as $n \to \infty$, for an appropriately chosen sequence of subspaces $\{Y_n \mid n \in \mathbb{N}\}$ of X?

Question 2.   What is the rate of convergence of the error of approximation as compared to the reference sequences $\{n^{-\alpha} \mid n \in \mathbb{N}\}$?

To investigate these questions, a definition is needed first.

Definition 1.63 (Approximation Order)   If the error of approximation $E_n(x) \in \mathcal{O}(n^{-\alpha})$ for some $\alpha \in \mathbb{R}_0^+$, then α is called the *order of approximation* (of $x \in X$ relative to Y_n).

Here, $\mathcal{O}$ is one of the Landau[18] symbols, formally defined as

$$f \in \mathcal{O}(g) \quad \Longleftrightarrow \quad 0 \le \limsup_{x \to a} \left|\frac{f(x)}{g(x)}\right| < \infty, \tag{1.20}$$

where f and g either denote two sequences of real numbers, in which case $x \in \mathbb{N}$ and $a = \infty$, or real functions of a real variable x in which case $a \in \overline{\mathbb{R}} := \mathbb{R} \cup \{-\infty\} \cup \{+\infty\}$.

Definition 1.64 (Approximation Spaces)   Suppose that (X, d) is a metric space and that $\{Y_n \mid n \in \mathbb{N}\} \subseteq X$ is a filtration of X. The linear spaces Y_n, $n \in \mathbb{N}$, are called *approximation spaces* for X.

In order to characterize approximation spaces for a given metric space (X, d), there are two types of theorems: *direct* or *Jackson*[19]*-type theorems* and

---

18 Edmund Georg Hermann Landau, 14 February 1877–19 February 1938. German mathematician who is known for his contributions to analytic number theory.
19 Durham Jackson, 1888–1946. American mathematician who worked in the theory of trigonometric and orthogonal polynomials and in approximation theory.

inverse or *Bernstein*[20]*-type theorems.* These can be described as follows. To this end, let Y be an approximation space of (X, d).

Jackson-type Theorems: $x \in \mathsf{Y} \implies E_n(x; \mathsf{Y}) \in \mathcal{O}(n^{-\alpha})$, for some $\alpha \in \mathbb{R}_0^+$;

Bernstein-type Theorems: $E_n(x; \mathsf{Y}) \in \mathcal{O}(n^{-\alpha})$, for some $\alpha \in \mathbb{R}_0^+ \implies x \in \mathsf{Y}$.

As an example of the versatility of direct and inverse theorems, we consider an important class of function spaces, namely the *Lipschitz*[21] *spaces.*

Assume that $\alpha \in (0, 1]$, X compact, and $(\mathsf{X}, \| \, \|)$ a normed linear space. Suppose that $f : \mathsf{X} \to \mathbb{R}$ is a function satisfying

$$|f(x) - f(y)| \le c \, \|x - y\|^{\alpha}, \quad \forall x, y \in \mathsf{X}, \tag{1.21}$$

for some positive constant $c \in \mathbb{R}$. Then f is called *Hölder continuous with exponent α.* For $\alpha = 1$, f is called *Lipschitz continuous.* The smallest constant c for which (1.21) holds, is called the *Lipschitz constant for f* and is denoted by $\mathrm{Lip}(f)$.

The set of all Hölder continuous functions with exponent α will be denoted by

$$\mathrm{Lip}^{\alpha}(\mathsf{X}) := \mathrm{Lip}^{\alpha}(\mathsf{X}, \mathbb{R})$$

$$:= \left\{ f : \mathsf{X} \to \mathbb{R} \, | \, \exists c \in \mathbb{R}^+ \, \forall x, y \in \mathsf{X} : |f(x) - f(y)| \le c \, \|x - y\|^{\alpha} \right\}$$

The set $\mathrm{Lip}^{\alpha}(\mathsf{X})$ can be made into a semi-normed linear space via the semi-norm

$$|f|_{\mathrm{Lip}^{\alpha}} := \sup \left\{ \frac{|f(x) - f(y)|}{\|x - y\|^{\alpha}} : x, y \in \mathsf{X}, x \ne y. \right\}$$

Note that $|f|_{\mathrm{Lip}^{\alpha}}$ is the smallest of all positive numbers for which (1.21) holds and that $|f|_{\mathrm{Lip}^{\alpha}} = 0$ iff $f = \mathrm{const}$. This latter observation suggests to define a *norm* on the linear spaces $\mathrm{Lip}^{\alpha}(\mathsf{X})$ by setting $\| \, \|_{\mathrm{Lip}^{\alpha}} : \mathrm{Lip}^{\alpha}(\mathsf{X}) \to \mathbb{R}$,

$$\|f\|_{\mathrm{Lip}^{\alpha}} := \|f\|_{\infty, \mathsf{X}} + |f|_{\mathrm{Lip}^{\alpha}}.$$

It can be shown that the normed linear spaces $(\mathrm{Lip}^{\alpha}(\mathsf{X}), \| \, \|_{\mathrm{Lip}^{\alpha}})$ are Banach spaces. (The reader is encouraged to verify this statement.)

It is also easy to verify that every Hölder continuous function is uniformly continuous. The Lipschitz spaces provide therefore a means of resolving the space of continuous functions into subspaces with increasing regularity. More precisely,

$$\mathrm{Lip}^{\beta}(\mathsf{X}) \subseteq \mathrm{Lip}^{\alpha}(\mathsf{X}) \subsetneq C(\mathsf{X}), \quad \text{for all } 0 < \alpha \le \beta \le 1. \tag{1.22}$$

---

20 SERGEI NATANOVICH BERNSTEIN, 5 March 1880–26 October 1968. Ukrainian mathematician who solved Hilbert's 19th Problem. He worked in probability theory, constructive function theory, and the mathematical foundations of genetics.
21 RUDOLF OTTO SIGISMUND LIPSCHITZ, 14 May 1832–7 October 1903. German mathematician who contributed to mathematical analysis, differential geometry, and classical mechanics.

(The reader is encouraged to verify this statement.) If we denote the set of all *bounded functions* $f : \mathbf{X} \to \mathbb{R}$ by $B(\mathbf{X})$, we have the following set inclusions:

$$\mathrm{Lip}^1(\mathbf{X}) \subsetneq \mathrm{Lip}^\alpha(\mathbf{X}) \subsetneq \mathrm{Lip}^0(\mathbf{X}) := B(\mathbf{X}),$$

for all $0 < \alpha < 1$. Because of the above inclusions, the spaces $\mathrm{Lip}^\alpha(\mathbf{X})$ are also denoted by $C^{0,\alpha}(\mathbf{X})$.

Example 1.65 The function $\sqrt{\ } : [0, 1] \to \mathbb{R}$, $x \mapsto \sqrt{x}$, is not Lipschitz continuous, but Hölder continuous for all exponents $0 < \alpha \le 1/2$. (The reader is encouraged to verify this statement.)

Example 1.66 The space $\mathrm{Lip}^\alpha(\mathbf{X})$ with $\alpha > 1$ consists only of the constant functions on $\mathbf{X}$. (The reader is encouraged to verify this statement.)

The inclusions in (1.22) show that the sequence of linear spaces $\{\mathrm{Lip}^\alpha(\mathbf{X}) \,|\, \alpha \in (0, 1]\}$ forms a (continuous) filtration of the space $C(\mathbf{X})$ where larger indices indicate higher regularity. However, also note that $\mathrm{Lip}^1(\mathbf{X}) \ne C(\mathbf{X})$. (See Example 1.65 for a validation of this claim.)

1.4.7 The Weierstrass Approximation Theorem

The above considerations have shown that a filtration of a metric space $(\mathbf{X}, d)$ may be such that its "last" element (with respect to the ordering defining the filtration) is not equal to, or more precisely, cannot be identified with $\mathbf{X}$. Better filtrations are those for which $\bigcup\{\mathbf{Y}_n \,|\, n \in \mathbb{N}\}$ is dense in $\mathbf{X}$ (with respect to the metric topology). The problem essentially reduces to the following setting.

Given an element of a Banach space $(\mathbf{X}, \|\ \|)$, determine a subspace $\mathbf{Y}$ of $\mathbf{X}$ with the property that $\mathbf{Y}$ is dense in $\mathbf{X}$ with respect to the norm topology, i.e.,

$$\forall \varepsilon > 0 \ \forall x \in \mathbf{X} \ \exists y \in \mathbf{Y}: \quad \|x - y\| < \varepsilon.$$

One of the first and also one of the most important approximation theorems in this setting is due to Weierstrass.

Theorem 1.67 (Weierstrass Approximation Theorem) Let $J \subseteq \mathbb{R}$ be a compact interval and $f \in C(J)$, endowed with the Chebyshev norm $\|\ \|_{\infty, J}$. Then $\bigcup\{\Pi^n \,|\, n \in \mathbb{N}\}$ is dense in $C(J)$. In other words,

$$\forall \varepsilon > 0 \ \forall f \in C(J); \ \exists (n = n(\varepsilon) \in \mathbb{N} \wedge p_n \in \Pi^n(J)) : \|f - p_n\|_{\infty, J} < \varepsilon.$$

Proof We prove this theorem for $J := [-1, 1]$ using the Landau kernels L_ν of order ν, which are defined as follows.

$$L_\nu(x) := c_\nu (1 - x^2)^\nu, \quad x \in [-1, 1] \tag{1.23}$$

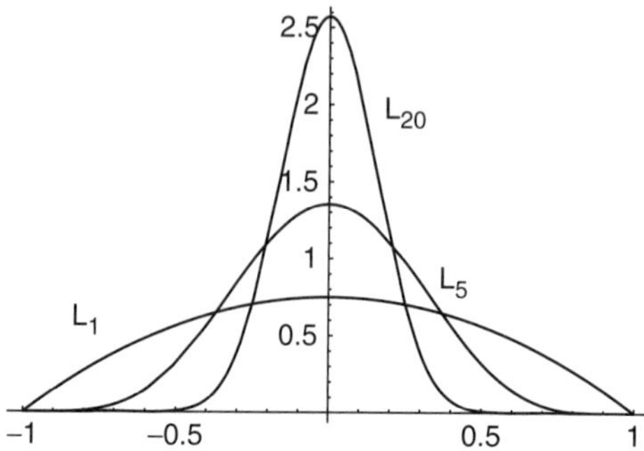

Figure 1.5   The Landau kernels L_1, L_5 and L_{20}.

where $v \in \mathbb{N}$ and c_v is chosen so that $\int_{-1}^{+1} L_v(x)dx = 1$. An explicit expression for the constant c_v is given by

$$c_v = \frac{\Gamma(v + \frac{3}{2})}{\sqrt{\pi}\, \Gamma(v + 1)}.$$

Using the fact that $\dfrac{\Gamma(n^\alpha)}{\Gamma(n^\beta)} \in \mathcal{O}(n^{\alpha - \beta})$, $\alpha, \beta \in \mathbb{R}$, (Cf. for instance [1].) we see that

$$c_v \in \mathcal{O}(\sqrt{v}). \tag{1.24}$$

Figure 1.5 shows the graphs of the Landau kernels L_1, L_5 and L_{20}.

For the purposes of this proof and for later needs, we introduce the *convolution* between two functions. To this end, let $f, g : K \subseteq \mathbb{R} \to \mathbb{R}$ and define

$$(f * g)(x) := \int_K f(x - t)g(t)\, dt,$$

where f and g are set equal to zero outside K. By a change of variables, it is easy to see that $f * g = g * f$.

The Landau kernels have the property that their convolution with a continuous function yields a polynomial, as described in the lemma below.

Lemma 1.68 $L_v * f$ is a polynomial in Π^{2v+1}, for all functions $f \in C[-1, 1]$.

Proof   First, we note that $L_v \in \Pi^{2v+1}$ and, therefore, $L_v(x - t)$ can be written in the form $L_v(x - t) = a_0(t) + a_1(t)x^2 + \cdots + a_v(t)x^{2v}$, for some polynomials $a_0(t), \ldots, a_v(t)$. Thus,

$$(L_v * f)(x) = \int_{-1}^{+1} L_v(x - t)f(t)\, dt$$

$$= \int_{-1}^{+1} \left[a_0(t) + a_1(t)\, x^2 + \cdots + a_v(t)\, x^{2v}\right] f(t)\, dt,$$

producing a polynomial in x of degree $\leq 2v$. $\square$

The next lemma shows that the polynomials $L_\nu * f$ provide approximations to $f \in C[-1, 1]$ and, indeed, converge to f.

Lemma 1.69 The sequence $\{L_\nu * f \mid \nu \in \mathbb{N}\} \subseteq \Pi^{2\nu+1}$ converges in the Chebyshev norm to $f \in C[-1, 1]$.

Proof Let $x \in [a, b]$ and consider

$$f(x) - (L_\nu * f)(x) = \int_{-1}^{+1} [f(x)L_\nu(t) - f(x-t)L_\nu(t)]dt$$

$$= \int_{-1}^{+1} [f(x) - f(x-t)]L_\nu(t)dt.$$

Since the Landau kernels L_ν become more concentrated around $x = 0$ when the index ν increases, we write the last integral above as

$$\int_{-1}^{+1} [f(x) - f(x-t)]L_\nu(t)dt$$

$$= \int_{-1}^{-\delta} [f(x) - f(x-t)]L_\nu(t)dt$$

$$+ \int_{-\delta}^{+\delta} [f(x) - f(x-t)]L_\nu(t)dt$$

$$+ \int_{+\delta}^{+1} [f(x) - f(x-t)]L_\nu(t)dt, \quad \text{for some } \delta > 0.$$

The idea now is to use the uniform continuity of f on $[-1, 1]$ to find such a $\delta > 0$ that works for all $x \in [-1, 1]$ and to show that for large ν all three integrals can be made less than some given $\varepsilon > 0$. To estimate the first and third integral above, one needs to employ (1.24). We leave the details to the reader. $\square$

Combining Lemmas 1.68 and 1.69 yields the Weierstrass Approximation Theorem. $\square$

We can use the Weierstrass Approximation Theorem to give another characterization of Lipschitz spaces. For this purpose, let $X := [a, b]$ and $f \in (C[a, b], \| \ \|_{\infty,[a,b]})$. We consider approximations of $C[a, b]$ by elements from Π^n noting that $\{\Pi^n \mid n \in \mathbb{N}\}$ is a filtration of $C[a, b]$. The error of best approximation of $f \in C[a, b]$ by $p_n \in \Pi^n$ is given by

$$E_n(f; \Pi^n) = \inf\{\|f - p_n\|_{\infty,[a,b]} \mid p_n \in \Pi^n\},$$

and by the Weierstrass Approximation Theorem we know that there exists for any given $\varepsilon > 0$ an $n = n(\varepsilon) \in \mathbb{N}$ so that $E_n(f; \Pi^n) < \varepsilon$. We can characterize the Lipschitz spaces by the rate at which the error of approximation goes to zero. The precise statement of this interesting fact is as follows:

$$\mathrm{Lip}^{\alpha}[a, b] = \{f : [a, b] :\to \mathbb{R} \mid E_n(f; \Pi^n) \in \mathcal{O}(n^{-\alpha})\}. \qquad (1.25)$$

The above equality actually means that there exist positive constants c_1 and c_2 so that for all $f \in \mathrm{Lip}^{\alpha}[a, b]$

$$c_1 \sup\{n^{\alpha} E_n(f; \Pi^n) \mid n \in \mathbb{N}\} \leq \|f\|_{\mathrm{Lip}^{\alpha}} \leq c_2 \sup\{n^{\alpha} E_n(f; \Pi^n) \mid n \in \mathbb{N}\}.$$

For, the first inequality states a direct or Jackson-type estimate and the second inequality an inverse or Bernstein-type estimate: If $\|f\|_{\mathrm{Lip}^{\alpha}} < \infty$, i.e., $f \in \mathrm{Lip}^{\alpha}[a, b]$, then the error of approximation decays like $n^{-\alpha}$, and conversely, if the error of approximation decays like $n^{-\alpha}$, i.e., $\sup\{n^{\alpha} E_n(f; \Pi^n) \mid n \in \mathbb{N}\} < \infty$, then the norm $\|f\|_{\mathrm{Lip}^{\alpha}} < \infty$, i.e., $f \in \mathrm{Lip}^{\alpha}[a, b]$. Both arguments show that the right-hand side of (1.25) can be identified with $\mathrm{Lip}^{\alpha}[a, b]$.

The example above illustrates a common type of argumentation using the direct and inverse estimates to establish connections between the rate of decay of the error of approximation and the approximation spaces themselves. For more details and an in-depth discussion, we refer the interested reader to the references given at the beginning of this chapter.

1.4.8 Approximation by Step Functions

Here we show that the metric space $(C[a, b], \|\,\|_{\infty,[a,b]})$ can be approximated by a sequence of spaces V_n, $n \in \mathbb{N}$, whose elements are *step functions*.

Definition 1.70 (Step Function)   Let $[a, b]$ be a nonempty interval in $\mathbb{R}$ and let $\{A_k \mid k = 1, \ldots, n\}$ be a partition of $[a, b]$ by intervals, i.e., $A_k = [\alpha_k, \beta_k]$, $a \leq \alpha_k < \beta_k \leq b$, $k = 1, \ldots, n$, $\bigcup_{k=1}^{n} A_k = [a, b]$, and $A_k \cap A_l = \emptyset$, for $k \neq l$. A function $s : [a, b] \to \mathbb{R}$ of the form

$$s(x) := s_n(x) := \sum_{k=1}^{n} a_k \, \chi_{A_k}(x),$$

where $a_k \in \mathbb{R}$, $k = 1, \ldots, n$, is called a *step function*.

We now assume, without loss of generality, that $a = 0$ and $b = 1$, and construct a nested sequence of linear spaces $\{\mathsf{V}_n \mid n \in \mathbb{N}\}$ as follows. Let $n \in \mathbb{N}$ and let $A_{nk} := [\frac{k-1}{2^n}, \frac{k}{2^n})$, $k = 1, \ldots, 2^n - 1$, and $A_n := [\frac{2^n-1}{2^n}, 1]$. Define sets

$$\mathsf{V}_n := \left\{ s_n := \sum_{k=1}^{2^n} a_k \, \chi_{A_{nk}} \in \mathrm{Map}([0, 1], \mathbb{R}) \,\middle|\, a_k \in \mathbb{R} \right\}, \qquad n \in \mathbb{N}. \qquad (1.26)$$

It follows immediately from (1.26) that the sets V_n are linear spaces (under the usual addition and scalar multiplication of functions) and nested: $\mathsf{V}_n \subset \mathsf{V}_{n+1}$, $n \in \mathbb{N}$.

Theorem 1.71   Let $f \in C[0, 1]$. Then f is the uniform limit of a sequence of step functions $s_n \in V_n$, $n \in \mathbb{N}$.

Proof   Let $n \in \mathbb{N}$ and choose an arbitrary $\xi_{nk} \in A_{nk}$, $k = 1, \ldots, 2^n$. Set

$$s_n(x) := \sum_{k=1}^{2^n} f(\xi_{nk}) \, \chi_{A_{nk}}(x), \quad n \in \mathbb{N}.$$

The fact that f is uniformly continuous on $[0, 1]$ now yields the result. (The reader is encouraged to verify this statement.) $\square$

1.4.9 Moduli of Continuity and Smoothness

Many approximation-theoretic results are stated in terms of quantities called *moduli of continuity* and *moduli of smoothness*. The modulus of continuity, respectively, the modulus of smoothness, allows for a very precise characterization of the local continuity, respectively, the local regularity of a function. Here, we introduce these quantities briefly and state some of their properties. As an application, we present the famous theorem of Jackson, regarding the error of the approximation of C^k-functions by polynomials.

Definition 1.72 (Modulus of Continuity)   Let $[a, b] \subseteq \mathbb{R}$ be a nonempty interval and suppose that $f \in \mathrm{Map}([a, b], \mathbb{R})$. For a given $h > 0$, the expression

$$\omega(f, h) := \sup_{0 < t < h} \sup\{|f(x + t) - f(x)| \,|\, x, x + t \in [a, b]\}$$

is called the *modulus of continuity* of f.

The modulus of continuity of a function f enjoys the following properties.

Proposition 1.73   Let $f \in \mathrm{Map}([a, b], \mathbb{R})$ and let $\omega(f, h)$, $h > 0$, be its modulus of continuity. Then

1. $\omega(f, \bullet)$ is a nonnegative and nondecreasing function of h;

2. If $f \in C[a, b]$, then $\omega(f, \bullet)$ is a continuous function of h;

3. $\omega(f, h)$ is subadditive in its first argument: $\omega(f + g, h) \leq \omega(f, h) + \omega(g, h)$;

4. $f \in C[a, b]$ implies $\omega(f, h) \leq 2\|f\|_{\infty, [a,b]}$;

5. Let $n \in \mathbb{N}$. Then $\omega(f, nh) \leq n\omega(f, h)$;

6. Suppose that $\lambda \in \mathbb{R}$. Then, $\omega(f, \lambda h) \leq (\lambda + 1)\omega(f, h)$;

7. If $f \in C^1[a, b]$, then $\omega(f, h) \leq h\|Df\|_{\infty, [a,b]}$.

Proof Exercise! □

In particular, we have the following characterizations of continuity and Hölder continuity.

$$f \in C[a, b] \quad \Longleftrightarrow \quad \forall \varepsilon > 0 \; \exists h > 0 : \omega(f, h) < \varepsilon \tag{1.27}$$

$$f \in \mathrm{Lip}^{\alpha}[a, b] \quad \Longleftrightarrow \quad \omega(f, h) \leq c\, h^{\alpha}, \text{ some } c \in \mathbb{R}^{+}. \tag{1.28}$$

The next theorem is one of the cornerstones of modern approximation theory.

Theorem 1.74 (Jackson) Let $k, n \in \mathbb{N}$, where $n > k + 1$ and suppose that $f \in C^{k}[a, b]$. Then

$$E_n(f; \Pi^n) \leq \left(\frac{6(3e)^k}{1 + k}\right) \left(\frac{b - a}{n - 1}\right)^k \omega\left(D^k f, \frac{b - a}{2(n - 1 - k)}\right) \in \mathcal{O}(n^{-k}). \tag{1.29}$$

Here, e denotes the Euler [22] constant.

Proof See, for instance, [129]. □

For our later purposes, we also need to introduce the *modulus of smoothness* of a function f. To this end, the *forward difference operator* has to be defined first.

Definition 1.75 (Forward Difference Operator) Assume that $f \in \mathrm{Map}(\mathbb{R}, \mathbb{R})$. The *$k$th-order forward difference operator* $\Delta_t^k : \mathrm{Map}(\mathbb{R}, \mathbb{R}) \to \mathrm{Map}(\mathbb{R}, \mathbb{R})$ with step size $t > 0$ is recursively given by

$$(\Delta_t f)(x) := f(x + t) - f(x)$$
$$\Delta_t^k f := \Delta_t(\Delta_t^{k-1} f),$$

with $\Delta_t^1 := \Delta_t$.

The forward difference operator Δ_t^k of order k has the following explicit representations.

$$(\Delta_t^k f)(x) = \sum_{i=0}^{k} (-1)^{k-i} \binom{k}{i} f(x + it); \tag{1.30}$$

---

22 LEONHARD EULER, 15 April 1707–18 September 1783. Swiss mathematician who made enormous contributions to a wide range of mathematical and physical theories. Amongst these are mathematical analysis, number theory, geometry, graph theory, mechanics, hydrodynamics, and optics.

$$(\Delta_{nt}^k f)(x) = \sum_{i_1=0}^{n-1} \cdots \sum_{i_k=0}^{n-1} (\Delta_t^k f)(x + i_1 t + \cdots + i_k t); \tag{1.31}$$

$$(\Delta_t^k f)(x) = \int_0^h \cdots \int_0^h (D^k f)(x + u_1 + \cdots + u_k)\, du_1 \cdots du_k. \tag{1.32}$$

The above representations can be proven by induction on k. (The reader is encouraged to verify these formulae.)

Definition 1.76 (Modulus of Smoothness) Suppose that $f \in \mathrm{Map}([a, b], \mathbb{R})$ is defined on a nonempty interval in $\mathbb{R}$ and $h > 0$. The *kth-order modulus of smoothness* of f is defined by

$$\omega_k(f, h) := \sup_{0 < t < h} \sup\{|\Delta_t^k f| \mid x, x + kt \in [a, b]\},$$

with $\omega(f, h) := \omega_1(f, h)$.

The modulus of smoothness enjoys properties similar to those of the modulus of continuity. These are

1. If $f \in C[a, b]$, then $\omega_k(f, \bullet)$ is a continuous, nonnegative, and nondecreasing function of h;

2. For all $n \in \mathbb{N}$, $\omega_k(f, nh) \le n^k \omega_k(f, h)$;

3. For all $\lambda \in \mathbb{R}$, $\omega_k(f, \lambda h) \le (\lambda + 1)^k \omega_k(f, h)$;

4. If $f \in C^k[a, b]$, then $\omega_k(f, h) \le h^k \|D^k f\|_{\infty, [a,b]}$.

The reader is encouraged to provide the proofs of these properties.

Exercises

1. Establish the inequalities in (1.1).

2. Justify Eqn. (1.2).

3. Use known properties of $\mathbb{R}$ and of continuous functions on a compact set, to prove that $(C(K), \| \;\|_{\infty, K})$ is a Banach space.

4. Investigate how one can define a norm on $C^k[a, b]$ using the semi-norms p_m, $m = 0, 1, \ldots, k$, defined in (1.3).

5. Find an example of a function $f \in C(a, b)$ that is not bounded and not uniformly continuous on (a, b), hence has no bounded and uniformly continuous extension to $[a, b]$.

6. Let $1 < p < \infty$. Show that $\| \;\|_{L^p} : C(\Omega) \to \mathbb{R}$ defines a norm on $C(\Omega)$. (*Hint: Use the Minkowski Inequality!*)

7. Exhibit a sequence $\{f_\nu \mid \nu \in \mathbb{N}\} \subsetneq C(\Omega)$ that does not converge in the L^p-norm to an element of $C(\Omega)$.

8. Show that there exists discontinuous linear functionals. This then implies that $X' \neq X^*$.(*Hint: Let $X := C^1[0, 1]$ endowed with the Chebyshev norm. Denote by $\Lambda : C^1[0, 1] \to \mathbb{R}$, the functional that evaluates the derivative of $f \in C^1[0, 1]$ at $x = 0$: $\Lambda(f) := (Df)(0)$. Consider the sequence $\{f_\nu \mid \nu \in \mathbb{N}\}$ of functions in $C^1[0, 1]$, where $f_\nu(x) := \frac{\sin \nu^2 x}{\nu}$, and show that $\Lambda f_\nu \xrightarrow[\nu \to \infty]{} \infty$ although $f_\nu \xrightarrow[\nu \to \infty]{} 0$.)*

9. Prove Theorem 1.11.

10. Establish *Young's* [23] *Inequality:* Suppose that $1 < p, q < \infty$ with $\frac{1}{p} + \frac{1}{q} = 1$. Further suppose that $A, B \in \mathbb{R}_0^+$. Then

$$
A B \leq \frac{A^p}{p} + \frac{B^q}{q},
$$

with equality iff $A^p = B^q$. (*Hint: Take $f(x) := x^{p-1}$ and consider the area between the graph of f above the interval $[0, A]$ on the x-axis, and the area enclosed by the interval $[0, B]$ on the y-axis and graph f.)*

11. Prove the Hölder Inequality. (*Hint: The cases $p = 1$ and $p = \infty$ are straight-forward. For $1 < p, q < \infty$, use Young's Inequality with $A := |f(x)|/\|f\|_{L^p, \Omega}$ and $B := |g(x)|/\|f\|_{L^q, \Omega}$, $x \in \Omega$.)*

12. Show the validity of the Minkowski Inequality. (*Hint: The cases $p = 1$ and $p = \infty$ are trivial. For $1 < p < \infty$, first show that $|f + g| \leq 2^p(|f|^p + |g|^p)$. Then use this result and the Hölder Inequality to establish the Minkowski Inequality.)*

13. Prove Theorem 1.22.

14. Show that the ternary Cantor Set is a null set.

15. Let $k \in \mathbb{N}$ and $m \in \mathbb{N}_0$. Show that $\Pi^k \subsetneq C^m(\mathbb{R})$ and that $\{1, x, \ldots, x^{k-1}\}$ is a basis for Π^k.

16. Prove Theorem 1.26. [*Hint: Suppose $x \neq x_\nu$, $\nu = 0, 1, \ldots, n$. Define a function $g := f - p - K \prod_{\nu=0}^{n}(x - x_\nu)$, where $K \in \mathbb{R}$. Use Rolle's [24] Theorem applied to g and its derivatives to determine the constant K.]*

17. Prove Lemma 1.27.

18. Prove Proposition 1.34.

19. Prove the Banach Fixed Point Theorem. (*Hint: Uniqueness follows from the fact that $0 \leq \lambda < 1$. To show existence of a fixed point x^*, prove that*

---

23 WILLIAM HENRY YOUNG, 20 October 1863–7 July 1942. British mathematician who worked in measure theory and Fourier series. He is best known for his major contributions to the theory of functions of several complex variables.
24 MICHEL ROLLE, 21 April 1652–8 November 1719. French mathematician who contributed to analysis, geometry, and the theory of diophantine equations.

$\{d(f^m(x_0), f^n(x_0)) \mid m, n \in \mathbb{N})\}$ *is a Cauchy sequence and use the continuity of* f. *Here* f^n *denotes the n-fold composition of* f *with itself:*
$$f^n := \underbrace{f \circ \cdots \circ f}_{n-times}.)$$

20. Show that the spaces $(L^p[a, b], \| \ \|_{L^p})$ are strictly convex for $1 < p < \infty$.

21. For each of the spaces $(L^1[a, b], \| \ \|_{L^1})$ and $(C[a, b], \| \ \|_{\infty, [a,b]})$, find an example that shows that the space is not strictly convex.

22. Establish the claim made in Example 1.50.

23. Establish the claim made in Example 1.51.

24. Establish the claims made in Example 1.53.

25. Establish the claims made in Example 1.54.

26. Establish the claim made in Example 1.55.

27. Show that the set $\{e^{in\bullet} \mid n \in \mathbb{Z}\}$ of complex exponentials forms an orthonormal basis of $L^2(0, 2\pi)$.

28. Given $(\mathbb{R}^3, \langle \ , \ \rangle)$ as a Hilbert space with the inner product defined as Example 1.53, compute the best linear approximation of $x := (5, -1, 3)^\top$ from the plane $2x + y - z = 1$.

29. Work out Example 1.65.

30. Prove the assertion in Example 1.66.

31. Complete the proof of Lemma 1.69.

32. Let $\{a := x_0 < x_1 < \cdots < x_n =: b\}$ be a knot set. Show that

    (a) $[x_0, \ldots, x_n]p = 0, \ \forall p \in \Pi^n$;
    (b) $[x_0, \ldots, x_n]p = \text{const.}, \ \forall p \in \Pi^{n+1}$.

33. Let $\{a := x_0 < x_1 < \cdots < x_n =: b\}$ be a knot set and $f : [a, b] \to \mathbb{R}$ a function. Prove that

$$[x_0, \ldots, x_n]f = \sum_{v=0}^{n} \frac{f(x_v)}{(D\omega)(x_v)},$$

    where $D := \frac{d}{dx}$ and $\omega(x) := \prod_{v=0}^{n}(x - x_v)$.

34. Let $\{a := x_0 < x_1 < \cdots < x_n =: b\}$ be a knot set with $x_v = x_0 + v$, $v = 1, \ldots, n$, and let $f \in \text{Map}([a, b], \mathbb{R})$. Prove the following explicit representation of the divided difference operator:

$$[x_0, \ldots, x_n]f = \frac{1}{n!} \sum_{v=0}^{n} (-1)^{v-1} \binom{n}{v} f(x_v).$$

35. Let $\{a := x_0 < x_1 < \cdots < x_n =: b\}$ be a knot set and $f \in C^n[a, b]$. Denote by Σ^n the n-simplex

$$\Sigma^n := \left\{ (u_0, u_1, \ldots, u_n) \in \mathbb{R}^n \,\middle|\, \forall i = 0, 1, \ldots, n : u_i \geq 0 \text{ and } \sum_{i=0}^{n} u_i = 1 \right\},$$

and, for any $g \in C(\Sigma^n)$, define

$$\int_{\Sigma^n} g(u)\,du := \int_0^1 \int_0^{1-t_1} \cdots \int_0^{1-t_1-\cdots-t_{n-1}}$$
$$\times\, g(1 - t_1 - \cdots - t_n, t_1, \ldots, t_n)\,dt_1 \cdots dt_n. \tag{1.33}$$

Prove the Hermite-Genocchi[25,26] formula

$$[x_0, \ldots, x_n]f = \int_{\Sigma^n} (D^n f)(u_0 x_0 + \cdots + u_n x_n)\,du. \tag{1.34}$$

(Hint: Use induction over n. Note that for $n = 1$ on has:

$$f(x_0) - f(x_1) = \int_0^1 (D_{x_0-x_1}f)(x_0 + t(x_1 - x_0))\,dt$$

$$= (x_0 - x_1) \int_0^1 (Df)((1-t)x_0 + tx_1)\,dt$$

$$= (x_0 - x_1) \int_{\Sigma^1} (Df)(u_0 x_0 + u_1 x_1)\,du.$$

Here, $D_{x_0-x_1}f := \langle Df, x_0 - x_1 \rangle$ denotes the directional derivative of a differentiable function f in the direction of $x_0 - x_1$).

36. Use the Hermite-Genocchi formula (1.34) to show that $\mathrm{vol}(\Sigma^n) = (n!)^{-1}$.

37. Show that $\mathrm{Lip}^\beta(\mathbf{X}) \subseteq \mathrm{Lip}^\alpha(\mathbf{X})$, for all $0 < \alpha \leq \beta \leq 1$.

38. Show that the function $\sqrt{} : [0, 1] \to \mathbb{R}$, $x \mapsto \sqrt{x}$, is not Lipschitz continuous, but Hölder continuous for all exponents $0 < \alpha \leq 1/2$. From this conclude that $\mathrm{Lip}^1(\mathbf{X}) \neq C(\mathbf{X})$.

39. Prove the claim made in Example 1.66.

40. Suppose that $[a, b] \subseteq \mathbb{R}$ is a nonempty interval and $f \in C^1[a, b]$. Show that the Lipschitz constant of f is given by $\sup\{(Df)(a + \vartheta(b - a)) \,|\, 0 < \vartheta < 1\}$. As a consequence one obtains that $C^1[a, b] \subseteq \mathrm{Lip}^1[a, b]$. Give an example showing that this inclusion is strict.

---

25 CHARLES HERMITE, 24 December 1822–14 January 1901. French mathematician who worked in number theory and algebra, orthogonal polynomials and elliptic functions. He proved that e is transcendental.

26 ANGELO GENOCCHI, 5 March 1817–7 March 1889. Italian mathematician whose main area of research was number theory.

41. Suppose that $(X, \langle\,,\,\rangle)$ is a pre-Hilbert space and Y a finite-dimensional subspace of X. Let $\{y_1, \ldots, y_n\}$ be a basis of Y. Show that the orthogonality condition in Theorem 1.61 provides a means of obtaining the best approximation $y^* = \sum_{i=1}^{n} c_i y_i \in Y$ for $x \in X$. Compute this best approximation.

42. Complete the proof of Theorem 1.71.

43. Prove Proposition 1.73.

44. Prove the assertions made in (1.27) and (1.28).

45. Establish the representations (1.30), (1.31), and (1.32). (*Hint: For the first, use the binomial identity* $\binom{k+1}{i} = \binom{k}{i-1} + \binom{k}{i}$.)

46. Show that if $f \in C^k[a, b]$, then (1.31) implies (1.32). (*Hint: Set $\eta := nh$ in (1.31) and let $n \to \infty$.*)

47. Show Properties $1 - 4$ of the modulus of smoothness. $\square$

2

Splines

This chapter is an introduction to the theory of splines and their applications. We present some aspects of this theory and discuss several properties of spline functions. In particular, we will define spline spaces and show that B-splines provide a natural basis for such spline spaces. The properties of B-splines are discussed and special cases, such as cardinal B-splines, fundamental cardinal splines, and Hermite-splines presented. We also consider repeated knot sequences and introduce the associated piecewise polynomial spaces.

Next, we investigate interpolation by splines and prove the uniqueness of spline interpolation for properly chosen knot sets and derive error estimates for the interpolation. In connection with these issues, we also show the minimality property of splines. Then, we consider approximation of functions by splines, in particular B-splines, and compute the approximation errors.

An alternative approach to defining polynomial splines, namely via certain linear differential operators, and the extension to exponential splines is introduced next. For this purpose, the concept of distributional derivative needed to be presented. It is shown that exponential splines form a filtration of $L^2(\mathbb{R})$ and an error estimate for asymptotic approximation is stated. The generalization to so-called $\mathcal{L}$-splines, where $\mathcal{L}$ is a linear differential operator, is introduced next. In the last section, after defining wavelets and introducing the concept of multiresolution analysis, it is shown that a certain class of splines forms a multiresolution analysis of $L^2(\mathbb{R})$.

Only a small portion of spline theory can be presented in this text. The interested reader is referred to the naturally incomplete list of references [26, 41, 44, 126, 135, 137, 140, 148, 159] to the spline literature.

2.1 Definitions

Let $n \in \mathbb{N}$ and let $a := x_0 < x_1 < \ldots < x_n =: b$ be points in $\mathbb{R}$. The set of points $X_n := \{x_\nu \mid \nu = 0, 1, \ldots, n\}$ is called a *knot set* and its elements *knots*. The knots $x_1, \ldots, x_{n-1}$ are named *interior knots* and the knots x_0 and x_n *boundary knots*.

Definition 2.1 (Spline and Spline Space) Let $k, n \in \mathbb{N}$ and let X_n be a knot set. A function $s : [a, b] \to \mathbb{R}$ is called a (*polynomial*) *spline of degree* $k - 1$ or of *order* k if it satisfies the following two conditions.

(1) $s \in C^{k-2}[a, b]$;

(2) $\forall \nu = 0, 1, \ldots, n - 1 \; \forall x \in [x_\nu, x_{\nu+1}) : s \in \Pi^k[a, b]$.

The linear space of all k-th order polynomial splines or, equivalently, the linear space of all polynomial splines of degree $< k$, with knot set X_n is denoted by $S^k(X_n)$.

If $k := 1$, then $C^{-1}[a, b]$ is taken as the linear space of piecewise continuous functions $f \in \mathrm{Map}([a, b], \mathbb{R})$.

Note that if we reformulate the two conditions above by defining $\Pi^k(X_n)$ to be the linear space of all functions whose restriction to each interval $(x_\nu, x_{\nu+1})$, $\nu = 0, 1, \ldots, n - 1$, is an element of $\Pi^k[a, b]$, then we obtain the following equivalent definition of spline space, namely,

$$S^k(X_n) = \Pi^k(X_n) \cap C^{k-2}[a, b].$$

Remark 2.2 Some authors use the polynomial degree $(k - 1)$ or even the degree of smoothness $(k - 2)$ instead of the polynomial order (k) to index spline spaces.

Remark 2.3 Condition (2) in Definition 2.1 can also be stated as follows. Let $D := \dfrac{d}{dx}$ be the univariate differential operator. Then

$$(2) \quad \Longleftrightarrow \quad \forall \nu = 0, 1, \ldots, n - 1 : s \text{ is a solution of } D^k y = 0 \text{ on } [x_\nu, x_{\nu+1})$$

$$\Longleftrightarrow \quad \forall \nu = 0, 1, \ldots, n - 1 : s \in \ker D^k \text{ on } [x_\nu, x_{\nu+1}).$$

We remark that each polynomial $p \in \Pi^k$ is automatically a spline of order k for any knot set X_n.

Let $-\infty < a < b < +\infty$. Every knot set X_n with $x_0 := a$ and $x_n := b$ induces *a partition* of the interval $[a, b]$. We denote this partition again with X_n. More precisely,

$$[a, b) = \bigcup_{\nu=0}^{n-1} [x_\nu, x_{\nu+1}).$$

2.2 A Basis for $S^k(X_n)$

The definition of spline implies that $S^k(X_n) \subseteq C^{k-2}[a, b]$ for all k, $n \in \mathbb{N}$, where $\subseteq$ denotes the vector space inclusion. Since every polynomial $p \in \Pi^k$ is also a spline of order k, we arrive at the following inclusions valid for any knot set X_n with first element $x_0 = a$ and last element $x_n = b$.

$$\forall k, n \in \mathbb{N}: \quad \Pi^k[a, b] \subseteq S^k(X_n) \subseteq C^{k-2}[a, b].$$

To prove that the inclusion on the left is indeed strict, we need to exhibit functions that are in $S^k(X_n)$ but not in Π^k. To this end, define functions

$$q_{vk} : [a, b] \to \mathbb{R}$$

$$q_{vk}(x) := (x - x_v)_+^{k-1}, \quad v = 0, 1, \ldots, n-1,$$

where

$$x_+ := \max\{x, 0\} \quad \text{and} \quad x_+^m := (x_+)^m, \ \forall m \in \mathbb{N}.$$

This definition implies that $q_{vk} \in C^{k-2}[a, b]$ and

$$\forall v = 0, 1, \ldots, n-1: \quad q_{vk}|_{[x_v, x_{v+1})} \in \Pi^k[a, b] \quad \text{but} \quad q_{vk}|_{[a,b]} \notin \Pi^k[a, b].$$

The functions q_{vk} are called *truncated power functions*. In Figure 2.1 the graphs of $q_{0,2}$ and $q_{2,4}$ are displayed.

Theorem 2.4 The set $\mathcal{B}^0 := \{1, x, \ldots, x^{k-1}\} \cup \{q_{vk} \mid v = 1, \ldots, n-1\}$ constitutes a basis for the spline space $S^k(X_n)$ for any knot set X_n. The dimension of $S^k(X_n)$ is therefore $\dim S^k(X_n) = n + k - 1$.

Proof Let X_n be a fixed knot set and $s \in S^k(X_n)$. We will show inductively that for each interval of the form $[x_0, x_m]$, $m = 1, \ldots, n$, there exist numbers $a_0, a_1, \ldots, a_{k-1} \in \mathbb{R}$ and $b_1, \ldots, b_{m-1} \in \mathbb{R}$ so that $s|_{[x_0, x_m]}$ has the unique representation

$$s(x) = \sum_{\mu=0}^{k-1} a_\mu x^\mu + \sum_{v=1}^{m-1} b_v (x - x_v)_+^{k-1}. \tag{2.1}$$

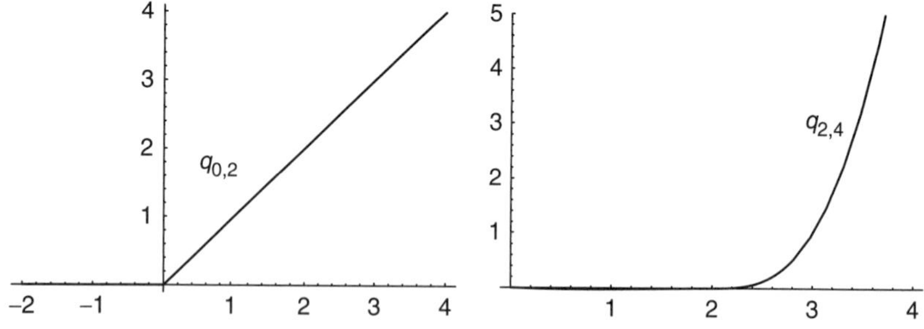

Figure 2.1 The graphs of two truncated power functions.

For $m := 1$, we have that $s|_{[x_0, x_1]} \in \Pi^k$ giving the unique representation

$$s(x) = \sum_{\mu=0}^{k-1} a_\mu x^\mu + \sum_{\nu=1}^{0} b_\nu (x - x_\nu)_+^{k-1},$$

where the second sum is identical to zero since $\sum_{\emptyset} := 0$.

For $m \in \{2, \ldots, n\}$ and $x \in [x_0, x_m]$, we assume that we have a representation of the form (2.1). Define

$$r(x) := s(x) - \sum_{\mu=0}^{k-1} a_\mu x^\mu - \sum_{\nu=1}^{m-1} b_\nu (x - x_\nu)_+^{k-1}. \tag{2.2}$$

Then $r \in C^{k-2}[x_0, x_{m+1}]$ and $r \equiv 0$ on $[x_0, x_m]$. For $x \in [x_m, x_{m+1}]$, $r \in \Pi^k$, but this is equivalent to saying that r is the solution of the linear differential equation $D^k y = 0$ with initial conditions $y(x_m) = (Dy)(x_m) = \cdots = (D^{k-2}y)(x_m) = 0$. The uniquely determined solution of this initial value problem is given by

$$r(x) = -b_m (x - x_m)_+^{k-1},$$

for some constant $b_m \in \mathbb{R}$ and for $x \in [x_0, x_{m+1}]$. This, together with (2.2), then proves the validity of the representation for $m + 1$. $\qquad\square$

The basis $\mathcal{B}^0$, however, has some definite disadvantages.

(i) The basis functions exhibit no symmetry.

(ii) The basis functions have infinite support.

(iii) There is no apparent relationship between the geometry and the coefficients in the representation (2.1).

(iv) It can be shown that the basis $\mathcal{B}^0$ is numerically not stable, i.e., the coefficients in (2.1) may be small but the function values large.

For these reasons, one searches for a better basis for the spline space $S^k(X_n)$. One such basis will be constructed in the next section.

2.3 B-Splines

In this section, we introduce a more satisfactory basis for the spline space $S^k(X_n)$. The functions in this new basis have compact support, are numerically stable, and allow an efficient computation of coefficient functionals.

To this end, let $k := 1$ and let X_n be an arbitrary knot set. For $\nu \in \{0, 1, \ldots, n-1\}$, define the piecewise constant function $B_{\nu,1,X_n} : \mathbb{R} \to \mathbb{R}$ by

$$B_{\nu,1,X_n} := \begin{cases} 1, & x \in [x_\nu, x_{\nu+1}); \\ 0, & \text{elsewhere,} \end{cases}$$

and, for $\nu \in \{0, 1, n-2\}$, the piecewise linear function $B_{\nu,2,X_n} : \mathbb{R} \to \mathbb{R}$ by

$$
B_{\nu,2,X_n} := \begin{cases} \dfrac{x - x_\nu}{x_{\nu+1} - x_\nu}, & x \in [x_\nu, x_{\nu+1}); \\[2ex] \dfrac{x_{\nu+2} - x}{x_{\nu+2} - x_{\nu+1}}, & x \in [x_{\nu+1}, x_{\nu+2}); \\[2ex] 0, & \text{elsewhere.} \end{cases}
$$

Now compute

$$
[x_\nu, x_{\nu+1}, x_{\nu+2}](\bullet - x)_+ = \frac{[x_{\nu+1}, x_{\nu+2}](\bullet - x_+) - x_\nu, x_{\nu+1}](\bullet - x_+)}{x_{\nu+2} - x_\nu}
$$

$$
= \frac{\frac{(x_{\nu+2}-x)_+ - (x_{\nu+1}-x)_+}{x_{\nu-2}-x_{\nu-1}} - \frac{(x_{\nu+1}-x)_+ - (x_\nu-x)_+}{x_{\nu+1}-x_\nu}}{x_{\nu+2} - x_\nu}
$$

$$
= \begin{cases} \dfrac{x - x_\nu}{(x_{\nu+2} - x_\nu)(x_{\nu+1} - x_\nu)}, & x \in [x_\nu, x_{\nu+1}); \\[2ex] \dfrac{x_{\nu+2} - x}{(x_{\nu+2} - x_\nu)(x_{\nu+2} - x_{\nu+1})}, & x \in [x_{\nu+1}, x_{\nu+2}). \end{cases}
$$

Therefore, we can write

$$
B_{2,k,X_n} = (x_{\nu+2} - x_\nu)[x_\nu, x_{\nu+1}, x_{\nu+2}](\bullet - x)_+.
$$

This simple calculation justifies the next definition.

Definition 2.5 (Curry–Schoenberg[1,2], 1966)   Let $k, n \in \mathbb{N}$ and let

$$
X := \{a := x_0 \leq x_1 \leq \cdots \leq x_{n+k} =: b\}, \tag{2.3}
$$

with $x_\nu < x_{\nu+k}$, $\nu = 0, 1, \ldots, n$, be a knot set. The ν-*th B-Spline of order k* is defined by

$$
B_{\nu k, X} := (x_{\nu+k} - x_\nu)[x_\nu, \ldots, x_{\nu+k}](\bullet - x)_+^{k-1}.
$$

Using the fact that $[x_\nu, \ldots, x_{\nu+k}]f = \dfrac{[x_{\nu+1}, \ldots, x_{\nu+k}]f - [x_\nu, \ldots, x_{\nu+k-1}]f}{x_{\nu+k} - x_\nu}$, the B-splines $B_{\nu k, X}$ can also be expressed as

$$
B_{\nu k, X} = [x_{\nu+1}, \ldots, x_{\nu+k}](\bullet - x)_+^{k-1} - [x_\nu, \ldots, x_{\nu+k-1}](\bullet - x)_+^{k-1}. \tag{2.4}
$$

---

1 HASKELL CURRY, 12 September 1900–1 September 1982. American mathematician and logician who mostly worked in combinatoric logic.
2 ISAAC JACOB SCHOENBERG, 21 April 1903–21 February 1990. Romanian-American mathematician contributing to approximation theory and who is considered the founder of spline theory.

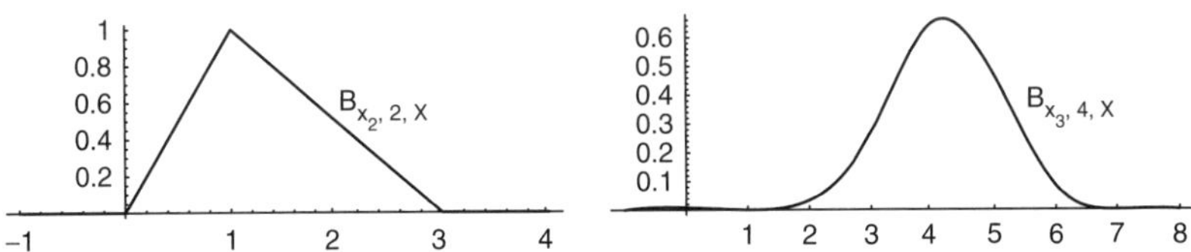

Figure 2.2   The graph of the linear B-splines $B_{x_2,2,X}$ and the cubic B-spline $B_{x_3,4,X}$.

There exists an alternate family of B-splines defined via

$$M_{vk,X} := k[x_v, \ldots, x_{v+k}](\bullet - x)_+^{k-1} = \frac{k}{x_{v+k} - x_v} B_{vk,X}. \tag{2.5}$$

Example 2.6   In Figure 2.2, we depict the graphs of two B-splines defined on the knot set $X := \{-2, 0, 1, 3, 4, 6, 7, 9\}$.

Remark 2.7   In the spline literature, the B-splines $B_{vk,X}$ are sometimes written $N_{vk,X}$ and referred to as normalized B-splines.

2.3.1 Properties of B-Splines

In this section, several properties of the B-splines are presented and the spline representation theorem is proved.

Elements of $S^k(X)$

For all $k, n \in \mathbb{N}$ and all knot sets X defined by (2.3), we have that $B_{vk,X}, M_{vk,X} \in S^k(X)$. We prove this assertion by using the following explicit representation of divided differences:

$$[x_v, \ldots, x_{v+k}]f = \sum_{\mu=v}^{v+k} \frac{f(x_\mu)}{(D\omega)(x_\mu)}, \tag{2.6}$$

where $\omega(x) := \prod_{\mu=v}^{v+k}(x - x_\mu)$ and D denotes the derivative with respect to x.

Eqn. (2.6) applied to $f := (\bullet - x)_+^{k-1}$ now yields

$$B_{vk,X} = (x_{v+k} - x_v) \sum_{\mu=v}^{v+k} \frac{(x_\mu - x)_+^{k-1}}{(D\omega)(x_\mu)}$$

and

$$M_{vk,X} = k \sum_{\mu=v}^{v+k} \frac{(x_\mu - x)_+^{k-1}}{(D\omega)(x_\mu)}.$$

As the truncated power functions are elements of $S^k(X)$ the inclusions are proved.

Peano Kernels for Divided Differences

Let X be defined as in (2.3) and let $f \in C^k[a, b]$. Then

$$[x_v, \ldots, x_{v+k}]f = \frac{1}{k!} \int_a^b M_{vk,X}(t)(D^k f)(t)\, dt. \qquad (2.7)$$

In other words, the B-spline $M_{vk,X}$ is the *Peano*[3] *kernel* of the integral representation of the divided difference operator $[x_v, \ldots, x_{v+k}](\bullet)$: $C^k[a, b] \to \mathbb{R}$.

Proof The Taylor expansion for f up to order $k - 1$ at $y \in (a, b)$ gives

$$f(x) = \sum_{\varkappa=0}^{k-1} \frac{(D^\varkappa f)(y)}{\varkappa!}(x - y)^\varkappa + \int_a^b (x - t)_+^{k-1} \frac{(D^k f)(t)}{(k-1)!}.$$

Now apply the divided difference operator $[x_v, \ldots, x_{v+k}]$ to both sides of this equation and take into account that $[x_v, \ldots, x_{v+k}](\bullet - y)^\mu = 0$ for $\mu < k$ (The reader is encouraged to verify this statement.). The definition of $M_{vk,X}$ then gives the result. $\square$

Partition of Unity

For all knot sets X of the form (2.3) we have that

$$\int_a^b M_{vk,X}(x)\,dx = 1, \qquad (2.8)$$

and

$$\sum_{v=1}^n B_{vk,X}(x) \equiv 1, \qquad \forall x \in [x_k, x_n]. \qquad (2.9)$$

Proof To show (2.8), apply (2.6) to the function $f(x) := x^k$ and use that $[x_v, \ldots, x_{v+k}]f = 1$. (The reader is encouraged to verify this statement.)

To prove (2.9), we use the representation (2.4) for $B_{vk,X}$ and sum over v. This yields a telescoping sum leaving only the first and last term:

$$\sum_{v=1}^n B_{vk,X} = [x_{n+1}, \ldots, x_{n+k}](\bullet - x)_+^{k-1} - [x_1, \ldots, x_k](\bullet - x)_+^{k-1}.$$

---

3 GUISEPPE PEANO, 27 August 1858–20 April 1932. Italian mathematician who is known for his work in mathematical logic, the axiomatization of the natural numbers (development of the Peano axioms), and first-order differential equations.

Now, if $x \geq x_k$ then $(t - x)_+^{k-1} = 0$ for $t = x_1, \ldots, x_k$, and, by Theorem 1.22, we have that $[x_1, \ldots, x_k](\bullet - x)_+^{k-1} = 0$. On the other hand, if $x \leq x_{n+1}$, then $(t - x)_+^{k-1} = (t - x)^{k-1}$ for t equal to $x_{n+1}, \ldots, x_{n+k}$. Theorem 1.22 and the fact that $[x_{n+1}, \ldots, x_{n+k}](\bullet - x)^{k-1} = 1$ gives the result. $\square$

Compactness of the Support

$$\operatorname{supp} M_{\nu k, X} = \operatorname{supp} B_{\nu k, X} = [x_\nu, x_{\nu+k}]. \tag{2.10}$$

Proof   The proof uses again the uniqueness of the interpolating polynomial and the annihilation properties of the divided difference operator. If $x \geq x_{\nu+k}$, then $(t - x)_+^{k-1} = 0$, for $t = x_\nu, \ldots, x_{\nu+k}$. Therefore, $[x_\nu, \ldots, x_{\nu+k}](\bullet - x)_+^{k-1} = 0$. However, if $x \leq x_\nu$, then $(t - x)_+^{k-1} = (t - x)^{k-1}$ for $t = x_\nu, \ldots, x_{\nu+k}$ implying that $[x_\nu, \ldots, x_{\nu+k}](\bullet - x)_+^{k-1} = 0$. $\square$

Recursion Formulae

The B-splines $B_{\nu k, X}$ and $M_{\nu k, X}$ satisfy recursion formulae, which will prove important later on and which are sometimes used to define the B-splines. For all $1 < k \in \mathbb{N}$, we have that

$$B_{\nu k, X} = \frac{x - x_\nu}{x_{\nu+k-1} - x_\nu} B_{\nu, k-1, X} + \frac{x_{\nu+k} - x}{x_{\nu+k} - x_{\nu+1}} B_{\nu+1, k, X} \tag{2.11}$$

and

$$M_{\nu k, X} = \frac{k}{k-1} \left[\frac{x - x_\nu}{x_{\nu+k} - x_\nu} M_{\nu, k-1, X} + \frac{x_{\nu+k} - x}{x_{\nu+k} - x_\nu} M_{\nu+1, k, X} \right]. \tag{2.12}$$

Proof   We will prove only the second recursion formula since then the first will be implied by the relationship between $B_{\nu k, X}$ and $M_{\nu k, X}$. To this end, notice that the following holds true for all $2 < k \in \mathbb{N}$: $(\bullet - x)_+^{k-1} = (\bullet - x)(\bullet - x)_+^{k-2}$.

Suppose that $f : [x_\nu, x_{\nu+k}] \to \mathbb{R}$ is a function. Define polynomials $p_0, p_1 \in \Pi^k$ with the property that p_0 interpolates the function f at the knots $\{x_\nu, \ldots, x_{\nu+k-1}\}$ and that p_1 interpolates the function f at the knots $\{x_{\nu+1}, \ldots, x_{\nu+k}\}$. Furthermore, denote by $p \in \Pi^{k+1}$ that polynomial which interpolates $(\bullet - x)f$ on the knot set $\{x_\nu, x_{\nu+1}, \ldots, x_{\nu+k-1}, x_{\nu+k}\}$. Now, note that by the uniqueness property of polynomial interpolation,

$$t - x = \left(\frac{x - x_\nu}{x_{\nu+k} - x_\nu} \right)(t - x_{\nu+k}) + \left(\frac{x_{\nu+k} - x}{x_{\nu+k} - x_\nu} \right)(t - x_\nu),$$

since both sides agree when $t = x_\nu$ and $t = x_{\nu+k}$. But then, using the same reasoning, we have that

$$p(x) = \left(\frac{x - x_\nu}{x_{\nu+k} - x_\nu} \right)(t - x_{\nu+k}) p_0(x) + \left(\frac{x_{\nu+k} - x}{x_{\nu+k} - x_\nu} \right)(t - x_\nu) p_1(x).$$

Since the divided differences are the coefficients for the polynomial p, written in the form (1.12), we obtain that

$$[x_v, \ldots, x_{v+k}](\bullet - x)f = \left(\frac{x - x_v}{x_{v+k} - x_v}\right)[x_v, \ldots, x_{v+k-1}]f$$

$$+ \left(\frac{x_{v+k} - x}{x_{v+k} - x_v}\right)[x_{v+1}, \ldots, x_{v+k}]f.$$

The recursion formula for $M_{vk,X}$ now follows from the above equation by setting $f := (\bullet - x)_+^{k-2}$. $\square$

Positivity

Using the recursion formulae for $B_{vk,X}$ and $M_{vk,X}$, together with the fact that $B_{v,1,X} > 0$ and $M_{v,1,X} > 0$, and that the coefficients in the recursion formulae are positive, we immediately obtain the next result.

$$\forall k \in \mathbb{N}: \quad \begin{cases} B_{vk,X} > 0 \\ M_{vk,X} > 0 \end{cases} \quad \text{on } (x_v, x_{v+k}).$$

Lee's Formula and Marsden's Identity

In order to prove that an appropriately chosen collection of B-splines $B_{vk,X}$ or $M_{vk,X}$ provides a basis for the spline space $S^k(X)$, we need two results that are of importance by themselves. To this end, let $k, n \in \mathbb{N}$ and let X be defined as in (2.3). For $v = 1, \ldots, n + k$, define polynomials $\psi_{vk} \in \Pi^k$ by

$$\psi_{vk}(x) := \frac{1}{(k-1)!} \prod_{\mu=v+1}^{v+k-1} (x - x_\mu).$$

For fixed x, denote by $L(\bullet, x) \in \Pi^k$ those polynomials that interpolate the truncated power function $(\bullet - x)_+^{k-1}$ on the knot set $\{x_v, \ldots, x_{v+k}\}$.

Lee's Formula. Let $k, n \in \mathbb{N}$. Then

$$\psi_{vk}(t)B_{vk,X}(x) = L_{v+1,k}(t, x) - L_{v,k}(t, x),$$

for all knot sets X and for all $t, x \in \mathbb{R}$.

Proof   Fix $x \in \mathbb{R}$. The difference $\Delta L_{vk}(\bullet, x) := L_{v+1,k}(\bullet, x) - L_{v,k}(\bullet, x)$ is an element of Π^k and $\Delta L_{vk}(x_{v+\ell}, x) = 0, \forall \ell = 1, \ldots, k - 1$. Since the polynomial ψ_{vk} vanishes on the same set of knots, $\Delta L_{vk}(\bullet, x)$ must be a constant multiple of ψ_{vk}, i.e., there exists $\gamma(x)$ so that

$$\Delta L_{vk}(\bullet, x) = \gamma(x)\,\psi_{vk}.$$

Comparing the leading coefficients in the above equality, we see that $\gamma(x)$ is the leading coefficient of the difference of the leading coefficients of the polynomials $L_{v+1,k}(\bullet, x)$ and $L_{v,k}(\bullet, x)$. By assumption, the polynomials $L_{v+1,k}(\bullet, x)$

and $L_{v,k}(\bullet, x)$ interpolate the truncated power function $(\bullet - x)_+^{k-1}$ and these leading coefficients are therefore the divided differences $[x_{v+1}, \ldots, x_{v+k}]$ $(\bullet - x)_+^{k-1}$ and $[x_v, \ldots, x_{v+k-1}](\bullet - x)_+^{k-1}$, respectively. Hence,

$$\gamma(x) = [x_{v+1}, \ldots, x_{v+k}](\bullet - x)_+^{k-1} - [x_v, \ldots, x_{v+k-1}](\bullet - x)_+^{k-1}$$

$$= (x_{v+k-x_v})[x_v, \ldots, x_{v+k}](\bullet - x)_+^{k-1}$$

$$= B_{vk,X}. \qquad \square$$

Marsden's Identity. Let $k, n \in \mathbb{N}$. Then

$$(t - x)^{k-1} = \sum_{v=1}^{n} \psi_{vk}(t) B_{vk,X}(x),$$

for all knot sets X, for all $t \in \mathbb{R}$, and for all $x \in [x_k, x_{n+1}]$.

Proof Summing Lee's formula over v and observing that the sum involving $\Delta L_{vk}(\bullet, x)$ is telescoping, we obtain

$$\sum_{v=1}^{n} \psi_{vk}(t) B_{vk,X}(x) = \sum_{v=1}^{n} \Delta L_{vk}(t, x) = L_{n+1,k}(t, x) - L_{1,k}(t, x).$$

Note that the polynomial $L_{1,k}(\bullet, x)$ interpolates the truncated power function $(\bullet - x)_+^{k-1}$ on the knot set $X^1 := \{x_1, \ldots, x_k\}$ and thus $(t - x)_+^{k-1} = 0$, for all $x \geq x_k$. By the uniqueness property of interpolating polynomials, $L_{1,k}(\bullet, x) \equiv 0$. Similarly, we obtain that for all $x \leq x_{n+1}$, $(t - x)_+^{k-1} = (\bullet - x)^{k-1}$ on the knot set $X^{n+1} := \{x_{n+1}, \ldots, x_{n+k}\}$. Using the fact that $L_{n+1,k}(\bullet, x)$ is the unique interpolating polynomial for the truncated power function $(\bullet - x)_+^{k-1}$ on X^{n+1}, we have that $L_{n+1,k}(\bullet, x) \equiv (\bullet - x)_+^{k-1}$. This proves Marsden's Identity. $\square$

2.3.2 The Basis Property of B-Splines

Here we prove that B-splines form a basis for the spline space $S^k(X)$, where X is an appropriately chosen knot set. To obtain this result, we need two propositions.

Proposition 2.8   For all $k, n \in \mathbb{N}$ and all knot sets X,

$$\Pi^k[x_k, x_{n+1}] \subseteq \text{span}\left\{B_{vk,X} \mid v = 1, \ldots, n\right\}.$$

(Here, $\Pi^k[x_k, x_{n+1}]$ denotes the linear space of all polynomials in Π^k restricted to the interval $[x_k, x_{n+1}]$.)

Proof   Let $m \in \mathbb{N}$ with $m \leq k$ and let $D_t := \dfrac{d}{dt}$ be the differential operator with respect to t. Divide Marsden's Identity by $(k - 1)!$ and then apply the operator D_t^{m-1} to the resulting equation. This yields

$$D_t^{m-1}\left[\frac{(t-x)^{k-1}}{(k-1)!}\right] = \frac{(t-x)^{k-m}}{(k-m)!} = \sum_{v=1}^{n}\frac{(D_t^{m-1}\psi_{vk})(t)}{(k-1)!}B_{vk,X}.$$

In other words, $\forall m \in \{1, \ldots, k\}$ and $t := a$,

$$\frac{(x-a)^{k-m}}{(k-m)!} = \sum_{v=1}^{n}\left[(-1)^{k-m}\frac{(D_t^{m-1}\psi_{vk})(a)}{(k-1)!}\right]B_{vk,X}. \qquad (2.13)$$

Hence, the set of polynomials $\left\{\left.\dfrac{(x-a)^{k-m}}{(k-m)!}\right| m = 1, \ldots, k\right\} \subseteq \Pi^k[x_k, x_{n+1}]$

belongs to the span of $\{B_{vk,X} \mid v = 1, \ldots, n\}$. $\square$

Remark 2.9 For $m := k$, one obtains from (2.13) the partition of unity property of the B-splines $B_{vk,X}$, since in this case $D_t^{k-1}\psi_{vk} \equiv 1$.

Remark 2.10 Setting $m := k-1$ and $k := 2$ in (2.13), gives the so-called *property of knot averages* for linear polynomials. For this choice of k and m, (2.13) implies that

$$x - t = \sum_{v=1}^{n}\left[\frac{x_{v+1} + \cdots + x_{v+k-1}}{k-1} - t\right]B_{v,2,X}(x),$$

and, therefore,

$$\forall p \in \Pi^2: \quad p = \sum_{v=1}^{n}p(x_{vk}^*)B_{v,2,X},$$

where the *Grenville sites* x_{vk}^* are defined by $x_{vk}^* := \dfrac{x_{v+1} + \cdots + x_{v+k-1}}{k-1}$.

Proposition 2.11 Assume that all knots $x_\mu \in X$ are distinct. Then the truncated power function $(\bullet - x_\mu)_+^{k-1}$ belongs to span $\{B_{vk,X} \mid v = 1, \ldots, n\}$.

Proof In (2.13), set $t := x_\mu$ and $m := 1$. If $x_v < x_\mu < x_{v+k}$, then $\psi_{vk}(x_\mu) = 0$. Thus,

$$\frac{(x-x_\mu)^{k-1}}{(k-1)!} = \sum_{v=1}^{n}(-1)^{k-1}\psi_{vk}(x_\mu)B_{vk,X}(x)$$

$$= \sum_{\mu \leq v}(-1)^{k-1}\psi_{vk}(x_\mu)B_{vk,X}(x)$$

$$+ \sum_{\mu \geq v+k}(-1)^{k-1}\psi_{vk}(x_\mu)B_{vk,X}(x).$$

The support properties of the B-splines $B_{vk,X}$ now imply that the sum on the left is zero, for all $x \leq x_\mu$. The sum on the right is zero for all $x \geq x_\mu$, since the support of $B_{vk,X}$ lies to the left of the knot x_μ when $v + k \leq \mu$. Thus, only the first sum contributes for $x > x_\mu$, and we obtain

$$(x - x_\mu)_+^{k-1} = \sum_{v=\mu}^{n} (-1)^{k-1}(k-1)! \psi_{vk}(x_\mu) B_{vk,X}(x) \times$$

$$\in \text{span}\left\{ B_{vk,X} \mid v = 1, \ldots, n \right\},$$

which concludes the proof. $\square$

These two propositions now give the next theorem which shows that B-splines provide a natural basis for splines.

Theorem 2.12 (Representation Theorem for Splines)   Let $k, N \in \mathbb{N}$, and let Ξ and X be knot sets defined by

$$\Xi := \{a := \xi_0 < \xi_1 \quad < \ldots < \xi_N < \xi_{N+1} =: b\},$$

$$X := \{x_1 \leq \cdots \leq \ a = x_k < x_{k+1} < \cdots < x_n < x_{n+1} = b \leq x_{n+2} \leq \cdots \leq x_{n+k}\}.$$

Then the set $\{B_{vk,X} \mid v = 1, \ldots, n\}$ is a basis for the spline space $S^k(\Xi)$. In other words, every spline s defined over the knot set Ξ has on $[a, b]$ a representation of the form

$$s = \sum_{v=1}^{n} c_v \, B_{vk,X}, \quad c_v \in \mathbb{R}.$$

Remark 2.13   Before we prove this theorem, a few comments are in order. The particular choice of the knot set X is motivated by the fact that we need to take into account the support of each B-spline when representing splines over the knot set Ξ. To obtain nonzero values of a spline s at the boundary knots $\xi_0 = a$ and $\xi_{N+1} = b$, one needs to use B-splines whose support contains these two knots. This, however, means that we have to introduce a total of $(k - 1)$ new knots to the left of ξ_0 and to the right of ξ_{N+1}.

Comparing the knot sets Ξ and X, one deduces that $n = k + N$. This relation can also be explained in the following manner. Given the knot set Ξ and the associated partition of $[a, b]$ into $(N + 1)$ subintervals, there is a totality of $(N + 1)k$ free parameters. As we require that each spline s is to be in $C^k[a, b]$, we have $(k - 1)$ smoothness conditions at each interior knot ξ_v, $v = 1, \ldots, N$, namely,

$$\text{jump}_{\xi_v} f^{(m)} := f^{(m)}(\xi_v+) - f^{(m)}(\xi_v-) = 0, \quad m = 0, 1, \ldots, k-2.$$

Thus,

$$\begin{array}{ccccc}
\text{Number of Parameter} & - & \text{Number of Smoothness Conditions} & = & \dim S^k(\Xi) \\
(N+1)k & - & N(k-1) & & = k + N = n.
\end{array}$$

Proof   We have shown so far that, besides a normalizing factor, each element of the basis $\mathcal{B}^0$ is in the span of $\{B_{\nu k,X} \mid \nu = 1, \dots, n\}$ for an appropriately chosen knot set. More precisely, Propositions 2.8 and 2.11 have shown that the functions

$$p_j := \frac{(\bullet - a)^{k-j}}{(k-j)!}, \quad j = 1, \dots, k,$$

and

$$q_{mk} := \frac{(\bullet - x_\mu)_+^{k-1}}{(k-1)!}, \quad m = 1, \dots, N,$$

are elements of span $\{B_{\nu k,X} \mid \nu = 1, \dots, n\}$. By Theorem 1.26, we know that the set $\{p_j \mid j = 1, \dots, k\} \cup \{q_{mk} \mid m = 1, \dots, N\}$ forms a basis for $S^k(\Xi)$. Moreover, since $\dim S^k(\Xi) = k + N = n = \operatorname{card} \mathcal{B}$, where $\mathcal{B} := \{B_{\nu k,X} \mid \nu = 1, \dots, n\}$, $\mathcal{B}$ is a basis for $S^k(\Xi)$. $\square$

Remark 2.14   In the above proof a standard argument from linear algebra was used to infer that a given set of vectors is a basis. This argument proceeds as follows. Suppose that E is a linear space and $\{f_1, \dots, f_n\}$ is a basis for a linear subspace F of E. Suppose further that $\{g_1, \dots, g_m\}$ is a sequence of vectors with the property that each f_i, $i = 1, \dots, n$, is contained in $\mathsf{G} := \operatorname{span}\{g_j \mid j = 1, \dots, m\}$. Then, $\mathsf{F} \subseteq \mathsf{G}$ and $n = \dim \mathsf{F} \leq \dim \mathsf{G} \leq m$. If in addition $n = m$, then we necessarily have that $\mathsf{F} = \mathsf{G}$ and $\dim \mathsf{G} = n$. Hence, the sequence $\{g_1, \dots, g_m\}$ is a minimal spanning set, and thus also linearly independent.

The next result is an immediate corollary of the above theorem.

Corollary 2.15   The set $\{B_{\nu k,X} \mid \nu = 1, \dots, n\}$ of B-splines is linearly independent.

Remark 2.16   The B-splines $B_{\nu k,X}$ are sometimes also written as

$$B_{\nu k}(\bullet \mid X) \quad \text{or} \quad B_{\nu k}(\bullet \mid x_1, \dots, x_n).$$

2.3.3 Derivatives and Integrals of B-Splines

Let $X := \{x_1, \dots, x_{n+k}\}$ be a knot set consisting of nondecreasing knots and let $\{B_{\nu k} \mid \nu = 1, \dots, n\}$ be the set of B-splines defined on this knot set. Then, on the interval $[x_1, x_n]$, we have that

$$D\left(\sum_{\nu=2-k}^{n-1} c_\nu B_{\nu k}\right) = (k-1)\sum_{\nu=3-k}^{n-1} \frac{c_\nu - c_{\nu-1}}{x_{\nu+k-1} - x_\nu} B_{\nu,k-1}. \tag{2.14}$$

Moreover, for every $x \in [x_1, x_n]$, the definite integral of a linear combination of B-splines is given by

$$\int_{x_1}^{x} \sum_{\nu=1}^{n} c_\nu B_{\nu k}(t)\,dt = \sum_{\nu=1}^{n-1} \left(\sum_{\mu=1}^{\nu} \frac{c_\mu(x_{\mu+k} - x_\mu)}{k} \right) B_{\nu,k+1}(x). \tag{2.15}$$

We encourage the reader to prove formulae (2.14) and (2.15) using induction on k.

2.4 Cardinal B-Splines

We now consider a special class of splines whose knot set X is infinite, but whose knots are uniformly distributed. By a simple rescaling, we can always achieve that the knot set X equals $\mathbb{Z}$.

Definition 2.17 (Cardinal Spline Space)   Let $k \in \mathbb{N}$ and let X be the infinite knot set $\mathbb{Z}$. The family of all functions $f \in C^{k-2}(\mathbb{R})$ with the property that $f|_{[\nu,\nu+1)} \in \Pi^k, \forall \nu \in \mathbb{Z}$, is a linear subspace of $C^{k-2}(\mathbb{R})$ and is called the *cardinal spline space of order k* (over the knot set $\mathbb{Z}$.) It is denoted by $S^k(\mathbb{Z})$.

We are also interested in the following classes of subspaces of $S^k(\mathbb{Z})$. Let $N \in \mathbb{N}$ and define

$$S_N^k(\mathbb{Z}) := \left\{ g \in \mathrm{Map}(\mathbb{R}, \mathbb{R}) \,|\, \exists f \in S^k(\mathbb{Z}) \text{ so that } g \equiv f|_{[-N,N]} \right\}.$$

The next proposition follows immediately from Theorem 2.4.

Proposition 2.18   The set $\{1, x, \ldots, x^{k-1}\} \cup \{(\bullet - \nu)_+^{k-1} \,|\, \nu = -N+1, \ldots, N-1\}$ forms a basis for $S_N^k(\mathbb{Z})$. Therefore, $\dim S_N^k(\mathbb{Z}) = k + 2N - 1$.

To obtain a basis for $S_N^k(\mathbb{Z})$ that consists only of one type of polynomial, we replace the functions $x \mapsto x^\varkappa$ by the truncated power functions $x \mapsto (x + N + \varkappa)_+^{k-1}$, for $\varkappa = 0, 1, \ldots, k-1$. Now, define

$$\mathcal{G}_N^k := \left\{ (\bullet - \nu)_+^{k-1} \,|\, \nu = -N - k + 1, \ldots, N - 1 \right\}.$$

Then, $\mathrm{card}\, \mathcal{G}_N^k = k + 2N - 1 = \dim S_N^k(\mathbb{Z})$ and $\mathcal{G}_N^k$ constitutes a basis for $S_N^k(\mathbb{Z})$. (The reader should prove this!)

The family $\mathcal{G}_N^k$ has the following advantages over the previously considered basis.

1. Each function $(\bullet - \nu)_+^{k-1}$ vanishes to the left of the knot ν, $\nu = -N - k + 1, \ldots, N - 1$;

2. *All* basis functions are generated by the integer translates of *one* function, namely, $x \mapsto x_+^{k-1}$.

We obviously have the following equality in the sense of sets:

$$S^k(\mathbb{Z}) = \bigcup_{N \in \mathbb{N}} S_N^k(\mathbb{Z}).$$

Definition 2.19 (Cardinal B-Spline)   The v-*th cardinal B-spline of order k* for the knot set $\mathbb{Z}$ is defined by

$$B_{vk,\mathbb{Z}} := k[v, \ldots, v+k](\bullet - x)_+^{k-1}.$$

Since the knots are integers, we immediately obtain $M_{vk,\mathbb{Z}} = B_{vk,\mathbb{Z}}$ and $\operatorname{supp} B_{vk,\mathbb{Z}} = [v, v+k]$. (See (2.5) and (2.10).)

There exist interesting relationships between divided differences, difference operators, and cardinal B-splines. These relationships will be exhibited next.

Definition 2.20 (Backward Difference Operator)   Let $k \in \mathbb{N}$ and suppose that $f \in \operatorname{Map}(\mathbb{R}, \mathbb{R})$. The *backward difference operator* $\nabla^k : \operatorname{Map}(\mathbb{R}, \mathbb{R}) \to \operatorname{Map}(\mathbb{R}, \mathbb{R})$ *of order k* is recursively defined by

$$(\nabla f)(x) := f(x) - f(x-1),$$

$$(\nabla^k f) := \nabla(\nabla^{k-1} f), \quad 1 < k \in \mathbb{N}.$$

The following properties of the backward difference operator ∇ are easily established and left as exercises for the reader.

$$\forall k \in \mathbb{N} \, \forall p \in \Pi^k : \nabla^k p = 0. \tag{2.16}$$

$$(\nabla^k f)(x) = \sum_{v=0}^{k} (-1)^v \binom{k}{v} f(x-v), \quad x \in \mathbb{R}. \tag{2.17}$$

$$\forall k \in \mathbb{N} : (\Delta^k f)(x) = (\nabla^k f)(x+k), \quad x \in \mathbb{R}. \tag{2.18}$$

Combining the definition of cardinal B-spline and properties (2.17) and (2.18), we obtain an explicit representation of $B_{vk} := B_{vk,\mathbb{Z}}$ of the form

$$B_{vk} = \frac{1}{(k-1)!} \sum_{\mu=v}^{v+k} (-1)^{v-\mu} \binom{k}{\mu - v} (\bullet - \mu)_+^{k-1}$$

$$= B_{0,k}(\bullet - v) =: B_k(\bullet - v).$$

In other words, all cardinal B-splines B_{vk} are integer translates (by v) of B_k. When studying the properties of cardinal B-splines it thus suffices to consider only those of B_k. For reference purposes, we state that

$$\forall k \in \mathbb{N} : \operatorname{supp} B_k = [0, k].$$

Next, we show that the cardinal B-splines form a basis for the cardinal spline space $S^k(\mathbb{Z})$. For this purpose, let

$$\mathcal{B}_k := \{B_k(\bullet - v)\,|\,v \in \mathbb{Z}\} \quad \text{and}$$

$$\mathcal{B}_{k,N} := \{B_k(\bullet - v)\,|\,v = -N - k + 1, \ldots, N - 1\}.$$

By Proposition 2.11, $\mathcal{B}_{k,N}$ forms a basis of $S_N^k(\mathbb{Z})$. Moreover, $\mathcal{B}_k$ is a basis for the cardinal spline space $S^k(\mathbb{Z})$ in the sense that

$$f \in S^k(\mathbb{Z}) \quad \Longleftrightarrow \quad f(x) = \sum_{v \in \mathbb{Z}} c_v \, B_k(x - v), \quad c_v \in \mathbb{R}, \tag{2.19}$$

point-wise for all $x \in \mathbb{R}$.

Remark 2.21   The compactness of the support of B_k implies that for any fixed $x \in \mathbb{R}$ the above sum has at most a finite number of nonzero terms.

In the next theorem, we summarize several properties of cardinal B-splines some of which are special cases already known from B-splines.

Theorem 2.22   Let $k \in \mathbb{N}$ and let B_k be the cardinal B-spline of order k.

1. $\forall f \in C(\mathbb{R})$: $\displaystyle\int_{\mathbb{R}} f(x) B_k(x)\,dx = \int_{W^k} f(x_1 + \cdots x_k)\,dx_1 \cdots dx_k$, where
   $W^k := [0, 1]^k$.

2. $\forall f \in C^k(\mathbb{R})$: $\displaystyle\int_{\mathbb{R}} (D^k f)(x) B_k(x)\,dx = \sum_{v=0}^{k} (-1)^{k-v} \binom{k}{v} f(v).$

3. $B_k = \dfrac{1}{(k-1)!}\, \nabla^k x_+^{k-1}.$

4. $\forall x \in \mathbb{R}$: $\displaystyle\sum_{v \in \mathbb{Z}} B_k(x - v) = 1.$

5. $DB_k = \nabla B_k = B_{k-1} - B_{k-1}(\bullet - 1).$

6. $\forall x \in \mathbb{R}$: $B_k(x) = \dfrac{x}{k-1}\, B_{k-1}(x) + \dfrac{k-x}{k-1}\, B_{k-1}(x - 1).$

7. $\forall x \in \mathbb{R}$: $B_k(x + \frac{k}{2}) = B_k(\frac{k}{2} - x)$. (Symmetry with respect to the support center)

8. $\forall 1 < k \in \mathbb{N}$: $B_k(x) = (B_{k-1} * B_1)(x) = \displaystyle\int_0^1 B_{k-1}(x - t)\,dt =: \left(\underset{i=1}{\overset{k}{*}} B_1 \right)(x).$

Proof   Exercise! $\square$

2.5 The Fourier Transform of Cardinal B-Splines

The Fourier[4] transform is an indispensable tool in many areas of pure and applied mathematics, as well as physics and engineering. The idea of Fourier to analyze the properties of a function by decomposing it into fundamental building blocks is essential in such areas as signal processing and data and image analysis.

In this section, we present the definition of the continuous Fourier transform for L^1-functions and list some of its properties, mostly those that are of interest in relation to B-splines. For more details, we refer the interested reader to the vast literature on Fourier analysis, of which [32, 35, 54, 95, 101], and [147] is only a rather short and subjective excerpt.

At this point, we recall the definition of the Banach spaces $(L^p(\mathbb{R}, \mathbb{C}), \| \ \|_{L^p})$, $1 \leq p < \infty$: A function $f : \mathbb{R} \to \mathbb{C}$ is in $L^p(\mathbb{R}, \mathbb{C})$ iff $\int_{\mathbb{R}} |f(x)|^p dx < \infty$. Here, we differ from our previous setting (Cf. Definition 1.8) in that we consider complex-valued functions, in which case the expression $|f(x)|$ is to be interpreted as $|f(x)| := \sqrt{f(x)\overline{f(x)}}$, where $^{-}$ denotes complex conjugation.

Definition 2.23 (Fourier Transform on $L^1(\mathbb{R}, \mathbb{C})$) The mapping $\mathscr{F} : L^1(\mathbb{R}, \mathbb{C}) \to C(\mathbb{R}, \mathbb{C})$ defined by

$$\mathscr{F}(f)(\omega) := \widehat{f}(\omega) := \frac{1}{\sqrt{2\pi}} \int_{\mathbb{R}} e^{-i\omega x} f(x) dx \qquad (2.20)$$

is called the *continuous Fourier transform* of f.

Remark 2.24 The definition above asserts that $\widehat{f} \in \mathrm{Map}(\mathbb{R}, \mathbb{C})$ is continuous. This, however, follows readily from (2.20). (The reader is encouraged to verify this statement.)

We like to extend the definition of Fourier transform to the space $L^2(\mathbb{R}, \mathbb{C})$, which being a Hilbert space, plays a pivotal role in applied mathematics. In order to do this, we need the so-called *Plancherel*[5] *Theorem*. We present this theorem in its simplest form without proof and refer the reader for the most general statement and its proof to the literature.

---

4 Jean Baptiste Joseph Fourier, 21 March 1768–16 May 1830. French mathematician and physicist who is best known for his mathematical studies of heat conduction and its solution in terms of infinite trigonometric series.

5 Michel Plancherel, 18 January 1885–4 March 1967. Swiss mathematician who worked in algebra, harmonic analysis, and mathematical physics.

Theorem 2.25 (Plancherel's Theorem) Suppose that $f \in L^1(\mathbb{R}, \mathbb{C}) \cap L^2(\mathbb{R}, \mathbb{C})$. Then the Fourier transform $\mathscr{F}(f) = \widehat{f}$ of f is in $L^2(\mathbb{R}, \mathbb{C})$ and the mapping $\mathscr{F}$ can be uniquely extended to a linear isometry on $L^2(\mathbb{R}, \mathbb{C})$.

Notationally, we will not distinguish between the mapping $\mathscr{F}$ and its extension to $L^2(\mathbb{R}, \mathbb{C})$. It will be clear from the context onto which functions the mapping $\mathscr{F}$ acts.

In particular, Plancherel's theorem states that

$$\|f\|_{L^2} = \|\widehat{f}\|_{L^2},$$

an equality usually referred to as *Parseval's*[6] *Identity*. Moreover, the mapping $\mathscr{F}$ is a unitary operator on $L^2(\mathbb{R}, \mathbb{C})$, i.e., the adjoint $\mathscr{F}^*$ of $\mathscr{F}$ is equal to the inverse $\mathscr{F}^{-1}$ of $\mathscr{F}$. Furthermore, this means that the operator $\mathscr{F}$ preserves inner products:

$$\int_{\mathbb{R}} f(x)\overline{g(x)}dx = \int_{\mathbb{R}} \mathscr{F}(f)(\omega)\overline{\mathscr{F}(g)(\omega)}d\omega = \int_{\mathbb{R}} \widehat{f}(\omega)\overline{\widehat{g}(\omega)}d\omega,$$

for all $f, g \in L^2(\mathbb{R}, \mathbb{C})$.

We remark that the Fourier transform $\widehat{f}$ of a function $f \in L^2(\mathbb{R}, \mathbb{C})$ evaluated at a point $\omega \in \mathbb{R}$ does in general not exist, i.e., the Fourier integral for a function in $L^2(\mathbb{R}, \mathbb{C})$ need not necessarily converge at every point $\omega \in \mathbb{R}$.

We now list a few properties of the Fourier transform whose proofs are left to the reader.

Proposition 2.26 Let $\alpha, \beta \in \mathbb{C}$ and let $f, g \in L^2(\mathbb{R}, \mathbb{C})$. Then the following hold.

1. Linearity: The mapping $\mathscr{F} : L^2(\mathbb{R}, \mathbb{C}) \to L^2(\mathbb{R}, \mathbb{C})$ is linear:
   $\mathscr{F}(\alpha f + \beta g) = \alpha \mathscr{F}(f) + \beta \mathscr{F}(g)$.

2. Modulation: Denote by $\tau_y : L^2(\mathbb{R}, \mathbb{C}) \to L^2(\mathbb{R}, \mathbb{C})$ the translation operator $(\tau_y f)(x) := f(x - y)$. Then, $\mathscr{F}(\tau_y f) = e^{iy} \mathscr{F}(f)$.

3. Scaling Property: Let $a \in \mathbb{R} \setminus \{0\}$ and let $\mathcal{D}_a : L^2(\mathbb{R}, \mathbb{C}) \to L^2(\mathbb{R}, \mathbb{C})$ denote the dilation operator $(\mathcal{D}_a f)(x) := f(ax)$. Then,
   $\mathscr{F}(\mathcal{D}_a f)(\omega) = |a|^{-1} \mathscr{F}(f)(a^{-1}\omega)$.

4. Fixed Point Property: The function $G(x) := e^{-x^2/2}$ is an eigenfunction of the operator $\mathscr{F}$ with eigenvalue one in the sense that $(\mathscr{F}G)(\omega) = e^{-\omega^2/2} = G(\omega)$.

---

6 MARC-ANTOINE PARSEVAL, 27 April 1755–16 August 1836. French mathematician who contributed to partial differential equations and who is most famous for what is known as the Parseval Identity.

5. Inverse Fourier Transform: Given the Fourier transform $\widehat{f} \in L^2(\mathbb{R}, \mathbb{C})$ of a function f, then f can be recovered from its Fourier transform via the formula

$$f(x) =: \mathscr{F}^{-1}(\widehat{f})(x) = \frac{1}{\sqrt{2\pi}} \int_{\mathbb{R}} e^{ix\omega} \widehat{f}(\omega) d\omega.$$

6. Fourier Transform of a Derivative: If f is differentiable with Fourier transform $\widehat{f}$, then the Fourier transform of its derivative Df is given by $i\omega\widehat{f}(\omega)$.

The next result is important for our purposes and we state it therefore separately. The proof is again left to the reader.

Theorem 2.27 (Convolution Theorem) Suppose that $f, g \in L^1(\mathbb{R}, \mathbb{C})$. Then the convolution $f * g$ is an element of $L^1(\mathbb{R}, \mathbb{C})$, and we have the following identity for the Fourier transforms:

$$\mathscr{F}(f * g) = \sqrt{2\pi}\, \mathscr{F}(f) \cdot \mathscr{F}(g). \tag{2.21}$$

Proof Exercise! $\square$

Remark 2.28 If $1 \leq p, q \leq \infty$, then the convolution of $f \in L^p(\mathbb{R}, \mathbb{C})$ with $g \in L^q(\mathbb{R}, \mathbb{C})$ exists and is in $L^r(\mathbb{R}, \mathbb{C})$, where $0 \leq \frac{1}{r} := \frac{1}{p} + \frac{1}{q} - 1$. This is part of W. H. Young's Theorem. (See, for instance [102], Theorem 4.2.)

Remark 2.29 Equality (2.21) holds in more general situations. It is true for f and g being tempered distributions, i.e., linear functionals on the space of all Schwartz[7] functions. Since Schwartz functions play an important role in the theory of distributions, we give their definition below.

Finally, we like to state without proof an important asymptotic result of the Fourier transform.

Lemma 2.30 (Riemann[8]-Lebesgue Lemma) Let $f \in L^1(\mathbb{R})$. Then $\lim\limits_{|\omega| \to \infty} |\widehat{f}(\omega)| = 0$. Moreover, $\|\widehat{f}\|_\infty \leq (2\pi)^{-1} \|f\|_1$.

The Riemann-Lebesgue lemma states that the Fourier transform of an L^1-function vanishes at infinity, but it does not give any indication about how fast this decay is. The next proposition provides an answer to this question.

---

7 LAURENT SCHWARTZ, 5 March 1915–4 July 2002. French mathematician who discovered the theory of distributions.
8 GEORG FRIEDRICH BERNHARD RIEMANN, 17 September 1826–20 July 1866. German mathematician who made major contributions to analysis, differential geometry, mathematical physics, and analytic number theory.

Proposition 2.31   Suppose that $f \in L^1(\mathbb{R}) \cap C^k(\mathbb{R})$. Then $\mathscr{F}(f) \in \mathcal{O}(\omega^{-k})$.

Definition 2.32 (Schwartz Function)   Let $\Omega \subseteq \mathbb{R}^n$, $n \in \mathbb{N}$, and let $\phi \in C^\infty(\Omega, \mathbb{R}^n)$. Then the *Schwartz space* $\mathscr{S}(\Omega, \mathbb{R}^n)$ or *space of rapidly decaying functions on Ω* is defined by

$$\mathscr{S}(\Omega, \mathbb{R}^n) := \{\phi \in C^\infty(\Omega, \mathbb{R}^n) \mid \forall \alpha, \beta \in \mathbb{N}_0^n : \sup\{\|\boldsymbol{x}^\alpha D^\beta \phi(\boldsymbol{x})\| < \infty\}\}$$

The elements of $\mathscr{S}(\Omega, \mathbb{R}^n)$ are called *Schwartz functions*.

As an example of a Schwartz function, we present

$$\phi(\boldsymbol{x}) := \boldsymbol{x}^\alpha \, e^{-a\|\boldsymbol{x}\|^2},$$

where $\alpha \in \mathbb{N}_0^n$ is a multi-index and $a > 0$ arbitrary. Moreover, any function with compact support in $\mathbb{R}^n$ is a Schwartz function on $\mathbb{R}^n$.

It can be shown that $\mathscr{S}(\Omega, \mathbb{R}^n) \subset L^p(\Omega, \mathbb{R}^n)$, for $p \in [1, \infty]$, and that $\mathscr{S}(\mathbb{R}^n) := \mathscr{S}(\mathbb{R}^n, \mathbb{R}^n)$ is dense in $L^2(\mathbb{R}^n)$. Finally, the Fourier transform $\mathscr{F}$ is a linear isomorphism $\mathscr{S}(\mathbb{R}^n) \to \mathscr{S}(\mathbb{R}^n)$. For details, we refer the interested reader to the literature given above.

We now compute the Fourier transform of cardinal B-splines using their characterization as a repeated convolution of the cardinal B-spline $B_1 = \chi_{[0,1]}$. (See item 8 in Theorem 2.22.) To this end, we need to compute the Fourier transform of B_1 first. A simple computation yields

$$\mathscr{F}(B_1)(\omega) = \widehat{B_1}(\omega) = \frac{1}{\sqrt{2\pi}} \int_{\mathbb{R}} e^{-ix\omega} \chi_{[0,1]}(x)\,dx = \frac{1}{\sqrt{2\pi}} \int_0^1 e^{-ix\omega}\,dx$$

$$= \frac{1}{\sqrt{2\pi}} e^{-i\omega/2} \left(\frac{e^{i\omega/2} - e^{-i\omega/2}}{i\omega} \right) = \frac{1}{\sqrt{2\pi}} e^{-i\omega/2} \frac{\sin \omega/2}{\omega/2}$$

$$=: \frac{1}{\sqrt{2\pi}} e^{-i\omega/2} \operatorname{sinc} \omega/2.$$

Using the Convolution Theorem 2.27 applied to $B_k = \overset{k}{\underset{i=1}{\ast}} B_1$, $k \in \mathbb{N}$, immediately produces

$$\widehat{B_k}(\omega) = (\sqrt{2\pi})^{k-1} \left(\widehat{B_1}(\omega)\right)^k = \frac{1}{\sqrt{2\pi}} e^{-ik\omega/2} \left(\frac{\sin \omega/2}{\omega/2} \right)^k$$

$$= \frac{1}{\sqrt{2\pi}} e^{-ik\omega/2} \operatorname{sinc}^k \omega/2.$$

Two easy consequences of the above result are the following. Firstly, we have the known fact that for all $k \in \mathbb{N}$,

$$\int_{\mathbb{R}} B_k(x)\,dx = \sqrt{2\pi} \, \widehat{B_k}(0) = 1,$$

and secondly, since the cardinal B-splines are compactly supported with support $[0, k]$, their Fourier transform $\widehat{B_k}$ decays polynomially fast:

$$|\widehat{B_k}(\omega)| \in \mathcal{O}(\omega^{-k}).$$

Employing the properties of the backward difference operator ∇, in particular (2.17) and (2.18), together with the identity in item 2 in Theorem 2.22, we can derive an interesting relationship between the Fourier transform of cardinal B-splines and the divided difference operator. For this purpose, recall that for any function $f \in C^k(\mathbb{R}, \mathbb{C})$

$$\int_{\mathbb{R}} B_k(x)(D^k f)(x)\, dx = (-1)^k (\nabla f)(0) = k![0, 1, \ldots, k] f.$$

Now substituting $f(x) := e^{-ix\omega}$ into the left-hand side, we obtain

$$(-i\omega)^k \int_{\mathbb{R}} B_k(x) e^{-ix\omega}\, dx = k![0, 1, \ldots, k]_x\, e^{-ix\omega},$$

where the subscript on the divided difference operator indicates that it operates on the variable x. Thus,

$$\widehat{B_k}(\omega) = \frac{1}{\sqrt{2\pi}}\, \frac{k!}{(-i\omega)^k}[0, 1, \ldots, k]_x\, e^{-ix\omega}.$$

Setting $t := -ix\omega$ and $x_\nu := \nu$, $\nu = 0, 1 \ldots, k$, and rewriting the divided difference yields

$$\widehat{B_k}(\omega) = \frac{k!}{\sqrt{2\pi}}\,[-i\omega x_0, -i\omega x_1, \ldots, -i\omega x_k]_t\, e^t.$$

The above identity can be used to define multivariate B-splines (see [86]), but here we will not pursue this topic further.

2.6 Cardinal Spline Interpolation

In this section, we consider an interpolation problem posed on the entire real line and for the knot sequence $\mathbb{Z}$, and present some basic results that will be important for our later purposes. References to the material presented here may be found in [137, 139] and to some extent also in [39].

The *cardinal spline interpolation problem* reads as follows.

CARDINAL INTERPOLATION PROBLEM: Given a sequence $Y := \{y_\nu \in \mathbb{K} \mid \nu \in \mathbb{Z}\}$, where $\mathbb{K} = \mathbb{R}$ or $\mathbb{K} = \mathbb{C}$, find an element $s \in S^k(\mathbb{Z})$ so that

$$s(\nu) = y_\nu, \quad \forall \nu \in \mathbb{Z}. \tag{2.22}$$

Problems of this type can be found for instance in the theory of time series, where data is sampled at regular intervals over a very long time period, or for periodically sampled data.

In order to state an existence and uniqueness theorem for cardinal interpolation, we need to introduce the following sequence set and function space reflecting the fact that we are dealing with a bi-infinite interpolation problem. Let

$$\mathcal{Y}^{\alpha} := \{ Y = \{ y_{\nu} \mid \nu \in \mathbb{Z} \} \mid y_{\nu} \in \mathcal{O}(|\nu|^{\alpha}) \}$$

and

$$S^{k,\alpha}(\mathbb{Z}) := \{ s \in S^{k}(\mathbb{Z}) \mid s(x) \in \mathcal{O}(|x|^{\alpha}) \}.$$

The next result, for whose proof we refer to [137] or [139], shows that the cardinal interpolation problem has a unique solution for even orders k.

Theorem 2.33 (Existence and Uniqueness of Cardinal Interpolation) Suppose that $k \in \mathbb{N}$ is even. If $Y \in \mathcal{Y}^{\alpha}$, then the cardinal interpolation problem (2.22) has a unique solution $s \in S^{k,\alpha}(\mathbb{Z})$. In case that $Y \in \ell^{p}(\mathbb{Z})$, $1 \le p \le \infty$, then (2.22) has a unique solution $s \in S^{k}(\mathbb{Z}) \cap L^{p}(\mathbb{R})$.

To construct the solution s to the cardinal interpolation problems, one considers the so-called *k-th order fundamental cardinal spline* L_{k}. The fundamental cardinal spline is the unique solution of the cardinal spline interpolation problem for $y_{\nu} := \delta_{0\nu}$, $\nu \in \mathbb{Z}$, i.e.,

$$L_{k}(\nu) = \delta_{0\nu}, \quad \nu \in \mathbb{Z}. \tag{2.23}$$

In [139], the following two results are proven.

Theorem 2.34 There exist two constants $A = A(k)$ and $\alpha = \alpha(k)$ so that

$$|L_{k}(x)| < A\, e^{-\alpha|x|}, \quad \forall x \in \mathbb{R}.$$

For orders $2k$, $k \in \mathbb{N}$, the unique solution s of the cardinal interpolation problem (2.22) is given in the form

$$s(x) = \sum_{\nu \in \mathbb{Z}} y_{\nu}\, L_{2k}(x - \nu) \quad x \in \mathbb{R}. \tag{2.24}$$

Moreover, the cardinal Lagrange interpolation formula (2.24) of order $2k$ converges absolutely and uniformly on finite intervals.

Example 2.35 For $k = 2$, one quickly obtains a closed expression for the fundamental cardinal spline. Notice that the shifted B-spline $B_{2}(\bullet + 1)$ satisfies the interpolation conditions (2.23) and, by uniqueness, must therefore be identical to L_{2}.

Our next goal is to derive a formula for the fundamental cardinal spline L_{k} with $k \ge 3$. To this end, we first write L_{k} as a linear combination of centered cardinal splines B_{k} (see Theorem 2.22, item 7):

$$L_{k}(x) = \sum_{\nu \in \mathbb{Z}} c_{\nu}^{k}\, B_{k}\left(x + \frac{k}{2} - \nu \right). \tag{2.25}$$

Secondly, we substitute the above expression for L_k into (2.23) and take the Fourier transform of both sides. This yields

$$\left[\sum_{\nu\in\mathbb{Z}} B_k\left(\frac{k}{2}+\nu\right)e^{-i\omega\nu}\right]\left[\sum_{\nu\in\mathbb{Z}} c_\nu^k e^{-i\omega\nu}\right] = 1. \tag{2.26}$$

Taking the Fourier transform of (2.25) gives

$$\widehat{L_k}(\omega) = \left[\sum_{\nu\in\mathbb{Z}} c_\nu^k e^{-i\omega\nu}\right]\widehat{B_k}\left(\bullet+\frac{k}{2}\right)(\omega),$$

where $\widehat{B_k}\left(\bullet+\frac{k}{2}\right)$ denotes the Fourier transform of $B_k\left(x+\frac{k}{2}\right)$. Eliminating the expression containing the coefficients c_ν^k from the latter two equations produces the sought-after formula for the computation of the fundamental cardinal spline:

$$\widehat{L_k}(\omega) = \frac{\widehat{B_k}\left(\bullet+\frac{k}{2}\right)(\omega)}{\displaystyle\sum_{\nu\in\mathbb{Z}} B_k\left(\frac{k}{2}+\nu\right)e^{-i\omega\nu}}. \tag{2.27}$$

Example 2.36   Although formula (2.27) provides a means of computing the fundamental cardinal spline, it is sometimes easier to use identity (2.26). We illustrate the latter approach by computing the fundamental cardinal spline L_4.

Setting $z := e^{-i\omega}$, $k = 4$, and using the fact that $B_4(1) = \frac{1}{6}$, $B_4(2) = \frac{2}{3}$, and $B_4(3) = \frac{1}{6}$, Equation (2.26) reduces to

$$\sum_{\nu\in\mathbb{Z}} c_\nu^k z^\nu = \frac{1}{\sum_{\nu\in\mathbb{Z}} B_4(\nu+2)\, z^\nu}$$

$$= \frac{1}{\frac{1}{6}z^{-1}+\frac{2}{3}+\frac{1}{6}z} = \frac{z}{\frac{1}{6}+\frac{2}{3}z+\frac{1}{6}z^2}.$$

Factoring the quadratic expression in z and employing a partial fraction decomposition yields

$$\sum_{\nu\in\mathbb{Z}} c_\nu^k z^\nu = \sqrt{3}\left(\frac{-2+\sqrt{3}}{z-(-2+\sqrt{3})} - \frac{-2-\sqrt{3}}{z-(-2-\sqrt{3})}\right).$$

Setting $a := -2+\sqrt{3}$ and $b := -2-\sqrt{3}$, we can rewrite the above expression as

$$\sum_{\nu\in\mathbb{Z}} c_\nu^k z^\nu = \sqrt{3}\left(\frac{\frac{a}{z}}{1-\frac{a}{z}}+\frac{1}{1-\frac{z}{b}}\right)$$

$$= \sqrt{3}\left[\sum_{\nu=1}^{\infty}\left(\frac{a}{z}\right)^\nu + \sum_{\nu=0}^{\infty}\left(\frac{z}{b}\right)^\nu\right]$$

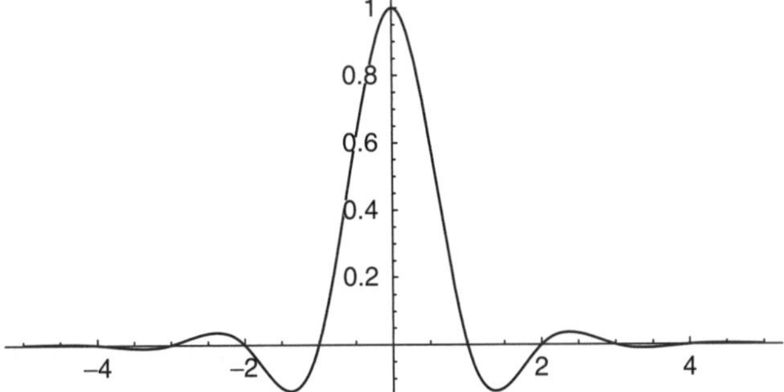

Figure 2.3   The graph of the fundamental cardinal spline L_4.

$$= \sqrt{3}\left[\sum_{v=-\infty}^{-1}\left(\frac{z}{a}\right)^v + \sum_{v=0}^{\infty}(az)^v\right]$$

$$= \sqrt{3}\sum_{v\in\mathbb{Z}}a^{|v|}z^v,$$

where we used that $ab = 1$. Comparing coefficients, we obtain that

$$c_v^4 = \sqrt{3}\,a^{|v|} = (-1)^v\sqrt{3}\,(2-\sqrt{3})^{|v|}, \quad v\in\mathbb{Z}.$$

The coefficients c_v^4, $v\in\mathbb{Z}$, are all nonzero, but decay exponentially fast. This, of course, implies that the fundamental cardinal spline of order 4 also decays exponentially fast, as expected by Theorem 2.34. The graph of L_4 is shown in Figure 2.3.

2.7 Repeated Knots

In this section, we investigate the case when knots coalesce, i.e., the setting when $\xi_v = \xi_{v+1} = \ldots = \xi_{v+m}$, for some $m\in\mathbb{N}$. We shall see that repeated knots are related to smoothness conditions and the order k of the underlying polynomial space Π^k.

To this end, let $N\in\mathbb{N}$ and let $\Xi := \{a := \xi_0 < \xi_1 < \ldots \xi_N < \xi_{N+1} =: b\} \subseteq \mathbb{R}$ be a knot set. The sequence $\boldsymbol{\xi} := (\xi_0, \xi_1, \ldots, \xi_{N+1}) \in \mathbb{R}^{\mathbb{Z}}$ is called the *ordered knot sequence associated with the knot set* Ξ. Conversely, every ordered knot sequence $\boldsymbol{\xi}$ determines a knot set Ξ.

Now, suppose that $\boldsymbol{\xi} := (\xi_0, \xi_1, \ldots, \xi_{N+1})$ is an ordered knot sequence with $a := \xi_0$ and $b := \xi_{N+1}$. Define

$$\Pi_{\boldsymbol{\xi}}^k[a, b] := \left\{f\in\mathrm{Map}([a, b], \mathbb{R}) \,|\, \forall v = 0, 1, \ldots, N : f|_{[\xi_v, \xi_{v+1})} \in \Pi^k[a, b]\right\}.$$

We will write $\Pi_{\boldsymbol{\xi}}^k := \Pi_{\boldsymbol{\xi}}^k[a, b]$ when the interval is understood. The set $\Pi_{\boldsymbol{\xi}}^k$ becomes a linear space under the usual definition of function addition and scalar multiplication of functions that contains Π^k. The linear space Π_ξ^k is called the *space of piecewise polynomial functions of order k on [a, b] with knot sequence ξ*. In order for Π_ξ^k to be well-defined, we impose that all functions in Π_ξ^k are right continuous. (For otherwise, one loses single valuedness at knots.)

2.7.1 Properties of the Spaces Π_ξ^k

The spaces Π_ξ^k enjoy the following properties whose proofs are left to the reader.

$$(1) \quad \Pi_\xi^k = \bigoplus_{i=1}^{k} \Pi^k. \tag{2.28}$$

$$(2) \quad \dim \Pi_\xi^k = (N+1)k. \tag{2.29}$$

$$(3) \quad S^k(\Xi) = \Pi_\xi^k \cap C^{k-2}[a, b]. \tag{2.30}$$

$$(4) \quad \Pi_\xi^k \subseteq C^{-1}[a, b]. \text{ (Here } \subseteq \text{ denotes vector space inclusion.)} \tag{2.31}$$

Figure 2.4 shows the graph of a piecewise polynomial function in Π_ξ^3 with knot sequence $\xi = (0, 1.5, 3.5, 4.5, 6)$.

2.7.2 Basis for Π_ξ^k

Let $f \in \Pi_\xi^k$. For $v = 0, 1, \ldots, N$ and $\mu = 1, \ldots, k$, define functionals $\Lambda_{v\mu} : \Pi_\xi^k \to \mathbb{R}$ by

$$\Lambda_{v\mu} f := \begin{cases} (D^{\mu-1}f)(\xi_0), & v = 0; \\ \text{jump}_{\xi_v} D^{\mu-1}f, & v = 1, \ldots, N. \end{cases} \tag{2.32}$$

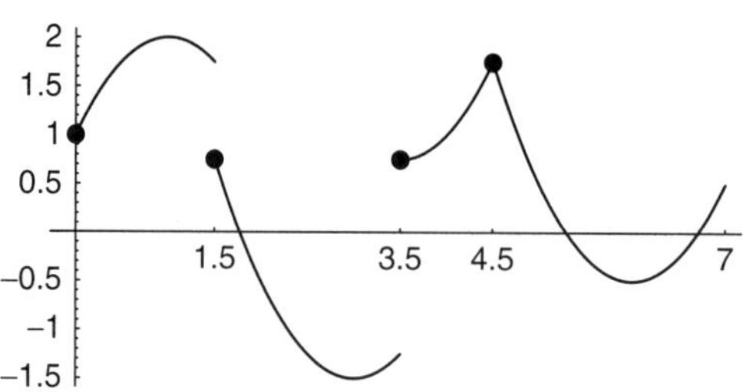

Figure 2.4 The graph of a piecewise polynomial function.

Corresponding to the functionals $\Lambda_{\nu\mu}$, we define functions $\varphi_{\nu\mu} : [a, b] \to \mathbb{R}$ via

$$
\varphi_{\nu\mu} := \begin{cases} \dfrac{(\bullet - \xi_0)^{\mu-1}}{(\mu - 1)!}, & \nu = 0; \\[2ex] \dfrac{(\bullet - \xi_\nu)_+^{\mu-1}}{(\mu - 1)!}, & \nu = 1, \ldots, N. \end{cases}
\tag{2.33}
$$

From the definition it follows that $\varphi_{\nu\mu} \in \Pi_{\xi}^k$, $\forall \nu \in \{0, 1, \ldots, N\}$ and $\forall \mu \in \{1, \ldots, k\}$. Moreover, it is straight-forward to show that $\forall \lambda, \nu \in \{0, 1, \ldots, N\}$ and $\forall \varkappa, \mu \in \{1, \ldots, k\}$:

$$
\Lambda_{\nu\mu}\varphi_{\lambda\varkappa} = \delta_{\nu\lambda}\delta_{\mu\varkappa}.
$$

Since the $\Lambda_{\nu\mu} \in (\Pi_{\xi}^k)^*$, where * denotes the *algebraic* dual space, it follows from linear algebra that the set

$$
\Phi := \{\varphi_{\nu\mu} \mid \nu = 0, 1, \ldots, N; \; \mu = 1, \ldots, k\}
$$

is linearly independent. In addition, card $\Phi = (N + 1)k = \dim \Pi_{\xi}^k$ and thus Φ forms a basis for Π_{ξ}^k. (The reader is encouraged to verify this statement.) We summarize these arguments in a proposition.

Proposition 2.37 Let $k, N \in \mathbb{N}$ and let $\boldsymbol{\xi} := (\xi_0, \xi_1, \ldots, \xi_{N+1})$ be an ordered knot sequence. Further, suppose that functionals $\Lambda_{\nu\mu}$ and associated functions $\varphi_{\nu\mu}$ are given as in (2.32) and (2.33), respectively. Then every $f \in \Pi_{\xi}^k$ can be uniquely represented in the form

$$
f = \sum_{\nu=0}^{N} \sum_{\mu=1}^{k} \Lambda_{\nu\mu}\,\varphi_{\nu\mu},
$$

or, more explicitly, by

$$
f = \sum_{\mu=1}^{k} \frac{(D^{\mu-1}f)(\xi_0)}{(\mu - 1)!} (\bullet - \xi_0)^{\mu-1} + \sum_{\nu=1}^{N} \sum_{\mu=1}^{k} \frac{\mathrm{jump}_{\xi_\nu} D^{\mu-1}f}{(\mu - 1)!} (\bullet - \xi_\nu)^{\mu-1}.
\tag{2.34}
$$

There are no smoothness conditions imposed on $f \in \Pi_{\xi}^k$ at the knots ξ_ν, $\nu = 1, \ldots, N$. One obtains linear subspaces of Π_{ξ}^k by requiring that the following conditions hold at these knots.

$$
\mathrm{jump}_{\xi_\nu} D^{\mu-1}f = 0, \quad \begin{cases} \forall \nu \in \{1, \ldots, N\}; \\ \forall \mu \in \{1, \ldots, m_\nu\}, \end{cases}
\tag{2.35}
$$

where $m_\nu = 1, \ldots, k - 1$.

Now, let $\boldsymbol{m} := (m_1, \ldots, m_N)^\top \in \mathbb{N}^N$ (Here $^\top$ denotes the *transpose* of a vector.) and assume that for fixed $1 < k \in \mathbb{N}$ such a vector whose components are elements from $\{1, \ldots, k-1\}$ is given. Then define

$$\Pi^k_{\boldsymbol{\xi m}} := \left\{ f \in \Pi^k_{\boldsymbol{\xi}} \,|\, \forall \nu = 1, \ldots, N;\ \forall \mu = 1, \ldots, m_\nu : \mathrm{jump}_{\xi_\nu} D^{\mu-1} f = 0 \right\}.$$

$$(2.36)$$

The linearity and homogeneity of the conditions (2.35) implies that $\Pi^k_{\boldsymbol{\xi m}}$ is a linear subspace of $\Pi^k_{\boldsymbol{\xi}}$. Under consideration of conditions (2.35), we obtain from (2.34) as a basis for the spaces $\Pi^k_{\boldsymbol{\xi m}}$ the set

$$\Psi := \{\varphi_{\nu\mu} \,|\, \mu = m_\nu + 1, \ldots, k;\ \nu = 0, 1, \ldots, N\},$$

with $m_0 := 0$. Therefore,

$$\dim \Pi^k_{\boldsymbol{\xi m}} = \sum_{\nu=0}^{N} (k - m_\nu) = (N+1)k - \sum_{\nu=1}^{N} m_\nu.$$

Note that for $m_\nu := k - 1, \forall \nu \in \{1, \ldots, N\}$, i.e., $\boldsymbol{m} := \boldsymbol{k} - \boldsymbol{1} \in \mathbb{R}^N$,

$$\Pi^k_{\boldsymbol{\xi}, k-1} = S^k(\Xi) = \mathrm{span}\{B_{\nu k} \,|\, \nu = 1, \ldots, k + N\}.$$

Since there are no continuity and smoothness conditions imposed on the functions in $\Pi^k_{\boldsymbol{\xi}}$, i.e., $\boldsymbol{m} = \boldsymbol{0} := (0, \ldots, 0)^\top \in \mathbb{N}_0^N$, it is useful to define

$$\Pi^k_{\boldsymbol{\xi}, 0} := \Pi^k_{\boldsymbol{\xi}}.$$

With this definition, we then have the following chain of linear spaces for any knot sequence $\boldsymbol{\xi}$ and associated knot set Ξ:

$$\Pi^k \subsetneq S^k(\Xi) = \Pi^k_{\boldsymbol{\xi}, k-1} \subseteq \Pi^k_{\boldsymbol{\xi m}} \subseteq \Pi^k_{\boldsymbol{\xi} 0} = \Pi^k_{\boldsymbol{\xi}} \subsetneq C^{-1}(\inf \boldsymbol{\xi}, \sup \boldsymbol{\xi}). \qquad (2.37)$$

Suppose now that $\boldsymbol{\xi}$ is a nondecreasing sequence of real numbers, i.e., $\boldsymbol{\xi} \in \mathbb{R}^{\mathbb{Z}}$ and $\xi_\nu \leq \xi_{\nu+1}, \nu \in \mathbb{Z}$. In the following, such a sequence $\boldsymbol{\xi}$ will be called a *knot sequence* and, if $\mathrm{card}\, \boldsymbol{\xi} < \infty$, a *finite knot sequence*. The terms ξ_ν in the sequence $\boldsymbol{\xi}$ are called *knots*. We write $\xi_\nu \in \boldsymbol{\xi}$ to denote this fact. Each knot sequence determines again a knot set $\Xi := \{\ldots \leq \xi_{\nu-1} \leq \xi_\nu \leq \xi_{\nu+1} \leq \ldots\} \subseteq \mathbb{R}$. For a fixed knot sequence $\boldsymbol{\xi}$, we introduce the function

$$\mathrm{mult} := \mathrm{mult}_{\boldsymbol{\xi}} : \mathbb{R} \to \mathbb{N}^\infty := \mathbb{N} \cup \{\infty\}$$

$$\xi_\nu \mapsto \mathrm{card}\{\mu \in \mathbb{N}_0 \,|\, \xi_\mu = \xi_\nu\},$$

and call it the *multiplicity of the knot ξ_ν in the sequence $\boldsymbol{\xi}$*. A knot ξ_ν whose multiplicity $\mathrm{mult}\, \xi_\nu > 1$ is called a *multiple knot* and a knot ξ_ν with $\mathrm{mult}\, \xi_\nu = 1$ is called a *simple knot*.

Example 2.38 Let $\boldsymbol{\xi}_1 := \{\xi_0, \xi_0, \xi_1, \xi_2, \xi_2, \xi_2, \xi_2\}$. Then $\mathrm{card}\, \boldsymbol{\xi} = 7$, $\mathrm{mult}\, \xi_0 = 2$, $\mathrm{mult}\, \xi_1 = 1$, and $\mathrm{mult}\, \xi_2 = 4$. The knot sequence $\boldsymbol{\xi}_1$ is obtained from the *ordered* knot sequence $\boldsymbol{\xi}_1^* := \{\xi_0, \xi_1, \xi_2\}$ by repeating the knot ξ_0 once and the knot ξ_2 thrice.

Example 2.39 Define $\boldsymbol{\xi}_2 := \{\cdots, 0, 0, 1, 1, \cdots\}$. Then $\operatorname{card}\boldsymbol{\xi} = \infty$ and $\operatorname{mult} 0 = \operatorname{mult} 1 = \infty$. The knot sequence $\boldsymbol{\xi}_2$ defines the so-called *Bernstein polynomials* on $[0, 1]$:

$$b_\nu^k(x) := \binom{k-1}{\nu} x^{k-1-\nu} (1-x)^\nu, \quad \nu = 0, 1, \ldots, k-1.$$

Let $\boldsymbol{\xi}$ be a knot sequence and let $\xi_\nu, \xi_{\nu+1} \in \boldsymbol{\xi}$ be such that $\inf \boldsymbol{\xi} < \xi_\nu < \xi_{\nu+1} < \sup \boldsymbol{\xi}$. Furthermore, for $k \in \mathbb{N}$, let $f \in C^k(\inf \boldsymbol{\xi}, \sup \boldsymbol{\xi})$. Consider the limit

$$\lim_{\xi_{\nu+1} \to \xi_\nu} \frac{f(\xi_\nu) - f(\xi_{\nu+1})}{\xi_\nu - \xi_{\nu+1}} = \lim_{\xi_{\nu+1} \to \xi_\nu} [\xi_\nu, \xi_{\xi_{\nu+1}}] f = (Df)(\xi_\nu).$$

In other words, if the knots ξ_ν and $\xi_{\nu+1}$ coalesce, then the value of f *and* Df are interpolated at ξ_ν. The above process generalizes to:

$$\forall \ell \in \mathbb{N} : \ell! \, [\xi_\nu, \ldots, \xi_{\nu+\ell}] f = (D^\ell f)(\xi_\nu), \tag{2.38}$$

if $\xi_\nu = \xi_{\nu+1} = \cdots = \xi_{\nu+\ell}$. In particular, it follows from this observation that the Taylor expansion of f up to order ℓ, namely,

$$\sum_{\lambda=0}^{\ell} \frac{(D^\lambda f)(\xi_\nu)}{\lambda!} (\bullet - \xi_\nu)^\lambda,$$

is an interpolating polynomial for f at the ℓ-times repeated knot ξ_ν.

A knot ξ_ν of multiplicity $\ell + 1$ loses for this reason ℓ degrees of freedom. Setting $t := \xi_\nu = \xi_{\nu+1} = \cdots = \xi_{\nu+\ell}$ in Marsden's Identity gives

$$(x - \xi_\nu)^{k-1} = \sum_{\nu=1}^{n} (-1)^{k-1} \psi_{\nu k}(\xi_\nu) B_{\nu k}(x)$$

and since $\psi_{\nu k} = \prod_{\mu=\nu+1}^{\nu+k-1} (\xi_\nu - \xi_\mu)$, only those terms with indices greater than $\nu + \ell$ or less than ν are nonzero. The proof of Proposition 2.11 now shows that

$$\frac{(\bullet - \xi_\nu)_+^{k-\ell}}{(k-\ell)!} = \sum_{\mu \geq \nu} \frac{(-1)^{k-\ell}(D^{\ell-1}\psi_{\nu k})(\xi_\nu)}{(k-1)!} B_{\nu k} \in \operatorname{span}\{B_{\nu k} \mid \nu = 1, \ldots, n\},$$

$\forall \nu \in \{1, \ldots, n\}$ and $\forall \ell \in \{1, \ldots, \operatorname{mult} \xi_\nu\}$. Note that $\dfrac{(\bullet - \xi_\nu)_+^{k-\ell}}{(k-\ell)!} \in C^{k-\ell-1}$ $(\inf \boldsymbol{\xi}, \sup \boldsymbol{\xi})$.

These observations now prove the next theorem.

Theorem 2.40 (Curry–Schoenberg) Let $k, N \in \mathbb{N}$ and let $\boldsymbol{\xi} := (\xi_0, \ldots, \xi_{N+1})$ be an ordered knot sequence. Furthermore, let $m \in \mathbb{N}^N$ with $m_i \in \{1, \ldots, k-1\}$, $i = 1, \ldots, N$, be given. Set

$$n := k + \sum_{i=1}^{N}(k - m_i) = k(N+1) - \sum_{i=1}^{N} m_i = \dim \Pi_{\boldsymbol{\xi}, m}^{k},$$

and define a new knot sequence $\boldsymbol{x} := (x_1, \ldots, x_{n+k})$ as follows:

$$x_1 \le x_2 \le \cdots \le x_k = \xi_0$$

$$x_{k + \sum_{\mu=1}^{\nu}(k - m_\mu - 1)} = \xi_\nu, \quad \nu = 1, \ldots, N,$$

$$x_{n+1} = \xi_{N+1} \le x_{n+2} \le \cdots \le x_{n+k}.$$

Then the set $\{B_{\nu k} \mid \nu = 1, \ldots, n\}$ forms a basis for $\Pi_{\boldsymbol{\xi}, m}^{k}$ on $[x_1, x_{n+1}]$.

Remark 2.41 Recall that m_i gives the number of smoothness conditions at the knot ξ_i, and by the above considerations, $k - m_i$ indicates the multiplicity of the knot ξ_i, $i = 1, \ldots, N$. Therefore, one has

$$k = (k - m_i) + m_i = \operatorname{mult} \xi_i + m_i, \quad \text{(at knot } \xi_i). \tag{2.39}$$

In words,

$$\text{Order} = \text{Multiplicity} + \text{Smoothness}.$$

It follows from Theorem 2.40 that the choice of the first k and last k knots is arbitrary and, since in general no information is available beyond the endpoints of the interval ξ_0, ξ_{N+1}, it is customary to set

$$x_1 = \cdots = x_k := \xi_0 \quad \text{and} \quad x_{n+1} = \cdots = x_{n+k} := \xi_{N+1},$$

i.e., no smoothness conditions are imposed at the boundary knots.

The following schematics brings the above-introduced concepts into relation. Let $k \in \mathbb{N}$ be fixed and let $\boldsymbol{\xi}$ be a given knot sequence with $a := \inf \boldsymbol{\xi}$ and $b := \sup \boldsymbol{\xi}$.

Π^k	$\subsetneq S^k(\Xi) = \Pi_{\boldsymbol{\xi}, k-1}^{k}$	$\subseteq \Pi_{\boldsymbol{\xi}, m}^{k}$	$\subseteq \Pi_{\boldsymbol{\xi}, 0}^{k} = \Pi_{\boldsymbol{\xi}}^{k}$	$\subsetneq C^{-1}[a,b]$
global basis	local basis	local basis	local basis	
maximal smoothness at knots	maximal smoothness at knots	some smoothness at knots	no smoothness at knots	no smoothness at knots
	minimal multiplicity at knots	some multiplicity at knots	maximal multipliciy at knots	
	$\operatorname{mult} \xi_i = 1$	$\max \operatorname{mult} \xi_i > 1$	$\operatorname{mult} \xi_i = k - 1$	

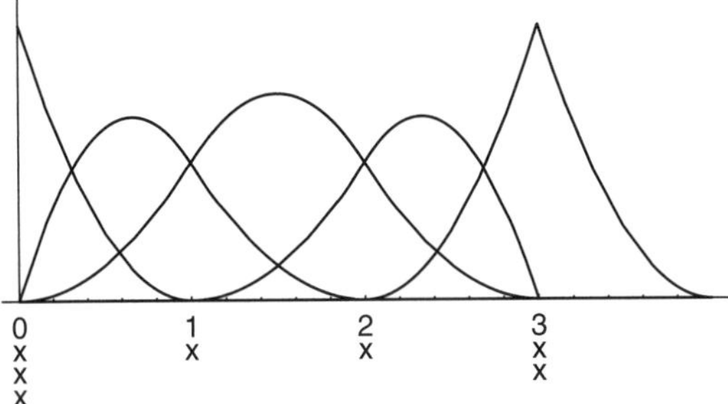

Figure 2.5   B-splines with repeated knots.

In Figure 2.5, some B-splines of order $k := 3$ are depicted for a sequence of knots with different multiplicities. The knot multiplicities are indicated by x.

Local Character of the B-Spline Basis for $\Pi^k_{\xi,m}$

Under the conditions of the Curry–Schoenberg Theorem, every $f \in \Pi^k_{\xi,m}$ allows a unique representation of the form

$$f = \sum_{v=1}^{n} c_v \, B_{vk}, \qquad c_v \in \mathbb{R}, \tag{2.40}$$

on the interval $[x_k, x_{n+1}]$.

Now, choose an $x \in [x_k, x_{n+1}]$. Then there exists a $\mu \in \{k, \ldots, n\}$ such that $x_\mu \le x \le x_{\mu+1}$. The compactness of the support of the B-splines B_{vk} then gives the *local representation of f at x* as

$$f(x) = \sum_{v=\mu-k+1}^{\mu} c_v \, B_{vk}(x),$$

i.e., a representation with a relatively small number $(= k)$ of nonzero coefficients c_v. This situation is summarized in Figure 2.6.

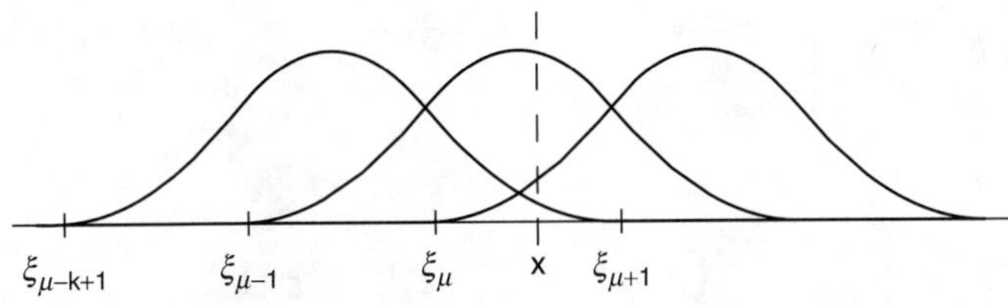

Figure 2.6   Local character of B-spline basis.

2.8 Hermite Splines

A special case of repeated knots is the following setting. Let $k, N \in \mathbb{N}$ and let $\boldsymbol{\xi} := (\xi_0, \xi_1, \ldots, \xi_{N+1})$ be an ordered knot sequence. Suppose that $\boldsymbol{m} := (r, \ldots, r) \in \mathbb{N}^N$ and $\text{mult}\,\xi_v = r$, $\forall v = 1, \ldots, N$. These assumptions imply that $k - r = r$, i.e., $k = 2r$. Letting $a := \xi_0$ and $b := \xi_{N+1}$, one has the inclusion

$$\Pi_{\boldsymbol{\xi},r}^{2r} \subseteq C^{2r-r-1}[a, b] = C^{r-1}[a, b].$$

An element $f \in \Pi_{\boldsymbol{\xi},r}^{2r}$ thus satisfies r smoothness conditions of the form

$$f(\xi_v), (Df)(\xi_v), \ldots, (D^{r-1}f)(\xi_v) \tag{2.41}$$

at each knot $\xi_v \in \boldsymbol{\xi}$. The functions in $\Pi_{\boldsymbol{\xi},r}^{2r}$ are called *Hermite Splines*.

Now, in addition, assume that $\text{mult}\,\xi_0 = \text{mult}\,\xi_{N+1} = r$. Recall that $\dim \Pi_{\boldsymbol{\xi},r}^{2r} = 2r + Nr = r(N + 2)$. The number of conditions at each knot ξ_v, $v = 0, 1, \ldots, N + 1$, is r, for a total of $r(N + 2)$ conditions. Thus, there exists a polynomial $p \in \Pi^{r(N+2)}$ so that for each function $f \in C^{r-1}$ the interpolation problem

$$(D^m p)(\xi_v) = (D^m f)(\xi_v), \quad \forall v \in \{0, 1, \ldots, N + 1\} \; \forall m \in \{0, 1, \ldots, r - 1\},$$

has a unique solution. This is an example of *Hermite Interpolation*.

Consider the setting when $r := 2$ and $\boldsymbol{\xi} := \mathbb{Z}$. The elements of $\Pi_{\mathbb{Z},2}^4$ are referred to as *cubic cardinal Hermite splines*. A basis for $\Pi_{\mathbb{Z},2}^4$ on the interval $[\xi_0, \xi_{N+1}]$ is given by the set of piecewise cubic polynomials $\{J_v, S_v \,|\, v = 1, \ldots, N\}$, where

$$J_1(x) := \begin{cases} 3(2 - x)^2 - 2(2 - x)^3, & x \in [1, 2], \\ 0, & \text{elsewhere;} \end{cases}$$

$$J_v(x) := \begin{cases} 3(x - v + 1)^2 - 2(x - v + 1)^3, & x \in [v - 1, v], \\ 3(v + 1 - x)^2 - 2(v + 1 - x)^3, & x \in [v, v + 1], \\ 0, & \text{elsewhere;} \end{cases}$$

$$J_N(x) := := \begin{cases} 3(x - N + 1)^2 - 2(x - N + 1)^3, & x \in [N - 1, N], \\ 0, & \text{elsewhere;} \end{cases}$$

and

$$S_1(x) := \begin{cases} (x - 2)^2 - (x - 2)^3, & x \in [1, 2], \\ 0, & \text{elsewhere;} \end{cases}$$

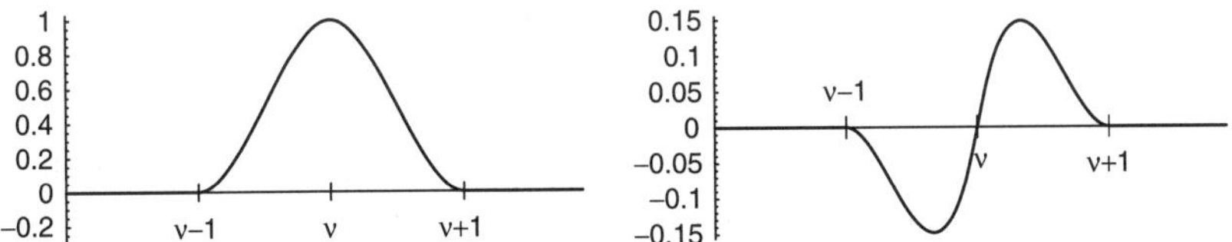

Figure 2.7   Examples of the basis functions J_v and S_v.

$$
S_v(x) := \begin{cases} -(x-v+1)^2 + (x-v+1)^3, & x \in [v-1, v], \\ (v+1-x)^2 - (v+1-x)^3, & x \in [v, v+1], \\ 0, & \text{elsewhere;} \end{cases}
$$

$$
S_N(x) := := \begin{cases} -(x-N+1)^2 + (x-N+1)^3, & x \in [N-1, N], \\ 0, & \text{elsewhere.} \end{cases}
$$

The basis functions J_v and S_v satisfy

$$
J_v(\mu) = \delta_{v\mu}, \qquad\qquad (DJ_v)(\mu) = 0,
$$
$$
S_v(\mu) = 0, \qquad\qquad |(DS_v)(\mu)| = \delta_{v\mu}.
$$

Moreover, every $f \in \Pi^4_{\mathbb{Z},2}[\xi_0, \xi_{N+1}]$ has a representation of the form

$$
f = \sum_{v=1}^{N} f(\xi_v)\, J_v + (Df)(\xi_v)\, S_v.
$$

The graphs of some representative basis functions J_v and S_v are depicted in
Figure 2.7

2.9 Interpolation with Splines

In this section, we consider the existence and uniqueness of the spline
interpolation problem and derive some error estimates. We also show that
splines satisfy a minimality property with respect to a certain semi-norm.

2.9.1 Preliminaries

Let $f \in C[a, b]$ be given and assume that the values of f are known on a
finite knot set $\Xi := \{a := \xi_1 < \cdots < \xi_n =: b\}$, $1 < n \in \mathbb{N}$. We seek a spline
$s \in \text{span}\{B_{vk} \mid v = 1, \ldots, n\}$ supported on a nondecreasing knot sequence
$\boldsymbol{x} := (x_1, \ldots, x_{n+k})$, where $x_v < x_{v+k}$, $\forall v \in \{1, \ldots, n\}$, so that

$$
\forall \mu \in \{1, \ldots, n\}: \quad s(\xi_\mu) = f(\xi_\mu).
$$

We abbreviate the above interpolation conditions as

$$s|_\Xi = f|_\Xi.$$

Since $s = \sum_{\nu=1}^n c_\nu B_{\nu k}$, for constants $c_\nu \in \mathbb{R}$, the above interpolation conditions are equivalent to the requirement to find a vector $c := (c_1, \ldots, c_n)^\top \in \mathbb{R}^n$, so that

$$\forall \mu \in \{1, \ldots, n\}: \quad \sum_{\nu=1}^n c_\nu B_{\nu k}(\xi_\mu) = f(\xi_\mu). \tag{2.42}$$

Introducing the vector $f := (f(\xi_1), \ldots, f(\xi_n))^\top \in \mathbb{R}^n$ and the matrix

$$\mathbf{B}(n, k; \Xi) := \begin{pmatrix} B_{1,k}(\xi_1) & \cdots & B_{nk}(\xi_1) \\ \vdots & \ddots & \vdots \\ B_{1,k}(\xi_n) & \cdots & B_{nk}(\xi_n) \end{pmatrix} \in \mathbb{R}^{n \times n},$$

we can express the equations in (2.42) in matrix form as

$$\mathbf{B}(n, k; \Xi)\, c = f. \tag{2.43}$$

In order to obtain a unique solution for the interpolation problem, the matrix $\mathbf{B}(n, k; \Xi)$ needs to be non-singular, i.e., invertible. The next theorem gives conditions for this to hold.

Theorem 2.42 (Schoenberg–Whitney[9])   The matrix $\mathbf{B}(n, k; \Xi)$ is invertible if and only if

$$\forall \mu \in \{1, \ldots, n\}: \quad B_{\mu k}(\xi_\mu) \neq 0, \tag{2.44}$$

or, equivalently,

$$\forall \mu \in \{1, \ldots, n\}: \quad x_\mu < \xi_\mu < x_{\mu+k}, \tag{2.45}$$

where $\xi_1 = x_1+$ and $\xi_n = x_n-$ is allowed.

Remark 2.43   We refer to the requirements (2.44) or (2.45) as the *Schoenberg–Whitney conditions*.

The theorem by Schoenberg and Whitney is a consequence of a much stronger result due to Karlin. To state this result, we need the following definition.

Definition 2.44 (Total Positivity)   A matrix $\mathbf{M}$ is called *totally positive* if the determinant of any of its square submatrices is positive.

Theorem 2.45 (Karlin)   The matrix $\mathbf{B}(n, k; \Xi)$ is totally positive.

---

9 Hassler Whitney, 23 March 1907–10 May 1989. American Mathematician who primarily worked in differential topology. He also contributed to graph theory, combinatorics, and integration theory.

As a corollary to the Schoenberg–Whitney theorem we obtain the result that the matrix $\mathbf{B}(n, k; \Xi)$ is banded.

Corollary 2.46 The matrix $\mathbf{B}(n, k; \Xi)$ has bandwidth less than k.

Proof The condition $B_{\nu k}(\xi_\mu) \neq 0$ is equivalent to $x_\nu < \xi_\mu < x_{\nu+k}$. Since $B_{\mu k}(\xi_\mu) \neq 0$, one has that on one hand $x_\nu < \xi_\mu < x_{\mu+k}$ and on the other hand $x_\mu < \xi_\mu < x_{\nu+k}$. These inequalities together with the fact that the support of a kth order B-spline ranges over $k + 1$ consecutive knot indices imply that $|\mu - \nu| < k$. $\square$

Now suppose that $x := \{x_1, \ldots, x_{n+k}\}$ is a nondecreasing knot sequence satisfying $x_\nu < x_{\nu+k}, \forall \nu \in \{1, \ldots, n\}$, that X is its associated knot set, and that $\{B_{\nu k, X} \mid \nu = 1, \ldots, n\}$ is the associated set of B-splines. Define

$$\mathcal{S}^k(X) := \mathrm{span}\{B_{\nu k, X} \mid \nu = 1, \ldots, n\}.$$

The Schoenberg–Whitney theorem has shown that every function $f \in C[a, b]$ defined on a finite knot set $\Xi := \{a := \xi_1 < \cdots < \xi_n =: b\}, 1 < n \in \mathbb{N}$, can be uniquely interpolated by a spline $s \in \mathcal{S}^k(X)$ provided that the knot set X is chosen so that $x_\mu < \xi_\mu < x_{\mu+k}, \forall \mu \in \{1, \ldots, n\}$. Such splines are called *interpolating splines* for f on Ξ.

As a side result, we obtain from the Schoenberg–Whitney theorem that the spaces $\mathcal{S}^k(X)$ and $S^k(\Xi)$ coincide on $[a, b]$:

$$\mathcal{S}^k(X)|_{[a,b]} = S^k(\Xi).$$

In the following, in order to ease the notation, we delete the explicit dependence of the B-splines from either the knot set X or the knot sequence x, when no confusion is expected.

Let $n \in \mathbb{N}$ and assume that $\Xi := \{\xi_1 < \xi_2 < \cdots < \xi_n\}$ is a knot set satisfying the Schoenberg–Whitney conditions. Then the above considerations show that there exists an *interpolation projector*

$$\pi_\Xi : C[a, b] \to \mathcal{S}^k(X)$$

$$f \mapsto \pi_\Xi f := \sum_{\nu=1}^{n} c_\nu B_{\nu k}, \quad \text{on } [a, b],$$

where $c := (c_1, \ldots, c_n)^\top$ is the uniquely determined solution of (2.42) for which

$$\pi_\Xi f|_\Xi = s|_\Xi = f|_\Xi.$$

The operator π_Ξ is a linear projection and satisfies Lebesgue's Inequality:

$$\|f - \pi_\Xi f\|_{\infty,[a,b]} = \|(f - s) - (\pi_\Xi f - s)\|_{\infty,[a,b]} \leq \|(I - \pi_\Xi)(f - s)\|_{\infty,[a,b]}$$

$$\leq (1 + \|\pi_\Xi\|)\|f - s\|_{\infty,[a,b]},$$

$$\tag{2.46}$$

$\forall s \in \mathcal{S}^k(X)$. To obtain the first inequality, we used the fact that π_Ξ is a projection. Equation (2.46) gives a first estimate on the accuracy of the interpolation of a continuous function in terms of a B-spline series.

Rewriting (2.46) slightly yields

$$\|f - s\|_{\infty,[a,b]} = \|(f - \pi_\Xi f) - (s - \pi_\Xi f)\|_{\infty,[a,b]} \le \|f - \pi_\Xi f\|_{\infty,[a,b]}$$
$$+ \|s - \pi_\Xi f\|_{\infty,[a,b]},$$

$\forall s \in \mathcal{S}^k(X)$, with equality if and only if $\|s - \pi_\Xi f\|_{\infty,[a,b]} = 0$. In other words, any other $s \in \mathcal{S}^k(X)$ gives a larger Chebyshev norm than the interpolation projection $\pi_\Xi f$.

To estimate the norm of the interpolation projector π_Ξ in (2.46), we proceed as follows.

$$\|\pi_\Xi f\|_{\infty,[a,b]} = \left\| \sum_{\nu=1}^{n} c_\nu\, B_{\nu k} \right\|_{\infty,[a,b]} \le \max\{|c_\nu|\} \left\| \sum_{\nu=1}^{n} B_{\nu k} \right\|_{\infty,[a,b]} \le \|\mathbf{c}\|_{\ell^\infty}.$$

(Since the B-splines are nonnegative and form a partition of unity.) The matrix equation (2.43) gives

$$\mathbf{c} = (\mathbf{B}(n, k; \Xi))^{-1} f,$$

and, therefore,

$$\|\mathbf{c}\|_{\ell^\infty} = \|(\mathbf{B}(n, k; \Xi))^{-1} f\|_{\ell^\infty} \le \|(\mathbf{B}(n, k; \Xi))^{-1}\|_{\ell^\infty} \cdot \|f\|_{\ell^\infty}$$

$$\le \|(\mathbf{B}(n, k; \Xi))^{-1}\|_{\ell^\infty} \cdot \|f\|_{\infty,[a,b]}.$$

Thus,

$$\|\pi_\Xi f\|_{\infty,[a,b]} \le \|(\mathbf{B}(n, k; \Xi))^{-1}\|_{\ell^\infty} \cdot \|f\|_{\infty,[a,b]}. \tag{2.47}$$

Defining the norm of the interpolation projector by

$$\|\pi_\Xi\|_{\infty,[a,b]} := \sup_{0 \ne f \in C[a,b]} \frac{\|\pi_\Xi f\|_{\infty,[a,b]}}{\|f\|_{\infty,[a,b]}},$$

and noting that (2.47) is trivially fulfilled for $f = 0$, we obtain the next proposition.

Proposition 2.47 $\|\pi_\Xi\|_{\infty,[a,b]} \le \|(\mathbf{B}(n, k; \Xi))^{-1}\|_{\ell^\infty}.$

We finally remark, that in the language of Section 1.4 the interpolation projector π_Ξ is a linear projection method.

2.9.2 Error Estimates for Spline Interpolation

Before we derive error estimates for spline interpolation, a few definitions are needed and a new class of function spaces has to be introduced. To this end, let $k \in \mathbb{N}_0$, let $n \in \mathbb{N}$, and let $\Omega \in \mathbb{R}^n$. We denote by $C^k(\Omega) := C^k(\Omega, \mathbb{R})$ the set of all functions $f : \Omega \to \mathbb{R}$ with the property that f and all its partial derivatives up to order k are continuous on Ω. To state these conditions more precisely, we will employ *multi-index notation*.

Let $m_1, \ldots, m_n \in \mathbb{N}_0$. The quantity $\boldsymbol{m} := (m_1, \ldots, m_n)$ is called a *multi-index*. For $\alpha \in \mathbb{R}$ and $\boldsymbol{x} \in \mathbb{R}^n$, we define the expressions $\boldsymbol{m}^\alpha$, $\boldsymbol{x}^{\boldsymbol{m}}$, and $\boldsymbol{m}!$ by

$$\boldsymbol{m}^\alpha := (m_1^\alpha, \ldots, m_n^\alpha), \quad \boldsymbol{x}^{\boldsymbol{m}} := (x_1^{m_1}, \ldots, x_n^{m_n}), \quad \boldsymbol{m}! := m_1! \cdots m_n!.$$

The length of a multi-index is defined as $|\boldsymbol{m}| := m_1 + \cdots m_n$. For $\alpha \in \mathbb{R}$ and multi-indices $\boldsymbol{k}$ and $\boldsymbol{m}$, we also define $\boldsymbol{k} \prec \alpha$ iff $k_i \prec \alpha$ and $\boldsymbol{k} \prec \boldsymbol{m}$ iff $k_i \prec m_i$, $\forall i \in \{1, \ldots, n\}$, where $\prec$ is any one of the binary relation symbols $<, \leq, >, \geq, =$. For $\alpha \in \mathbb{R}$ and a multi-index $\boldsymbol{k}$, we set $\alpha \boldsymbol{k} := (\alpha k_1, \ldots, \alpha k_n)$.

Remark 2.48   Note that the relations $\leq$ and $\geq$ defined on multi-indices are only *partial* orders. There exist multi-indices such as for instance $\boldsymbol{k} := (1, 0)$ and $\boldsymbol{m} := (0, 1)$ that cannot be compared.

With this notation, we can express any partial derivative of the form $\dfrac{\partial^{m_1 + \cdots m_n}}{\partial x_1^{m_1} \cdots \partial x_n^{m_n}}$, where $m_1, \ldots, m_n \in \{0, 1, \ldots, m\}$ with $m_1 + \ldots + m_n = m$, more compactly as $\dfrac{\partial^{|\boldsymbol{m}|}}{\partial \boldsymbol{x}^{\boldsymbol{m}}}$, or simply, as $\partial^{|\boldsymbol{m}|}$, if no confusion arises as to which variables is differentiated.

Example 2.49   For $n := 3$, the partial derivatives $\dfrac{\partial^3}{\partial x_1 \partial x_3^2}$ and $\dfrac{\partial^5}{\partial x_1^2 \partial x_2 \partial x_3^2}$ are in multi-index notation simply $\dfrac{\partial^{|\boldsymbol{m}|}}{\partial \boldsymbol{x}_{\boldsymbol{m}}}$, where $\boldsymbol{m} := (1, 0, 2)$ for the first and $\boldsymbol{m} := (2, 1, 2)$ for the second expression.

Equipped with multi-index notation, we state the definition of $C^k(\Omega)$ now more precisely as

$$C^k(\Omega) := \left\{ f \in \mathrm{Map}(\Omega, \mathbb{R}) \,|\, \partial^{|\boldsymbol{m}|} f \text{ continuous on } \Omega \text{ for } 0 \leq |\boldsymbol{m}| \leq k \right\}.$$

Similar to the procedure in Example 1.5, we define the space $C^k(\overline{\Omega})$ by

$$C^k(\overline{\Omega}) := \left\{ f \in C^k(\Omega) \,|\, \partial^{|\boldsymbol{m}|} f \text{ bounded and uniformly continuous on } \Omega \text{ for } \right.$$
$$\left. 0 \leq |\boldsymbol{m}| \leq k \right\}.$$

Now let $k \in \mathbb{N}$ and let $p \in [1, \infty)$. We define functionals $\| \ \|_{k,p} : C^k(\Omega) \to \mathbb{R}$ as follows.

$$\|f\|_{k,p} := \left(\sum_{0 \leq |\boldsymbol{m}| \leq k} \|D^{\boldsymbol{m}} f\|_{L^p}^p \right)^{1/p}.$$

The functionals $\| \; \|_{k,p}$ define norms on the linear spaces $C^k(\Omega)$. (The reader is encouraged to verify this statement.) The spaces $C^k(\Omega)$, however, are not complete with respect to these norms.

Definition 2.50 (Sobolev[10] Spaces)   Let $k \in \mathbb{N}_0$ and let $p \in [1, \infty)$. The *Sobolev spaces* $H^{k,p}(\Omega)$ are defined as the completion of $\{f \in C^k(\Omega) \,|\, \|f\|_{k,p} < \infty\}$ with respect to the norms $\| \; \|_{k,p}$.

Notice that for $k := 0$, the norms $\| \; \|_{0,p} = \| \; \|_{L^p}$ and thus $H^{0,p}(\Omega)$ can be identified with $L^p(\Omega)$. In addition, we have the following strict inclusion

$$H^{k,p}(\Omega) \subsetneq L^p(\Omega).$$

As an example of a function that is in $L^p(\Omega)$ but not in $H^{k,p}(\Omega)$, we set $\Omega := (-1, 1)$ and consider the function

$$\Theta : (-1, 1) \to \mathbb{R},$$

$$x \mapsto \begin{cases} 1, & x \in [0, 1) \\ 0, & x \in (-1, 0) \end{cases}.$$

Then $\Theta \in L^p(-1, 1)$, $p \in [1, \infty)$, but Θ is clearly not differentiable at $x = 0$, hence not in $H^{1,p}(-1, 1)$.

The next theorem is an immediate consequence of the definition of Sobolev spaces.

Theorem 2.51   The spaces $(H^{k,p}(\Omega), \| \; \|_{k,p})$ are Banach spaces.

Example 2.52   Let $\Omega := (-1, 1) \subset \mathbb{R}$, $k \in \mathbb{N}_0$, and consider $q : (-1, 1) \to \mathbb{R}$, $q(x) := x_+^k$. Then $q \in H^{k,p}(-1, 1)$, for $D^k q = k! \, \Theta \in L^p(-1, 1)$, $1 \le p < \infty$, where Θ is defined above, but $D^{k+1} q$ is not differentiable at $x = 0$.

Example 2.53   Now suppose that $\Omega := (0, 1) \times (0, 1) \subset \mathbb{R}^2$ and $f : \Omega \to \mathbb{R}$ is defined by

$$f(x, y) := \begin{cases} xy, & (x, y) \in (0, 1) \times (0, \tfrac{1}{2}), \\ x(1 - y), & (x, y) \in (0, 1) \times (\tfrac{1}{2}, 1), \\ x, & (x, y) \in (0, \tfrac{1}{2}) \times \{\tfrac{1}{2}\}, \\ \sqrt{\pi^e}, & (x, y) = (\tfrac{1}{2}, \tfrac{1}{2}), \\ x, & (x, y) \in (\tfrac{1}{2}, 1) \times \{1\}. \end{cases}$$

Then, $\partial f / \partial x, \partial f / \partial y \in L^p((0, 1) \times (0, 1))$, $p \in [1, \infty)$, and thus $f \in H^{1,p}((0, 1) \times (0, 1))$ as is easily verified.

---

10 Sᴇʀɢᴇɪ L'ᴠᴏᴠɪᴄʜ Sᴏʙᴏʟᴇᴠ, 6 October 1908–3 January 1989. Russian mathematician who worked in mathematical analysis and partial differential equations.

An important subclass of Sobolev spaces is obtained by taking $p := 2$. The spaces $H^{k,2}(\Omega)$ are then also denoted by $H^k(\Omega)$. Analog to the case of $L^2(\Omega)$, it is possible to define a bilinear form $\langle\,,\,\rangle_k : H^k(\Omega) \times H^k(\Omega) \to \mathbb{R}$ by

$$\langle f, g \rangle_k := \sum_{0 \le |m| \le k} \int_\Omega (D^m f)(x)(D^m g)(x)\,dx. \tag{2.48}$$

It is not difficult to verify that the bilinear form $\langle\,,\,\rangle_k$ is symmetric and positive definite, hence an inner product on $H^k(\Omega)$. Indeed, more can be said about the spaces $H^k(\Omega)$.

Theorem 2.54   The linear spaces $(H^k(\Omega), \langle\,,\,\rangle_k)$ are Hilbert spaces.

For the sequel, we also need the functional $|\,|_k := |\,|_{k,2} : H^k(\Omega) \to \mathbb{R}$ defined by

$$|f|_k := \sqrt{\sum_{|m|=k} \int_\Omega |(D^m f)(x)|^2 dx}.$$

Note that $|\,|_k$ is a semi-norm on the spaces $H^k(\Omega)$. Since in the semi-normed spaces $(H^k(\Omega), |\,|_k)$ the functions f and $f + p$, where $p \in \Pi^k$, have the same semi-norm, $|\,|_k$ may be interpreted as a norm modulo polynomials. The semi normed spaces $(H^k(\Omega), |\,|_k)$ are also *homogeneous* in the sense that $|f(\lambda \bullet)|_k = |\lambda|^k |f|_k$, for all $\lambda \in \mathbb{R}$.

For a bounded domain $\Omega \subset \mathbb{R}^n$, we also set

$$H^\infty(\Omega) := \bigcap_{k \in \mathbb{N}_0} H^k(\Omega).$$

We remark that $H^\infty(\Omega)$ is a linear space but not a Hilbert space. It becomes a Banach space when endowed with the norm $\|\,\|_{H^\infty} : H^\infty(\Omega) \to \mathbb{R}$:

$$\|f\|_{H^\infty} := \operatorname{ess\,sup}\{|f(x)| \,|\, x \in \Omega\},$$

where $\operatorname{ess\,sup}\{|f(x)| \,|\, x \in \Omega\} := \inf\{M \in \mathbb{R}^+ \,|\, \{x \in \Omega \,|\, |f(x)| > M\}$ is a null set$\}$.

For the sake of completeness, we next summarize some properties of Sobolev spaces.

Properties of Sobolev Spaces

In this section, we introduce the Sobolev spaces on $\mathbb{R}$ in a slightly more general setting and show that the definition here agrees with the one in the previous section for $p \in [1, \infty)$, but differs for $p = \infty$. For proofs or more details, we refer the interested reader to the literature on Sobolev spaces, in particular, to [2].

To this end, define $\overline{\mathbb{R}} := \mathbb{R} \cup \{\pm\infty\}$ and let $M \subseteq \overline{\mathbb{R}}$ be nonempty, closed, and connected.

Definition 2.55 (Absolute Continuity)   A function $f : M \to \mathbb{R}$ is called *absolutely continuous* if $\forall \varepsilon > 0$ $\exists \delta = \delta(\varepsilon)$ $\forall$ finite systems of nonempty intervals $[a_i, b_i] \subseteq M$ satisfying $[a_i, b_i) \cap [a_j, b_j) = \emptyset$, for $i \neq j$,

$$\sum_{i=1}^{m} |b_i - a_i| < \delta \quad \text{implies} \quad \sum_{i=1}^{m} |f(b_i) - f(a_i)| < \varepsilon.$$

We denote the set of all absolutely continuous functions on M by $AC(M)$ and, in addition, define for $k \in \mathbb{N}_0$

$$AC^k(M) := \{f \in \text{Map}(M, \mathbb{R}) \mid D^\ell f \text{ is absolutely continuous for } \ell = 0, 1, \ldots, k\},$$

with $AC^0(M) := AC(M)$.

Example 2.56   Let $M := [a, b] \subset \mathbb{R}$ be nonempty. Then, every absolutely continuous function on $[a, b]$ is also uniformly continuous, hence continuous on $[a, b]$. Moreover, every Lipschitz-continuous function is absolutely continuous. Thus, we have the following strict inclusions:

$$AC[a, b] \subsetneq C[a, b] \quad \text{and} \quad \text{Lip}^1[a, b] \subsetneq AC[a, b].$$

An example of a continuous function that is not absolutely continuous is given by choosing $a < 0 < b$ and setting

$$f(x) := \begin{cases} x \sin \frac{1}{x}, & x \in [a, b] \setminus \{0\}; \\ 0, & x = 0. \end{cases}$$

One important property of absolutely continuous functions is that they are almost everywhere differentiable. More precisely,

$$f \in \text{Map}(M, \mathbb{R}) \text{ is absolutely continuous iff } f(x) = \int_{x_0}^{x} f'(t)\,dt + f(x_0),$$

$x_0 \in M$ arbitrary.

We refer to the literature on real analysis for a proof of this statement.

Now, let $k \in \mathbb{N}$ and let $1 \leq p \leq \infty$. Define collections $W^{k,p}(\Omega)$ of functions on a nonempty domain $\Omega \subseteq \mathbb{R}$ by (see also [26])

$$W^{k,p}(\Omega) := \left\{ f \in \text{Map}(\Omega, \mathbb{R}) \mid f \in AC^{k-1}(\Omega), D^k F \text{ exists a.e. on } \Omega \text{ and} \right.$$

$$\left. D^k f \in L^p(\Omega). \right\}$$

The sets $W^{k,p}(\Omega)$ are easily seen to become linear spaces under the usual operations of function addition and scalar multiplication of functions. In addition, we define on $W^{k,p}(\Omega)$ the functionals $\| \ \|_{k,p} : W^{k,p}(\Omega) \to \mathbb{R}$,

$$\|f\|_{k,p} := \left(\sum_{\varkappa=0}^{k} \int_{\Omega} |D^\varkappa f(x)|^p \, dx \right)^{1/p}, \quad 1 \leq p < \infty,$$

and

$$\|f\|_{k,\infty} := \max\{\|D^\varkappa f\|_{L^\infty} \mid \varkappa = 0, 1, \ldots, k\}, \quad p = \infty.$$

It can be shown (cf., for instance, [2]) that the spaces $(W^{k,p}(\Omega), \| \ \|_{k,p})$ are Banach spaces for $k \in \mathbb{N}$ and $p \in [1, \infty]$.

In the case $p := 2$, the norm $\| \ \|_{k,2}$ is induced by the bilinear form $\langle \ , \ \rangle_{k,2} : W^{k,2}(\Omega) \times W^{k,2}(\Omega) \to \mathbb{R}$,

$$\langle f, g \rangle_{k,2} := \sum_{\varkappa=0}^{k} \int_\Omega (D^\varkappa f)(x)(D^\varkappa g)(x)\, dx,$$

and the spaces $(W^{k,2}(\Omega), \langle \ , \ \rangle_{k,2})$ are Hilbert spaces.

To discuss some properties of Sobolev spaces, we need to introduce the notion of *continuous embedding* between Banach spaces. To this end, let $(\mathsf{X}, \| \ \|_\mathsf{X})$ and $(\mathsf{Y}, \| \ \|_\mathsf{Y})$ be two Banach spaces with $\mathsf{X} \subset \mathsf{Y}$.

Definition 2.57 (Continuous Embedding between Banach Spaces) We refer to the Banach space $(\mathsf{X}, \| \ \|_\mathsf{X})$ as being *continuously embedded* in the Banach space $(\mathsf{Y}, \| \ \|_\mathsf{Y})$, written $\mathsf{X} \hookrightarrow \mathsf{Y}$, if the inclusion mapping $\iota : \mathsf{X} \to \mathsf{Y}$ is continuous, or, equivalently, if $\exists\, c \in \mathbb{R}^+$ such that $\forall f \in \mathsf{X}$

$$\|f\|_\mathsf{Y} \leq c \|f\|_\mathsf{X}.$$

The following embedding results hold for Sobolev spaces of functions $f : \Omega \to \mathbb{R}$, where $k \in \mathbb{N}$ and $p \in [1, \infty]$:

$$W^{k,p}(\Omega) \hookrightarrow L^p(\Omega);$$

$$W^{k,p}(\Omega) \hookrightarrow W^{k',p'}(\Omega), \quad \forall k \geq k';\ \forall p \geq p';$$

$$\forall\, kp = 1 : f \in W^{k,p}(\Omega),\ \mathrm{supp}\, f \ \text{compact} \implies f \in L^q(\Omega), \forall q \in [1, \infty).$$

The next result shows that for $1 \leq p < \infty$, the spaces $H^{k,p}(\Omega)$ and $W^{k,p}(\Omega)$ coincide.

Theorem 2.58 (Meyers–Serrin) If $p \in [1, \infty)$, then $H^{k,p}(\Omega) = W^{k,p}(\Omega)$, for any $\Omega \subseteq \mathbb{R}^n$.

Example 2.59 In the present setting, we have $\mathrm{AC}[a, b] = W^{1,1}[a, b]$ and $\mathrm{Lip}^1[a, b] = W^{1,\infty}[a, b]$.

Minimalization of the $| \ |_k$-Semi-Norm

In order to obtain the sought-after error estimates for spline interpolation, we will make use of the results stated in the theorem below.

Theorem 2.60 Assume that V is a linear space and $T : H^k[a, b] \to V$ a linear operator. Furthermore, for $v \in V$ and $s \in H^k[a, b]$ we suppose that $Ts = v$. Then the following statements are equivalent:

1. Among all $g \in H^k[a, b]$ satisfying $Tg = v$, s minimalizes the Sobolev semi-norm $|\ |_k$. In other words,

$$s = \mathrm{argmin}\left\{ |g|_k = |D^k g|_{L^2[a,b]} : Tg = v \right\}.$$

2. $\langle s, g - s \rangle_k = 0, \ \forall g \in H^k[a, b]$ fulfilling $Tg = v$.

3. $\langle s, f \rangle_k = 0, \ \forall f \in H^k[a, b]$ satisfying $Tf = 0$.

Remark 2.61 Here, we defined the argmin as follows. Let $f \in \mathrm{Map}(D, \mathbb{R})$. Then

$$\mathrm{argmin}\{f(x) \,|\, x \in D\} := \{x \in D \,|\, f(x) \text{ is minimal}\}.$$

Proof (3) implies (2): Follows immediately from $Ts = v = Tg$ by setting $f := g - s$. (2) implies (1): For any $g \in H^k[a, b]$ with $Tg = v$ and the given s, we have

$$|g|_k^2 = |g - s + s|_k^2 = \langle g - s + s, g - s + s \rangle_k = |g - s|_k^2 + |s|_k^2 + 2\langle s, g - s \rangle_k.$$

By (2), this last equality equals $|g - s|_k^2 + |s|_k^2 \geq |s|_k^2$, proving statement (1). (1) implies (3): For all $g \in H^k[a, b]$ with $Tg = v$, one has

$$|g|_k^2 = |g - s|_k^2 + |s|_k^2 + 2\langle s, g - s \rangle_k \geq |s|_k^2$$

by (1). Thus,

$$|g - s|_k^2 + 2\langle s, g - s \rangle_k \geq 0. \tag{2.49}$$

Now, set $g := s + cf$, where $c \in \mathbb{R}$ will be chosen later, and with $f \in H^k[a, b]$, $Tf = 0$, and $|f|_k = 1$. (If $f = 0$, then (3) is clearly satisfied.) Substituting g into (2.49) yields

$$2c\langle s, f \rangle_k + c^2 |f|_k^2 \geq 0. \tag{2.50}$$

Set $c := -\langle s, f \rangle_k$ then (2.50) becomes $-\langle s, f \rangle_k \leq 0$, and therefore $\langle s, f \rangle_k = 0$. $\square$

The following sequence of lemmata is needed for the proof of the minimality property of splines.

Lemma 2.62 Suppose that $f \in C[a, b]$ and let $\Xi := \{a := \xi_1 < \cdots < \xi_n =: b\}$ be a knot set. Set $\omega(x) := \prod_{\nu=1}^{n} (x - \xi_\nu)$. Then, the following relation holds for all $x \in [a, b]$:

$$f(x) - \omega(x) \sum_{\nu=1}^{n} \frac{f(\xi_\nu)}{(x - \xi_\nu)(D\omega)(\xi_\nu)} = \omega(x)\,[x, \xi_1, \ldots, \xi_n] f. \tag{2.51}$$

Proof By Exercise 26 in Chapter 1, we have with $\tilde{\omega}(\bar{x}) := \prod_{v=0}^{n}(\bar{x} - \xi_v)$ and

$$\omega(x) := \prod_{v=1}^{n}(\bar{x} - \xi_v)$$

$$[x, \xi_1, \ldots, \xi_n]f = \sum_{v=0}^{n} \frac{f(\xi_v)}{(D\tilde{\omega})(\xi_v)} = \frac{f(x)}{(D\tilde{\omega})(x)} + \sum_{v=1}^{n} \frac{f(\xi_v)}{(D\tilde{\omega})(\xi_v)}. \qquad (2.52)$$

Since $\tilde{\omega}(\bar{x}) = (\bar{x} - x)\omega(\bar{x})$, we have that $D\tilde{\omega} = \omega(\bar{x}) + (\bar{x} - x)(D\omega)(\bar{x})$, and thus $(D\tilde{\omega})(x) = \omega(x)$ and $(D\tilde{\omega})(\xi_v) = \omega(\xi_v) + (\xi_v - x)(D\omega)(\xi_v) = (\xi_v - x)(D\omega)(\xi_v)$. Substituting these identities into (2.52) and multiplying by $\omega(x)$ gives

$$\omega(x)[x, \xi_1, \ldots, \xi_n]f = f(x) + \omega(x) \sum_{v=1}^{n} \frac{f(\xi_v)}{(\xi_v - x)(D\omega)(\xi_v)},$$

which proves the lemma. $\quad\square$

Lemma 2.63 Let $H_\Xi^k[a, b] := \{f \in H^k[a, b] \,|\, f|_\Xi = 0\}$ and suppose that $f \in H_\Xi^k[a, b]$. Then

$$f(x) = \frac{\omega_{v+k-1}(x)}{k!} \int_{[\xi_v, \xi_{v+k})} M_k(t \,|\, x, \xi_v, \ldots, \xi_{v+k-1})(D^k f)(t)\,dt \qquad (2.53)$$

is a local representation of f for all $x \in [\xi_v, \xi_{v+k})$. Here, $\omega_{v+k-1}(x) = \prod_{\mu=v}^{v+k-1}(x - \xi_\mu)$.

Proof The statement follows from Lemma 2.62, since $f(\xi_v) = 0$, for all $v = 1, \ldots, n$, and the fact that the B-splines $M(\bullet \,|\, x, \xi_v, \ldots, \xi_{v+k-1})$ are the Peano kernels in the integral representation of the divided difference operators $[x, \xi_v, \ldots, \xi_{v+k-1}](\bullet)$. (See also (2.7).) $\quad\square$

Lemma 2.64 Suppose that $f \in H_\Xi^k[a, b]$. Then we have the following local estimate on the size of the L^∞-norm of f:

$$\|f\|_{L^\infty[\xi_v, \xi_{v+k})} \leq c_k \|\bar{\xi}\|^{k-\frac{1}{2}} \|D^k f\|_{L^2[\xi_v, \xi_{v+k})},$$

where the constant $c_k \in \mathbb{R}^+$ depends only on k and

$$\|\bar{\xi}\| := \max\{|\xi_{\mu+1} - \xi_\mu| \,|\, v = v, \ldots, v + k - 2\}.$$

Proof Equation (2.53) together with (see Lemma 1.27)

$$|\omega_{v+k-1}(x)| = \prod_{\mu=v}^{v+k-1} |x - \xi_\mu| \leq \frac{(k-1)!}{4} \|\bar{\xi}\|^k$$

implies that, for all $x \in [\xi_v, \xi_{v+k})$,

$$|f(x)| \leq \frac{|\omega_{v+k-1}(x)|}{k!} \int_{[\xi_v, \xi_{v+k})} |M_k(t \,|\, x, \xi_v, \ldots, \xi_{v+k-1})| \cdot |(D^k f)(t)| \, dt$$

$$\leq c'_k \|\overline{\boldsymbol{\xi}}\|^k \, \|M_k\|_{L^2[\xi_v, \xi_{v+k})} \, \|D^k f\|_{L^2[a,b]},$$

where, in the last step, we used the Cauchy–Schwarz inequality. Now, rewrite the product $|M_k(t \,|\, x, \xi_v, \ldots, \xi_{v+k-1})|^2$ as follows:

$$|M_k(t \,|\, x, \xi_v, \ldots, \xi_{v+k-1})|^2 = |M_k(t \,|\, x, \xi_v, \ldots, \xi_{v+k})| \cdot |M_k(t \,|\, x, \xi_v, \ldots, \xi_{v+k-1})|$$

$$= \frac{n}{\xi_{v+k-1} - \xi_v} |B_k(t \,|\, x, \xi_v, \ldots, \xi_{v+k-1})|$$

$$\cdot |M_k(t \,|\, x, \xi_v, \ldots, \xi_{v+k-1})|,$$

and apply the Hölder inequality to the product

$$|B_k(t \,|\, x, \xi_v, \ldots, \xi_{v+k-1})| \cdot |M_k(t \,|\, x, \xi_v, \ldots, \xi_{v+k-1})|$$

to obtain

$$\|M_k\|_{L^2[\xi_v, \xi_{v+k})}\|^2 \leq \left\| \left(\frac{k}{\xi_{v+k-1} - \xi_v} \right) B_k \right\|_{L^\infty[\xi_v, \xi_{v+k})} \|M_k\|_{L^1[\xi_v, \xi_{v+k})}.$$

The estimate

$$|\xi_{v+k-1} - \xi_v| \geq \sum_{\mu=v}^{v+k-2} |\xi_{\mu+1} - \xi_\mu| \geq \|\overline{\boldsymbol{\xi}}\|,$$

gives, using the partition of unity properties of the B-splines B and M,

$$\|M_k\|_{L^2[\xi_v, \xi_{v+k})}\|^2 \leq \left(\frac{k}{\|\overline{\boldsymbol{\xi}}\|} \right) \|B_k\|_{L^\infty[\xi_v, \xi_{v+k})} \|M_k\|_{L^1[\xi_v, \xi_{v+k})} \leq k \|\overline{\boldsymbol{\xi}}\|^{-1},$$

and, thus,

$$|f(x)| \leq c_k \|\overline{\boldsymbol{\xi}}\|^{k - \frac{1}{2}} \|D^k f\|_{L^2[\xi_v, \xi_{v+k})}.$$

Taking the supremum over all $x \in [\xi_v, \xi_{v+k})$ yields the estimate given in the lemma. $\square$

Lemma 2.65 Assume that $h \in H^{2k-m}[a, b]$, for $m = 1, \ldots, k$, and that $g \in H^k[a, b]$. Then the bilinear form $\langle \, , \, \rangle_k : H^k[a, b] \times H^k[a, b] \to \mathbb{R}$ defined in (2.48) can be expressed in the form

$$\langle g, h \rangle_k = (-1)^{k-m} \int_a^b (D^{2k-m} h)(t) \, (D^m g)(t) \, dt$$

$$+ \sum_{j=1}^{k-m} (-1)^{j-1} \big[(D^{k+j-1} h)(b)(D^{k-j} g)(b)$$

$$- (D^{k+j-1} h)(a)(D^{k-j} g)(a) \big].$$

Proof   Use integration by parts and induction on m. Exercise! $\square$

Now, fix $x \in [a, b]$ and define functionals $R_i^j(x) : H^k[a, b] \to \mathbb{R}^{j-i}$ by

$$R_i^j(x)f := ((D^i f)(x), \ldots, (D^{j-1} f)(x))^\top,$$

where $0 \leq i < j \leq k+1$, and, for $m = 0, 1, \ldots, k$, functionals $T_m : H^k[a, b] \to \mathbb{R}^{nm}$ by

$$T_m f := (R_0^m(\xi_1), \ldots, R_0^m(\xi_n))^\top.$$

Note that the functionals T_m have the form of the mappings in Theorem 2.60. Condition (iii) in this theorem can then be expressed as

$$0 = (-1)^{k-m} \int_a^b (D^{2k-m} h)(t)\,(D^m g)(t)\, dt$$

$$+ \sum_{j=1}^{k-m} (-1)^{j-1} \left[(D^{k+j-1} h)(b)(D^{k-j} g)(b) - (D^{k+j-1} h)(a)(D^{k-j} g)(a) \right],$$

$$(2.54)$$

for all $f \in H^2[a, b]$ satisfying $T_m f = 0$.

The integral in Equation (2.54) vanishes if one requires that $D^{2k-m} h|_{[\xi_\nu, \xi_{\nu+k}]} \in \Pi^m$ or, equivalently, $h|_{[\xi_\nu, \xi_{\nu+k}]} \in \Pi^{2k}$. (The reader is encouraged to verify this statement.) Let us therefore assume now that we are given an $s \in S^{2k}(\Xi)$ and an $f \in H^k[a, b]$ satisfying $T_m f = 0$, $m = 1, \ldots, k$. Condition (iii) in Theorem 2.60 is then equivalent to the vanishing of the bilinear form

$$(s, f) \mapsto \sum_{j=1}^{k-m} (-1)^{j-1} \left[(D^{k+j-1} s)(b)(D^{k-j} f)(b) - (D^{k+j-1} s)(a)(D^{k-j} f)(a) \right].$$

Setting $m := 1$, one easily verifies that this bilinear form vanishes uniformly for all $f \in H^k[a, b]$ with $T_m f = 0$, $m = 1, \ldots, k$, if

$$(R_k^{2k-1} a)(s) = 0 = (R_k^{2k-1} b)(s). \qquad (2.55)$$

Splines of even order fulfilling condition (2.55) are called *natural splines*. The set of all natural splines associated with the knot set Ξ will be denoted by $\mathcal{N}^{2k}(\Xi)$:

$$\mathcal{N}^{2k}(\Xi) := \left\{ s \in S^{2k}(\Xi) \,|\, (R_k^{2k-1} a)(s) = 0 = (R_k^{2k-1} b)(s) \right\}.$$

It is a proper subset of $S^{2k}(\Xi)$.

Theorem 2.60 implies then that every natural spline with the property that $s|_\Xi = f|_\Xi$, minimizes the $|\ |_k$-semi-norm. We therefore arrive at the next theorem.

Theorem 2.66 (Minimal Semi-Norm Property of Splines) Let $\Xi := \{a := \xi_1 < \xi_2 < \cdots < \xi_{n-1} < \xi_n =: b\}$ be a knot set and assume that $s \in \mathcal{N}^{2k}(\Xi) \subseteq \mathcal{S}^{2k}(X)$ is an interpolating spline for the function $f \in H^k[a, b]$ on Ξ. Then s minimizes the Sobolev semi-norm $|\;|_k$ among all functions $g \in H^k[a, b]$ satisfying the interpolation condition $g|_\Xi = f|_\Xi$.

Proof Note that on $[a, b]$, we have $\mathcal{S}^{2k}(\Xi) = \mathcal{S}^{2k}(X)$. The claim now follows from the above considerations. $\square$

The next result is an immediate consequence of Theorems 2.60 and 2.66.

Corollary 2.67 $\mathcal{N}^{2k}(\Xi) \perp_{\langle\,,\,\rangle_k} H^k_\Xi[a, b]$.

Here, $\perp_{\langle\,,\,\rangle_k}$ denotes orthogonality with respect to the inner product $\langle\,,\,\rangle_k$.

Spline Interpolation Error

We are now ready to state the first of two theorems that give estimates on the spline interpolation error.

Theorem 2.68 Suppose that $f \in H^k[a, b]$ and that $s \in \mathcal{N}^{2k}(\Xi)$ is an interpolating spline for f on Ξ. Then the estimate

$$\|f - s\|_{L^\infty[a,b]} \le c_k \|\boldsymbol{\xi}\|^{k-\frac{1}{2}} \|D^k f\|_{L^2[a,b]} = c_k \|\boldsymbol{\xi}\|^{k-\frac{1}{2}} |f|_k$$

holds, where $c_k \in \mathbb{R}^+$ depends only on k.

Proof Notice that $f - s \in H^k_\Xi[a, b]$ and that, by Lemma 2.64, we have the inequalities

$$|f(x) - s(x)| \le c_k \|\overline{\boldsymbol{\xi}}\|^{k-\frac{1}{2}} \|D^k(f - s)\|_{L^2[\xi_v,\xi_{v+k}]} \le c_k \|\boldsymbol{\xi}\|^{k-\frac{1}{2}} \|D^k(f - s)\|_{L^2[a,b]}$$

$$= c_k \|\boldsymbol{\xi}\|^{k-\frac{1}{2}} |f - s|_k,$$

where $\|\boldsymbol{\xi}\| := \max\{|\xi_{v+1} - \xi_v| \,|\, v = 1, \ldots, n - 1\}$. Theorem 2.60 (ii) implies that

$$|f|^2_k = |f - s|^2_k + |s|^2_k \ge |f - s|^2_k,$$

and, thus,

$$|f(x) - s(x)| \le c_k \|\boldsymbol{\xi}\|^{k-\frac{1}{2}} |f|_k = c_k \|\boldsymbol{\xi}\|^{k-\frac{1}{2}} |D^k f|_{L^2[a,b]}. \quad \square$$

Now, suppose that $f \in H^k[a, b]$ and that $s \in \mathcal{N}^{2k}(\Xi)$ is an interpolating spline to f on Ξ. Then,

$$|f - s|_k = \langle f - s, f - s \rangle_k = \langle f - s, s \rangle_k = \int_a^b (D^k f)(t)(D^k[f - s])(t)\,dt$$

$$= (-1)^k \int_a^b (D^{2k}f)(t)(f - s)(t)\,dt$$

$$+ \sum_{j=1}^{k-1} (-1)^{j-1} \big[(D^{k+j-1}f)(b)(D^{k-j}[f - s])(b)$$

$$- (D^{k+j-1}f)(a)(D^{k-j}[f - s])(a) \big].$$

The second equality above follows from Theorem 2.66 and the last line is obtained through integration by parts. If one requires in addition that

$$(D^{k-j}[f - s])(b) = (D^{k-j}[f - s])(a) = 0, \quad \forall j = 1, \ldots, k,$$

or, equivalently,

$$a = x_1 = \cdots = x_k \quad \text{and} \quad x_{n+1} = \cdots = x_{n+k},$$

(since then $(D^i f)(b) = (D^i s)(b)$ and $(D^i f)(b) = (D^i s)(b)$, for all $i = 0, 1, \ldots, k-1$), then the sum in the last equality above vanishes, leaving

$$|f - s|_k = (-1)^k \int_a^b (D^{2k}f)(t) \cdot (f - s)(t)\,dt \leq \|f - s\|_{L^\infty[a,b]} \|D^{2k}f\|_{L^1[a,b]},$$

$$(2.56)$$

by an application of Hölder's Inequality to the above integral. In the proof of Theorem 2.68 we showed that

$$|f(x) - s(x)| = c_k \|\xi\|^{k-\frac{1}{2}} |f - s|_k$$

and, therefore,

$$|f(x) - s(x)|^2 \leq c_k' \|\xi\|^{2k-1} |f - s|_k^2 \leq c_k' \|\xi\|^{2k-1} \|f - s\|_{L^\infty[a,b]} \|D^{2k}f\|_{L^1[a,b]},$$

by Equation (2.56). Taking the supremum on the left-hand side over all $x \in [a, b]$ and cancelling the term $\|f - s\|_{L^\infty[a,b]}$ ($f \neq s$, since otherwise all conclusions are vacuous) produces

$$\|f - s\|_{L^\infty[a,b]} \leq c_k' \|\xi\|^{2k-1} \|D^{2k}f\|_{L^1[a,b]}.$$

Thus, we have arrived at a strengthening of the error estimate for spline interpolation under some additional conditions on the knot sequence x and the function f, which is now required to lie in the Sobolev space $W^{2k,1}[a, b]$. The next theorem is a summary of the current derivations.

Theorem 2.69 Assume that $f \in W^{2k,1}[a, b]$ and that $s \in \mathcal{N}^{2k}(\Xi)$ is an interpolating spline to f on Ξ. Further assume that f and s satisfy the boundary conditions

$$(D^j s)(a) = (D^j f)(a) \quad \text{and} \quad (D^j s)(b) = (D^j f)(b), \quad \forall j = 0, 1, \ldots, k - 1.$$

Then the L^∞-interpolation error satisfies the following inequality:

$$\|f - s\|_{L^\infty[a,b]} \leq C_k \|\boldsymbol{\xi}\|^{2k-1} \|D^{2k} f\|_{L^1[a,b]},$$

where the positive constant C_k depends only on k.

For later purposes and for the sake of completeness, we state without proof an error estimate for natural cubic spline interpolation in $C^k[a, b]$, where $k = 2, 3$, or 4.

Theorem 2.70 Let $s \in \mathcal{N}^4(\Xi)$ and suppose that $s|_\Xi = f|_\Xi$, where $f \in C^k[a, b]$ and $k \in \{2, 3, 4\}$. Then, for all $\xi_i \leq x \leq \xi_i$ and $\varkappa \in \{0, 1, 2\}$, the following pointwise estimate holds:

$$|D^\varkappa (s - f)(x)| \leq C_{k\varkappa} \|D^k f\|_\infty h^{k-\varkappa} + K_\varkappa \beta_\varkappa (2^{1-i} + 2^{1-n+i}) h, \qquad (2.57)$$

where we set $h := \max\{\xi_i - \xi_{i-1} \mid i = 0, 1, \ldots, n\}$. The values of the constants $C_{k\varkappa}$, K_k, and $\beta_\varkappa$ are as follows:

$$C_{20} = \frac{9}{8}, \qquad\qquad C_{21} = 4, \qquad\qquad C_{22} = 10,$$

$$C_{30} = \frac{71}{216}, \qquad\qquad C_{31} = \frac{31}{27}, \qquad\qquad C_{32} = 5$$

$$C_{40} = \frac{5}{384}, \qquad\qquad C_{41} = \frac{9 + \sqrt{3}}{316}, \qquad\qquad C_{42} = 5,$$

$$\beta_0 = \frac{\Delta \xi_i}{216}, \qquad\qquad \beta_1 = 1, \qquad\qquad \beta_2 = \frac{6}{\Delta \xi_i}$$

and

$$K_2 = \frac{5}{2} \|D^2 f\|_\infty + R, \quad K_3 = \|D^3 f\|_\infty h + \frac{1}{2} R, \quad K_4 = \frac{7}{24} \|D^4 f\|_\infty h^4 + \frac{1}{2} R.$$

Here, we set $\Delta \xi_i := \xi_{i+1} - \xi_i$, $i = 0, 1, \ldots, n - 1$, and $R := \max\{|D^2 (f - s)(a)|, |D^2(f - s)(b)|\}$.

Proof We refer to [78] for the proof of this theorem. $\quad\square$

2.10 Approximation with Splines

For this section, we assume that we are given a knot sequence $\boldsymbol{x} := (x_1, \ldots, x_{n+k})$ with $a := x_1 = \cdots = x_k$ and $x_{n+1} = \cdots = x_{n+k} =: b$ and the associated knot set X. Moreover, suppose that $f : [a, b] \subseteq \mathbb{R} \to \mathbb{R}$ is a function defined on a nonempty interval and $\Xi := \{\xi_1 \le \cdots \le \xi_n\}$ is a knot set contained in $[a, b]$.

Define a linear approximation method $T := T_{k,X} : C[a, b] \to \mathcal{S}^k(X)$ by

$$Tf := \sum_{\nu=1}^{n} f(\xi_\nu)\, B_{\nu k, X}, \quad \text{on } [a, b].$$

(From now on, we drop the indices k and X from the B-splines if no confusion is to be expected.) In case $f \equiv c$, $c \in \mathbb{R}$, then

$$Tc = \sum_{\nu=1}^{n} c\, B_\nu = c,$$

by the Partition of Unity Property (2.9) of the B-splines. In other words, the linear approximation method T reproduces constant functions exactly.

Now, suppose that $x_0 \in [a, b]$. Then, there exists a $\mu \in \{1, \ldots, n\}$ so that $x_0 \in [x_\mu, x_{\mu+1}] \subseteq [a, b]$. Thus,

$$(Tf)(x_0) = \sum_{\nu=\mu+1-k}^{\mu} f(\xi_\nu) B_\nu(x_0),$$

and, again by the Partition of Unity Property (2.9) of B-splines,

$$f(x_0) = f(x_0) \sum_{\nu=\mu+1-k}^{\mu} B_\nu(x_0) = \sum_{\nu=\mu+1-k}^{\mu} f(x_0) B_\nu(x_0).$$

Hence,

$$|(Tf)(x_0) - f(x_0)| \le \sum_{\nu=\mu+1-k}^{\mu} |f(\xi_\nu) - f(x_0)| B_\nu(x_0)$$

$$\le \max\{|f(\xi_\nu) - f(x_0)| \mid \nu = \mu + 1 - k, \ldots, \mu\}$$

The knots $\{\xi_\nu \mid \nu = 1, \ldots, n\}$ are now chosen according to

$$\xi_\nu := \begin{cases} x_{\nu+k/2}, & k \text{ even;} \\ \dfrac{x_{\nu+(k-1)/2} + x_{\nu+(k+1)/2}}{2}, & k \text{ odd.} \end{cases}$$

The above choice of knots then entails the estimate

$$|(Tf)(x_0) - f(x_0)| \le \max\{|f(x) - f(y)| \mid x, y \in [\xi_{\mu+1-k}, x_{\mu+1}] \cup [x_\mu, \xi_\mu]\}$$

$$\le \max\{|f(x) - f(y)| \mid x, y \in [a, b], \ |x - y| \le (k\,\|\boldsymbol{x}\|)/2\}$$

$$= \omega(f, (k\,\|\boldsymbol{x}\|)/2) \le \left(\frac{k+1}{2}\right) \omega(f, \|\boldsymbol{x}\|).$$

Taking the supremum over all $x_0 \in [a, b]$, we arrive at

$$\| Tf - f \|_{\infty,[a,b]} \leq \left(\frac{k+1}{2} \right) \omega(f, \|\boldsymbol{x}\|),$$

and, finally, at

$$E_k(f; \mathcal{S}^k(X)) \leq \left(\frac{k+1}{2} \right) \omega(f, \|\boldsymbol{x}\|). \tag{2.58}$$

Remark 2.71 Using more sophisticated estimates, Marsden [106] has shown that for $1 < k \in \mathbb{N}$,

$$E_k(f; \mathcal{S}^k(X)) \leq 2\omega\left(f, \min\left\{ \frac{b-a}{\sqrt{2(k-1)}}, \sqrt{k/12}\, \|\boldsymbol{x}\| \right\} \right).$$

Next, we derive an estimate for the approximation error of functions in $C^k[a, b]$ by splines from $\mathcal{S}^k(X)$. To this end, let $f \in C^k[a, b]$ and let $s \in \mathcal{S}^k(X)$, $1 < k \in \mathbb{N}$. We note that $\mathrm{dist}(f, \mathcal{S}^k(X)) = \mathrm{dist}(f - s, \mathcal{S}^k(X))$ and that the Theorem of Curry–Schoenberg 2.40 implies that

$$\mathcal{S}^{k-1}(X) = \{Ds \mid s \in \mathcal{S}^k(X)\} =: D\mathcal{S}^k(X).$$

The first of these two observations gives, together with (2.58),

$$E_k(f; \mathcal{S}^k(X)) = E_k(f - s; \mathcal{S}^k(X)) \leq \left(\frac{k+1}{2} \right) \omega(f - s, \|\boldsymbol{x}\|) =: c_k\,\omega(f - s, \|\boldsymbol{x}\|),$$

for all $s \in \mathcal{S}^k(X) \cap C[a, b]$.

Property 7 of the module of continuity in Proposition (1.73) implies

$$\omega(f - s, \|\boldsymbol{x}\|) \leq \|\boldsymbol{x}\| \, \|D(f - s)\|_\infty = \|\boldsymbol{x}\| \, \|Df - Ds\|_{\infty,[a,b]}.$$

Now, $Ds \in \mathcal{S}^{k-1}(X) \cap C[a, b]$, and, therefore,

$$E_k(f; \mathcal{S}^k(X)) \leq c_k \|\boldsymbol{x}\| \, \mathrm{dist}(Df, \mathcal{S}^{k-1}).$$

As in (2.58), one similarly obtains that

$$\mathrm{dist}(Df, \mathcal{S}^{k-1}(X)) \leq c_{k-1}\,\omega(Df, \|\boldsymbol{x}\|),$$

for a positive constant c_{k-1}, which produces the following estimate:

$$E_k(f; \mathcal{S}^k(X)) \leq (c_k c_{k-1}) \|\boldsymbol{x}\| \, \omega(Df, \|\boldsymbol{x}\|).$$

Repetition of this argument proves the next theorem.

Theorem 2.72 (Jackson-type Estimate) Let $k \in \mathbb{N}$ and let $f \in C^k[a, b]$. Moreover, let $\boldsymbol{x} := (x_1, \ldots, x_n)$ denote an arbitrary knot sequence with $a = x_1 = \cdots = x_k \leq x_{k+1} \leq \cdots \leq x_n \leq x_{n+1} = \cdots = x_{n+k} =: b$. Then the

following estimate for the interpolation error holds for $j = 0, 1, \ldots,$
$k - 1$:

$$E_k(f; \mathcal{S}^k(X)) \leq c_{k,j} \|\boldsymbol{x}\|^j \omega(D^j f, \|\boldsymbol{x}\|),$$

where $c_{k,j}$ denotes a positive constant depending only on j and k.

Corollary 2.73 $E_k(f; \mathcal{S}^k(X)) \leq c_k \|\boldsymbol{x}\|^k \|D^k f\|_{\infty,[a,b]}.$

Proof   To establish the statement, set $j = k - 1$ in Theorem 2.72 and use the
fact that $\omega(D^{k-1}f, \|\boldsymbol{x}\|) \leq \|\boldsymbol{x}\| \|D^k f\|_{\infty,[a,b]}$. $\square$

As a special case of the above setting, we consider the knot sequence
$\boldsymbol{x}$, whose knots are of the form $x_{k+i} := a + ih$, $i = 1, \ldots, n - k$, with $h :=$
$\dfrac{b - a}{n - k + 1}$. Then, $\|\boldsymbol{x}\| = h \in \mathcal{O}(n^{-1})$, and, therefore,

$$E_k(f; \mathcal{S}^k(X)) \in \mathcal{O}(n^{-k}).$$

2.11 L^2-Approximation with Splines

Since $(C^k[a, b], \| \; \|_{\infty,[a,b]})$ is not a strictly convex normed linear space,
one can not in general guarantee the existence of a unique best linear
approximation from the spline space $\mathcal{S}^k(X)$. In this section, we consider best
linear approximations from $\mathcal{S}^k(X)$ using a modified L^2-norm.

For this purpose, assume that $f \in C[a, b]$ is known on a knot set $\Xi := \{a :=$
$\xi_1 < \xi_2 < \cdots < \xi_n =: b\}$, $1 < n \in \mathbb{N}$. Moreover, let $X := \{x_\nu \mid \nu = 1, \ldots, n + k\}$
be a knot set and let $\mathcal{S}^k(X)$ be the associated linear space of B-splines defined
on X.

Define a bilinear form

$$(\, , \,)_{\Xi,w} : C[a, b] \times C[a, b] \to \mathbb{R},$$

$$(g, h) \longmapsto \sum_{i=1}^{n} w_i \, g(\xi_i) \, h(\xi_i),$$

where the *weight vector* $\boldsymbol{w} := (w_1, \ldots, w_n)^\top \in (\mathbb{R}^+)^n$ is chosen in such a way
that $(\, , \,)_{\Xi,w}$ represents an approximation of the integral $\int_a^b g(t)h(t)dt$. One
possible such choice for $\boldsymbol{w}$ is given by

$$w_i := \begin{cases} \dfrac{\Delta\xi_1}{2}, & i = 1; \\[2ex] \dfrac{\Delta\xi_{i-1}}{2} + \dfrac{\Delta\xi_i}{2}, & i = 2, \ldots, n - 1; \\[2ex] \dfrac{\Delta\xi_{n-1}}{2}, & i = n. \end{cases}$$

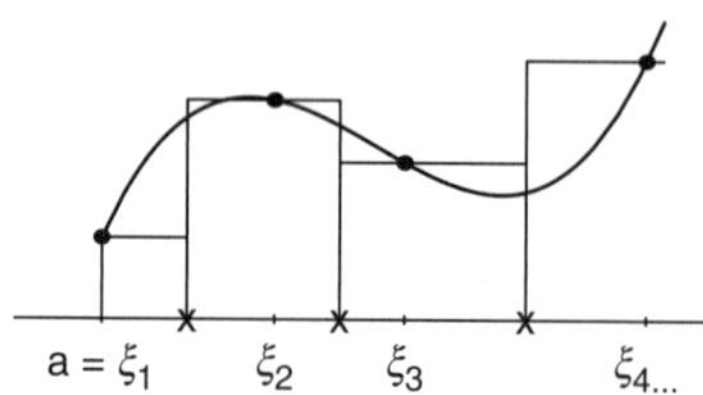

Figure 2.8 A choice of weights w.

(Here $\Delta\xi_\nu := \xi_{\nu+1} - \xi_\nu$, $\nu = 1, \dots, n$.) This particular choice is motivated by Figure 2.8.

Associated with the bilinear form $(,)_{\Xi,w}$ is the semi-norm $|\ |_{\Xi,w} : C[a, b] \to \mathbb{R}$, given by

$$|g|_{\Xi,w} := \sqrt{(g, g)_{\Xi,w}}.$$

Note that $g|_\Xi = 0$ does not imply $g \equiv 0$ on $[a, b]$ since card $\Xi < \infty$. In order for $|\ |_{\Xi,w}$ to become a norm on $C[a, b]$, one chooses Ξ such that $g|_\Xi = 0$ implies $g|_{[a,b]} \equiv 0$, i.e., one requires that the interpolation problem on Ξ has a unique solution. The Theorem of Schoenberg–Whitney immediately gives the next result.

Proposition 2.74 $|\ |_{\Xi,w}$ is a norm on $C[a, b]$ iff for $1 \leq j_1 < \cdots < j_n \leq n$: $\xi_i < x_{j_i} < \xi_{i+k}$, $\forall i \in \{1, \dots, n\}$.

Using Theorem 1.61, the best linear approximation to $f \in C[a, b]$, where $C[a, b]$ is now endowed with the *norm* $|\cdot|_{\Xi,w}$, can be computed as follows. On $[a, b]$, one has $s = \sum_{\nu=1}^{n} c_\nu B_{\nu k}$, for constants c_ν. Then, for all $\mu = 1, \dots, n$,

$$0 = (f - s, B_{\mu k})_{\Xi,w} = \left(f - \sum_{\nu=1}^{n} c_\nu B_{\nu k}, B_{\mu k} \right)_{\Xi,w}$$

$$= (f, B_{\mu k})_{\Xi,w} - \sum_{\nu=1}^{n} c_\nu \left(B_{\nu k}, B_{\mu k} \right)_{\Xi,w}.$$

By Karlin's Theorem, the matrix $\left((B_{\nu k}, B_{\mu k})_{\Xi,w}) \,|\, \mu, \nu = 1, \dots, n \right)$ is symmetric and totally positive, and by Corollary 2.46 we know that $\left(B_{\nu k}, B_{\mu k} \right)_{\Xi,w} = 0$, for $|\nu - \mu| > k$. Hence, the B-spline coefficients c_ν, $\nu = 1, \dots, n$, can be computed using Gaussian[11] elimination without pivoting. This then produces the best linear approximation in the L^2-sense.

---

11 JOHANN CARL FRIEDRICH GAUSS, 30 April 1777–23 February 1855. German mathematician, physicist and astronomer who is considered to be one of the greatest and most influential mathematicians of all times. He made major contributions to algebra, probability theory, Euclidean geometry, real and complex analysis, and geodesy.

2.12 Exponential Splines

In this section and the next, we introduce a new class of splines via certain classes of differential operators. Part of our approach here relies on the original essays [42, 90, 118, 134, 138, 152] and the articles by Micchelli and Schoenberg in [93]. First, however, we need to revisit polynomial splines and introduce a generalization of the concept of derivative of a function.

2.12.1 Polynomial Splines and Distributional Derivatives

At this point, we recall Remark 2.3, where we noted that a spline $s \in S^k(X_n)$ of order k over a knot set $X_n \subset \mathbb{R}$ of cardinality $n + 1$, when restricted to the half-open subintervals determined by the knots, lies in the kernel of the differential operator D^k:

$$D^k s|_{[x_\nu, x_{\nu+1})} = 0, \quad \nu = 0, 1, \ldots, n - 1.$$

However, the $(k - 1)$st-derivative at the knots x_ν, $\nu = 0, 1, \ldots, n$, does not exist but has a jump whose size is given by

$$\operatorname{jump}_\nu D^{k-1} s := \operatorname{jump}_{x_\nu} D^{k-1} s = D^{k-1}_+ s(x_\nu) - D^{k-1}_- s(x_\nu), \quad \nu = 0, 1, \ldots, n.$$

Here, $(D_\pm f)(x_0)$ denotes the right hand, respectively, left hand derivative of a function f at x_0:

$$(D_\pm f)(x_0) := \lim_{\varepsilon \to 0\pm} \frac{f(x_0 + \varepsilon) - f(x_0)}{\varepsilon}.$$

For illustrative purposes, let $s \in S^2(X_3)$, where $X_3 := \{1, 3, 4\}$ is a knot set on the real line $\mathbb{R}$ and $Y := \{1, 3, 2\}$. In Figure 2.9, we show s and its derivatives Ds and $D^2 s$ on the subintervals determined by X_3.

The question arises whether it is possible to assign a "derivative" to Ds at a jump or point of discontinuity. To provide an affirmative answer, we need to digress and generalize the concept of differentiation. To this end, denote by $\mathscr{D}(\mathbb{R})$ the set of all infinitely differentiable functions on the real line $\mathbb{R}$ with compact support:

$$\mathscr{D}(\mathbb{R}) := \mathscr{D}(\mathbb{R}, \mathbb{R}) := \{\varphi \in C^\infty(\mathbb{R}) \mid \operatorname{supp} \varphi \text{ compact}\}.$$

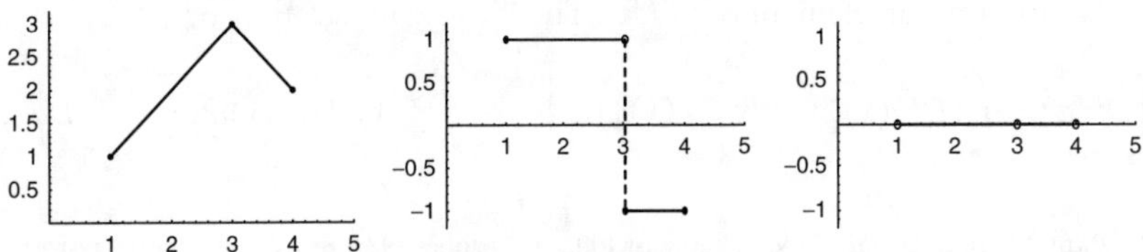

Figure 2.9 A linear spline s and its derivatives Ds and $D^2 s$.

The set $\mathscr{D}(\mathbb{R})$ is nonempty since it contains, for instance, the functions

$$\varphi_r(x) := \begin{cases} \exp\left(\frac{r^2}{x^2-r^2}\right), & |x| \le r; \\ 0, & \text{elsewhere} \end{cases} \tag{2.59}$$

for any $r \in \mathbb{R}^+$.

It is clear that $\mathscr{D}(\mathbb{R})$ becomes a real (complex) vector space under the usual operation of function addition and real (complex) scalar multiplication of functions. The elements of $\mathscr{D}(\mathbb{R})$, considered as a set or linear space, are also referred to as *test functions*.

We now consider linear functionals on $\mathscr{D}(\mathbb{R})$, i.e., elements of the *algebraic dual* $\mathscr{D}'(\mathbb{R})$ of $\mathscr{D}(\mathbb{R})$. One such linear functional is the so-called *Dirac* [12] *delta distribution* defined by

$$\delta : \mathscr{D}(\mathbb{R}) \to \mathbb{R}, \quad \varphi \mapsto \varphi(0).$$

In other words, the δ functional evaluates the test function φ at the point $x = 0$. One may generalize this property by defining the functional $\delta_a : \mathscr{D}(\mathbb{R}) \to \mathbb{R}$, $\delta_a(\varphi) := \varphi(a)$, for an arbitrary $a \in \mathbb{R}$.

Assume now that $f \in \mathrm{Map}(\mathbb{R}, \mathbb{R})$ is a *locally integrable function*, i.e., a function for which the integral $\int_K |f(x)|dx < \infty$, for all nonempty compact subsets $K \subset \mathbb{R}$. Then, setting

$$(f, \varphi) := \int_{\mathbb{R}} f(x)\varphi(x)dx, \tag{2.60}$$

we obtain the following estimate:

$$|L_f(\varphi)| := |(f, \varphi)| = \left| \int_K f(x)\varphi(x)dx \right| \le \int_K |f(x)||\varphi(x)|dx \le \|f\|_{K,\infty} \cdot \|\varphi\|_{\infty,K},$$

where $K \supset \mathrm{supp}\,\varphi$. In other words, the linear functional L_f is bounded and thus continuous. It is not hard to see that every integrable function f defines a bounded linear functional of the form (2.60). However, not every bounded linear functional can be obtained this way. A prominent example is the Dirac distribution δ.

Now suppose that $f \in C^1(\mathbb{R})$ is fixed. Consider the integral (f, φ) defined above. It is a linear functional $L_f := (f, \bullet) : \mathscr{D}(\mathbb{R}) \to \mathbb{R}$ on the vector space $\mathscr{D}(\mathbb{R})$ and thus an element of $\mathscr{D}'(\mathbb{R})$. Then, using integration-by-parts,

$$(Df, \varphi) = \int_{\mathbb{R}} (Df)(x)\varphi(x)dx = f(x)\varphi(x)\Big|_{-\infty}^{+\infty} - \int_{\mathbb{R}} f(x)(D\varphi)(x)dx = -(f, D\varphi),$$

---

12 PAUL ADRIEN MAURICE DIRAC, 8 August 1902–20 October 1984. British theoretical physicist and one of the founders of relativistic quantum theory. In 1933 he received the Nobel Prize in Physics.

where the last equality follows from the fact that φ has compact support. Similarly, one can prove that for all $k \in \mathbb{N}_0$ and any function $f \in C^k(\mathbb{R})$,

$$(D^k f, \varphi) = (-1)^k (f, D^k \varphi).$$

(The reader is encouraged to verify this statement.) Based on these observations, we define the *generalized derivative* or *distributional derivative* of a locally integrable function f as follows:

Definition 2.75 (Generalized Derivative or Distributional Derivative) Suppose that $f \in \mathrm{Map}(\mathbb{R}, \mathbb{R})$ is locally integrable. Then the *generalized* or *distributional derivative of* f is the linear functional $f' : \mathscr{D}(\mathbb{R}) \to \mathbb{R}$ such that

$$f'(\varphi) := -(f, D\varphi), \quad \forall \varphi \in \mathscr{D}(\mathbb{R}).$$

Similarly, we define the higher-order distributional derivatives $f^{(k)}$, $k \in \mathbb{N}$, of f as those linear functionals $f^{(k)} \in \mathrm{Map}(\mathscr{D}(\mathbb{R}), \mathbb{R})$ for which

$$f^{(k)}(\varphi) = (-1)^k (f, D^k \varphi), \quad \forall \varphi \in \mathscr{D}(\mathbb{R}). \tag{2.61}$$

Note that for a locally integrable function $f \in C^k(\mathbb{R})$, $f^{(k)} = D^k f$.

Example 2.76 Consider the Heaviside[13] function $H : \mathbb{R} \to \mathbb{R}$ defined by

$$H(x) := \begin{cases} 0, & x \in (-\infty, 0); \\ 1, & x \in [0, +\infty). \end{cases} \tag{2.62}$$

We compute the distributional derivative H' of the locally integrable function H as follows. Let $\varphi \in \mathscr{D}(\mathbb{R})$ be arbitrary. Then

$$H'(\varphi) = -\int_{\mathbb{R}} H(x)(D\varphi)(x)\,dx = -\int_0^\infty (D\varphi)(x)\,dx$$

$$= -\lim_{M \to +\infty} \varphi(M) + \varphi(0) = \varphi(0) = \delta(\varphi).$$

Thus,

$$H' = \delta.$$

Remark 2.77 Using the Dirac delta distribution, it is straight-forward to establish the following representation of the k-th derivative of a cardinal B-spline B_k:

$$D^k B_k = \sum_{\varkappa=0}^{k} (-1)^\varkappa \binom{k}{\varkappa} \delta(x - \varkappa). \tag{2.63}$$

(The reader is encouraged to verify this statement.)

---

13 OLIVER HEAVISIDE, 18 May 1850–3 February 1925. British mathematician and physicist who applied complex numbers to the study of electrical circuits and reformulated Maxwell's Equations in terms of the vector calculus.

Notice that if $[a, b]$ is a nonempty interval in $\mathbb{R}$, then

$$\chi_{[a,b]}(x) = H(x - a) - H(x - b), \quad x \in \mathbb{R},$$

which implies that any step function is a linear combination of shifted Heaviside functions. Moreover, the distributional derivative of $\chi_{[a,b]}$ is thus

$$\chi'_{[a,b]} = \delta_a - \delta_b.$$

With this observation, we can compute the distributional derivative of the constant spline $D^{k-1}s$, $s \in S^k(X_n)$, as follows. Let $\alpha_\nu := D^{k-2}s|_{[x_\nu,x_{\nu+1})}$, $\nu = 0, 1, \ldots, n - 1$. Then, since the values of the constants are the slopes of the linear spline over the partition of $[a, b]$ induced by the knot set X_n, we have

$$D^{k-1}s = \sum_{\nu=0}^{n-1} \alpha_\nu \chi_{[x_\nu,x_{\nu+1})},$$

and, therefore,

$$s^{(k)} = \sum_{\nu=0}^{n-1} \alpha_\nu (\delta_{x_\nu} - \delta_{x_{\nu+1}}) = \sum_{\nu=0}^{n} \beta_\nu \delta_{x_\nu}, \tag{2.64}$$

where

$$\beta_\nu := \begin{cases} \alpha_0 = \operatorname{jump}_0 D^{k-1}s, & \nu = 0; \\ \alpha_\nu - \alpha_{\nu-1} = \operatorname{jump}_\nu D^{k-1}s, & \nu = 1, \ldots, n - 1; \\ \alpha_{n-1} = \operatorname{jump}_n D^{k-1}s, & \nu = n. \end{cases}$$

We may use (2.64) to give an alternate definition of a polynomial spline of order k, $k \in \mathbb{N}$.

Definition 2.78 (Alternate Definition of Polynomial Spline)   A function $s : [a, b] \to \mathbb{R}$ is called a *polynomial spline of order* k, $k \in \mathbb{N}$, if its k-th (distributional) derivative is of the form

$$s^{(k)} = \sum_{\nu=0}^{n} \beta_\nu \delta_{x_\nu}, \tag{2.65}$$

for real numbers β_ν, $\nu = 0, 1 \ldots, n$.

The above arguments show that every polynomial spline as defined in Definition 2.1 satisfies the characterization given in Definition 2.78. For consistency, however, we also need to show that every expression of the form (2.65) is also a spline in the sense of Definition 2.1. To this end, we again consider the case $k := 2$ and the simple example of a linear spline given at the beginning of this section with $X_3 := \{x_0, x_1, x_2\} := \{1, 3, 4\}$ and $Y := \{y_0, y_1, y_2\} := \{1, 3, 2\}$. Then

$$s^{(2)} = \beta_1 \delta_{x_0} + \beta_2 \delta_{x_1} + \beta_3 \delta_{x_2},$$

and integration with respect to x gives

$$s' = \beta_1 H(\bullet - x_0) + \beta_2 H(\bullet - x_1) + \beta_3 H(\bullet - x_2) + c_1, \quad c_1 \in \mathbb{R},$$

using the fact that $H' = \delta$. It is straight-forward to verify that $(x_+)' = H(x)$ (The reader is encouraged to verify this statement.) and thus

$$s(x) = \beta_1(x - x_0)_+ + \beta_2(x - x_1)_+ + \beta_3(x - x_2)_+ + c_1 x + c_2, \quad c_1, c_2 \in \mathbb{R}.$$

Theorem 2.4, together with the requirement that s interpolate $X_3 \times Y$, now implies that $\beta_1 = \beta_3 = 0$, $\beta_2 = y_2 - 2y_1 + y_0$, $c_1 = y_1 - y_0$, and $c_2 = y_0$, producing the spline depicted in Figure 2.9.

We summarize these arguments in a theorem.

Theorem 2.79 A function $s : [a, b] \to \mathbb{R}$ is a polynomial spline of order k over a knot set $X_n = \{x_v \mid v = 0, 1, \ldots, n\}$, $n \in \mathbb{N}$, iff

$$s^{(k)} = \sum_{v=0}^{n} \beta_v \delta_{x_v},$$

for real numbers β_v, $v = 0, 1, \ldots, n$.

Proof The proof is a minor generalization of the procedure above for the case $k := 2$ and is left to the reader. $\square$

2.12.2 Linear Differential Operators and Exponential Splines

In this subsection, we generalize the ideas that led to Theorem 2.79 by considering more general differential operators. This generalization produces the class of *exponential splines*. Our construction of such splines is based on the presentation given in [152].

To this end, let $k \in \mathbb{N}_0$ and let $a_0, a_1, \ldots, a_{k-1}$ be real numbers. Furthermore, suppose that $[a, b] \subset \mathbb{R}$ is a nonempty interval.

Definition 2.80 (Linear Differential Operator) An operator $\mathcal{L} : C^k[a, b] \to C[a, b]$ is called a *k-th order linear differential operator* with real coefficients $a_0, a_1, \ldots, a_{k-1}$, iff it is of the form

$$\mathcal{L} := D^k + a_{k-1} D^{k-1} + \cdots + a_1 D + a_0 I =: \sum_{j=0}^{k} a_j D^j,$$

where D denotes the ordinary differential operator and I the identity operator on $C^k[a, b]$. For the purpose of consistency we defined $D^0 := I$.

Remark 2.81 The differential operator $\mathcal{L}$ depends on the k-tuple of coefficients $\boldsymbol{a} := (a_0, a_1, \ldots, a_{k-1})$. When necessary, we will express this dependence by writing $\mathcal{L}_{\boldsymbol{a}}$.

With each linear differential operator $\mathcal{L}$, we can associate its *symbol* given by

$$\mathcal{L}(s) := s^k + a_{k-1}s^{k-1} + \cdots + a_1 s + a_0, \tag{2.66}$$

where $s \in \mathbb{R}$. Since (2.66) is a polynomial of degree k it can be factored into linear polynomials by Gauss's Fundamental Theorem of Algebra. This yields

$$\mathcal{L}(s) = \prod_{\varkappa=1}^{k}(s - \alpha_{\varkappa}) = \prod_{\mu=1}^{l}(s - \alpha_{\mu})^{k_{\mu}}, \tag{2.67}$$

where l denotes the number of distinct roots $\alpha_{\mu} \in \mathbb{C}$ of the polynomial $\mathcal{L}(s)$ and k_{μ} their multiplicity. Clearly, $\sum_{\mu=1}^{l} k_{\mu} = k$. To this factorization of the symbol $\mathcal{L}(s)$ of the linear differential operator $\mathcal{L}$ we, formally, associate the factored form of $\mathcal{L}$:

$$\mathcal{L} = \prod_{\mu=1}^{l}(D - \alpha_{\mu}I)^{k_{\mu}}. \tag{2.68}$$

Here, $(D - \alpha I)^k$ is defined as the k-fold successive differentiation $\underbrace{(D - \alpha I) \cdots (D - \alpha I)}_{k-\text{times}}$.

Example 2.82 Let $k := 3$, and define a third order linear differential operator $\mathcal{L} : C^3(\mathbb{R}) \to C(\mathbb{R})$ by

$$\mathcal{L} := D^3 - 4D^2 + 5D - 2I.$$

Then the symbol is given by

$$\mathcal{L}(s) = s^3 - 4s^2 + 5s - 2 = (s - 1)^2(s - 2),$$

and the factored form of $\mathcal{L}$ thus by

$$\mathcal{L} = (D - I)^2(D - 2I).$$

It is a straight-forward computation to show that the function $y = y(x) = c_1 e^x + c_2 x e^x + c_3 e^{2x} \in C^3(\mathbb{R})$, with arbitrary $c_1, c_2, c_3 \in \mathbb{R}$, satisfies $\mathcal{L}y = 0$. In particular, the functions $y_1(x) := c_1 e^x + c_2 x e^x$ and $y_2(x) := c_3 e^{2x}$ are the solutions of $(D - I)^2 y_1 = 0$, respectively, $(D - 2I)y_2 = 0$. However, $(D - I)^2 y_2 \neq 0$ and $(D - 2I)y_1 \neq 0$. These observations are reflected in the following general result from the theory of linear differential operators with constant coefficients.

Proposition 2.83 Let $\mathcal{L}$ be a kth order linear differential operator with constant coefficients whose factorization is given by (2.68). Then every solution $y = y(x)$ of $\mathcal{L}y = 0$ is such that

$$y \in \mathcal{N}(\mathcal{L}) := \operatorname{span}\left\{ x^{m_{\mu}} e^{\alpha_{\mu}x} \mid m_{\mu} = 1, \ldots, k_{\mu} - 1; \ \mu = 1, \ldots, l \right\}.$$

In other words, $\mathcal{N}(\mathcal{L})$ is the null space of the linear operator $\mathcal{L}$. Moreover, the functions $y_\mu = y_\mu(x) := x^{m_\mu} e^{\alpha_\mu x}$ are the solutions of $(D - \alpha_\mu I)^{k_\mu} y_\mu = 0$, $\mu = 1, \ldots, l$.

Proof For a proof of this result, we refer to the literature on differential equations, for instance [88]. $\quad\square$

With most differential operator equations of the form $\mathcal{L}y = f$, where $f \in C[a, b]$, one associates boundary conditions of the form

$$\mathcal{B}_\nu y = 0, \quad \nu = 0, 1, \ldots, k - 1, \tag{2.69}$$

where the *boundary operators* $\mathcal{B}_\nu : C^\nu[a, b] \to \mathbb{R}$ are given by

$$\mathcal{B}_\nu y := \sum_{\mu=0}^{\nu} \left[a_{\nu\mu}(D^\mu y)(a) + b_{\nu\mu}(D^\mu y)(b) \right],$$

with sets of real numbers $\{a_{\nu\mu} \,|\, \mu, \nu = 0, 1, \ldots, k - 1\}$ and $\{b_{\nu\mu} \,|\, \mu, \nu = 0, 1, \ldots, k - 1\}$.

The set of functions $y \in C^k[a, b]$ satisfying $\mathcal{L}y = f$ together with (2.69) will be denoted by $C_B^k[a, b]$. The linearity of the operators $\mathcal{L}$ and $\mathcal{B}_\nu$, $\nu = 0, 1, \ldots, k - 1$, implies that $C_B^k[a, b]$ is a vector space over $\mathbb{R}$.

Now, if the operator $\mathcal{L} : C^k(a, b) \to \mathrm{im}\,\mathcal{L} \subseteq C[a, b]$, where $\mathrm{im}\,\mathcal{L} := \{f \in C[a, b] \,|\, f = \mathcal{L}y\}$ denotes the *image of* $\mathcal{L}$, is injective, i.e., when the only solution to $\mathcal{L}y = 0$ is the *null solution* $y \equiv 0$, then there exists a unique inverse of $\mathcal{L}$, denoted by $\mathcal{L}^{-1}$. It is shown in the theory of differential equations that this inverse can be written as the kernel function $G : C[a, b] \to C_B^k[a, b]$ of an integral operator, called a *Green's* [14] *function*. The solution of (2.69) can then be expressed in the form

$$y(x) = \int_\mathbb{R} G(x - x')f(x')dx' = (G * f)(x), \quad x \in [a, b].$$

Note that, at least formally, one has that

$$f(x) = (\mathcal{L}y)(x) = \int_\mathbb{R} (\mathcal{L}G)(x - x')f(x')dx'. \tag{2.70}$$

Therefore, again formally, $(\mathcal{L}G)(x - x') = \delta(x - x')$, and the integral in (2.70) may be interpreted as a "sum over point-values of the function f." These formal arguments can be made rigorous by considering distributions, and they provide an alternative definition of a Green's function, namely, as being that linear operator which is the right inverse to the linear differential operator $\mathcal{L}$ acting on distributions:

$$\mathcal{L}G = \delta. \tag{2.71}$$

For more details and proofs, we refer the interested reader to the literature on differential operators, for instance, [40] or [155].

---

14 GEORGE GREEN, 14 July 1793–31 March 1841. British mathematician and physicist who was one of the founders of potential theory and the theory of electromagnetism.

If we take the Fourier transform of the differential operator equation $\mathcal{L}y = f$, assuming that $y, f \in C[a, b] \cap L^1[a, b]$, we obtain

$$(i\omega)^k \widehat{y} + a_{k-1}(i\omega)^{k-1}\widehat{y} + \cdots + a_0 \widehat{y} = \widehat{f}. \tag{2.72}$$

Note that the left-hand side of the above equation is nothing but the symbol of $\mathcal{L}$ evaluated at $s = i\omega$. Thus, we may express (2.72) in the form

$$\mathcal{L}(i\omega)\widehat{y} = \widehat{f}.$$

Factoring the symbol as in (2.67), we obtain

$$\prod_{\mu=1}^{l}(i\omega - \alpha_\mu)^{k_\mu}\,\widehat{y} = \widehat{f}.$$

Taking the Fourier transform of (2.71) and using the fact that $\widehat{\delta} = 1$ (see Exercise 35), one can formally solve the resulting *algebraic system* for $\widehat{y}$:

$$\widehat{G}(\omega) = \prod_{\mu=1}^{l} \frac{1}{(i\omega - \alpha_\mu)^{k_\mu}}. \tag{2.73}$$

At this point, let us assume that *all roots* $\alpha_\mu, \mu = 1, \ldots, l$, are real.

In case that $\mathcal{L}$ is just the first order linear operator $\mathcal{L} = D - \alpha I$, then its Green's function G can be obtained from (2.73) by taking the inverse Fourier transform. Note that we can add any element of the null space $\mathcal{N}(\mathcal{L})$ of $\mathcal{L}$ and still satisfy (2.71). However, if we impose boundary conditions on G of the type considered above, then one obtains a unique solution. For instance, if we require that $G \equiv 0$ on $(-\infty, x_0)$, $x_0 \in [a, b]$, then a unique solution for a linear first order operator is of the form

$$G_\alpha(x) = H(x - x_0)\, e^{\alpha x},$$

where H is the Heaviside function (cf. (2.62)). (The reader is encouraged to verify this statement.) The general solution under the aforementioned boundary condition is then obtained via the convolution theorem and yields

$$G = G_{\alpha_1} * \cdots * G_{\alpha_k}.$$

In particular, taking into account the multiplicities of the roots $\alpha_\varkappa$, $\varkappa = 1, \ldots, k$, we thus get

$$G(x) = \sum_{\mu=1}^{l}\sum_{\varkappa=1}^{k_\mu} c_{\mu\varkappa}\, \frac{(x - x_0)_+^{\varkappa-1}}{(\varkappa - 1)!}\, e^{\alpha_\mu x}. \tag{2.74}$$

The coefficients $c_{\mu\varkappa}$ are determined by the partial fraction expansion of

$$\prod_{\mu=1}^{l}\frac{1}{(i\omega - \alpha_\mu)^{k_\mu}} = \sum_{\mu=1}^{l}\sum_{\varkappa=1}^{k_\mu}\frac{c_{\mu\varkappa}}{(i\omega - \alpha_\mu)^{\varkappa}}. \tag{2.75}$$

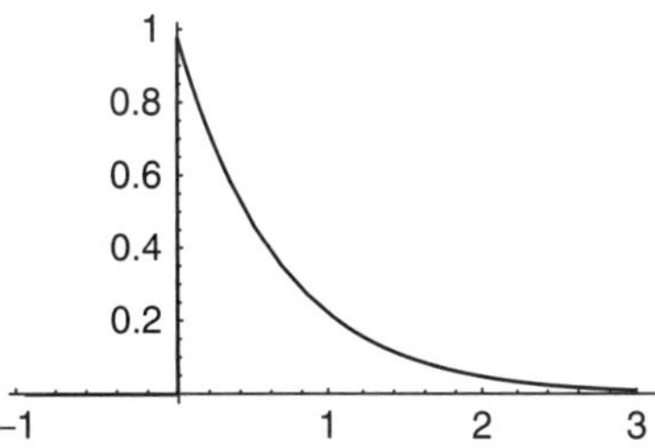

Figure 2.10 A causal solution of $\mathcal{L} = D + \frac{3}{2} I$.

Remark 2.84 It is no loss of generality, if we assume that $x_0 := 0$. In this case, one also refers to $G(x) = H(x) e^{\alpha x}$ as the causal solution. (See Figure 2.10.)

We now associate with every k-th order linear differential operator $\mathcal{L}$ with constant coefficients $a_0, a_1, \ldots, a_{k-1}$, a spline, called an *exponential spline*.

Definition 2.85 (Exponential Spline) Suppose that $\mathcal{L} := \mathcal{L}_a$ is a k-th order linear differential operator with constant real coefficient tuple $a := (a_0, a_1, \ldots, a_{k-1})$. Furthermore, suppose that the symbol of $\mathcal{L}$ has only real roots. Let $X := \{x_\nu \mid \nu = 0, 1, \ldots, N\}$, $N \in \mathbb{N}$, be a sequence of knots on the real line $\mathbb{R}$, with $x_0 := a$ and $x_N := b$. An *exponential spline* $E_a := E_{\mathcal{L}_a;X} : [a, b] \to \mathbb{R}$ associated with the differential operator $\mathcal{L}_a$ is any solution of the equation

$$\mathcal{L} E_a = \sum_{\nu=0}^{N} \beta_\nu \, \delta_{x_\nu}, \qquad (2.76)$$

where $\{\beta_\nu \mid \nu = 0, 1, \ldots, N\} \subseteq \mathbb{R}$. In case $X \subset \mathbb{Z}$, E_a is called a *cardinal exponential spline*.

Remark 2.86 The derivatives in (2.76) are to be interpreted as distributional derivatives.

Remark 2.87 In case $\mathcal{L} := D^k$, $k \in \mathbb{N}$, the exponential spline $E_{(0,\ldots,0)}$ is nothing but a k-th order polynomial spline $s = s_X$ supported on the knot set X.

It follows from the definition (2.76) of exponential spline, that E_a has discontinuities of order k at the knots x_ν, but when restricted to any interval of the form $(x_\nu, x_{\nu+1})$, $\nu \in \mathbb{Z}$, coincides with a smooth function from the null space $\mathcal{N}(\mathcal{L})$ of $\mathcal{L}$. For real roots α_μ of $\mathcal{L}$ all these functions are given by

$$E_a(x) = x^{m_\mu} e^{\alpha_\mu} \chi_{(x_\nu, x_{\nu+1})}, \quad m_\mu = 1, \ldots, k_\mu - 1; \ \mu = 1, \ldots, l; \ \sum_{\mu=1}^{l} = k.$$

Note that E_a is globally, i.e., on $[a, b]$, of class C^{k-2}. Since the Green's function is the inverse of the operator $\mathcal{L}$, the exponential spline is then explicitly given by

$$E_a(x) = \sum_{\nu=0}^{N} \beta_\nu \, G(x - x_\nu) + \phi(x), \tag{2.77}$$

where $\phi \in \mathrm{Map}(\mathbb{R}, \mathbb{R})$ is a (global) function from the null space $\mathcal{N}(\mathcal{L})$ of $\mathcal{L}$ that will be determined by boundary conditions of the type (2.69). Since an arbitrary element from $\mathcal{N}(\mathcal{L})$ contains $\dim \mathcal{N}(\mathcal{L}) = k$ free parameters, one can employ them together with the set $\{\beta_\nu \mid \nu = 0, 1, \ldots, N\}$ to develop parameter-dependent models. This is illustrated by the following simple example.

Example 2.88 Consider the differential operator $\mathcal{L} := D^2 - I = (D + I)(D - I)$ and the knot set $\{0, \frac{3}{2}, \frac{5}{2}\}$. The Green's function for $D + I$ is given by $G_{-1}(x) = H(x)\,e^{-x}$ and for $D - I$ by $G_1(x) = H(x)\,e^x$. Thus, the (causal) Green's function G for $\mathcal{L}$ is obtained by convolution of G_1 and G_2, and yields

$$G(x) = (G_{-1} * G_1)(x) = \int_{\mathbb{R}} H(x - t) H(t) e^{-(x-t)} e^t \, dt$$

$$= H(x)\,e^{-x} \int_0^x e^{2t}\, dt = H(x) \sinh x.$$

Hence, the exponential spline for the coefficient pair $a = (-1, 0)$ and given real numbers β_ν, $\nu = 0, 1, 2$, is on the interval $[0, \frac{5}{2}]$ of the form

$$E(x) := E_{(-1,0)}(x)$$

$$= \beta_0 H(x) \sinh(x) + \beta_1 H\left(x - \frac{3}{2}\right) \sinh\left(x - \frac{3}{2}\right)$$

$$+ \beta_2 H\left(x - \frac{5}{2}\right) \sinh\left(x - \frac{5}{2}\right) + \phi(x)$$

$$= \beta_0 H(x) \sinh(x) + \beta_1 H\left(x - \frac{3}{2}\right) \sinh\left(x - \frac{3}{2}\right) + c_1 e^{-x} + c_2 e^x.$$

Here, $\phi \in \mathcal{N}(D^2 - I)$ and, as such, must be of the form

$$\phi(x) = c_1 e^{-x} + c_2 e^x,$$

for $c_1, c_2 \in \mathbb{R}$. Imposing, for instance, the boundary conditions

$$(\mathcal{B}_0 E)(0) = E(0) := 1 \quad \text{and} \quad (\mathcal{B}_1 E)(0) = (DE)(0) := 0,$$

produces the unique solution

$$E(x) = \beta_0 H(x) \sinh(x) + \beta_1 H\left(x - \frac{3}{2}\right) \sinh\left(x - \frac{3}{2}\right) + \frac{1 + \beta_0}{2} e^{-x} + \frac{1 - \beta_0}{2} e^x$$

$$= \beta_0 [1 + H(x)] \sinh(x) + \beta_1 H\left(x - \frac{3}{2}\right) \sinh\left(x - \frac{3}{2}\right) + \cosh(x).$$

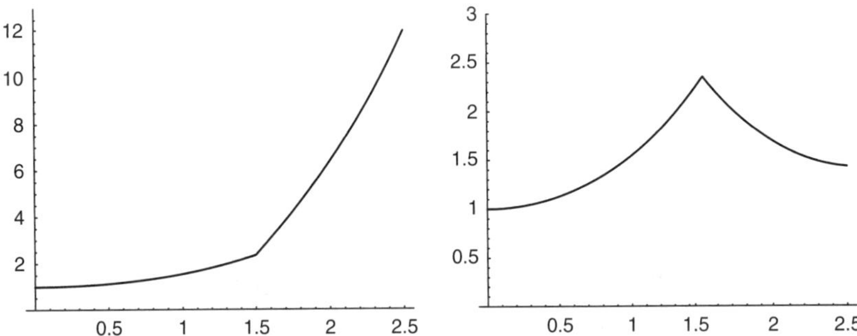

Figure 2.11   Two examples of exponential splines.

In Figure 2.11 two exponential splines for different values of β_0 and β_1 are displayed. The graph on the left hand side corresponds to $\beta_0 := -1$ and $\beta_1 := 5$, whereas the graph on the right hand side corresponds to $\beta_0 := 2$ and $\beta_1 := -4$. Note that both examples are of class C^0.

Remark 2.89   Since the hyperbolic functions sinh and cosh can be expressed as linear combinations of exponential functions with positive and negative argument, exponential splines can be rewritten in terms of hyperbolic functions alone and produce in this case so-called *hyperbolic splines* [94, 141, 160]. (See also Example 2.88 above.)

Remark 2.90   The reader is encouraged to show that (2.77) is nothing but the representation of a spline $s \in S^k(X_n)$ in the form (2.1) in case $\mathcal{L} = D^k$. In other words, the Green's functions $G(\bullet - x_\nu)$ generalize the truncated power functions $q_{\nu k}$ introduced in Section 2.2.

2.12.3 Exponential B-Splines

We have seen in Theorem 2.12 that every polynomial spline of order k can be represented as a linear combination of (polynomial) B-splines, and it is therefore natural to inquire whether there exists the analogue of a polynomial B-spline and a similar representation theorem in the case of exponential splines.

In this subsection, we will present the construction of exponential B-splines on integer knots and show that every exponential spline can be written as a linear combination of exponential B-splines. We base our arguments again on the presentation provided by [152].

For this purpose, we again assume that we are given a k-order linear differential operator $\mathcal{L}$ with constant real coefficient tuple $\boldsymbol{a} := (a_0, a_1, \ldots, a_{k-1})$ whose symbol $\mathcal{L}(s)$ has only real roots. Further we assume that our knot set X is identical to $\mathbb{Z}$. For the construction of exponential B-splines, we recall a possible construction of polynomial B-splines on the knot set $\mathbb{Z}$ via the method given by item (8) of Theorem 2.22.

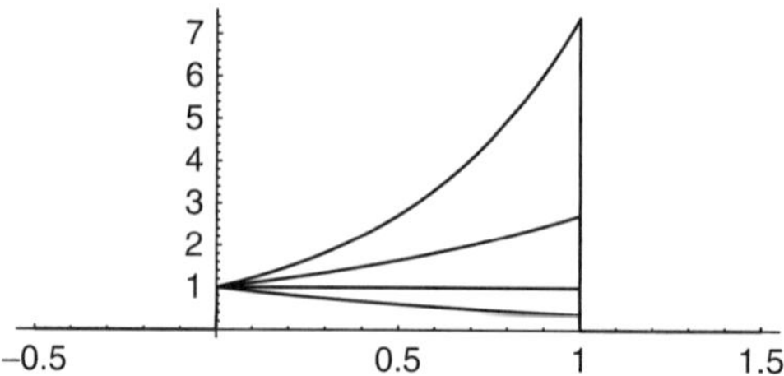

Figure 2.12 First order exponential B-splines β_α for $\alpha = 2, 1, 0, -1$.

To this end, we start with the simplest possible case, namely, $k := 1$, and consider the Green's function G_α associated with the first order linear operator $\mathcal{L} := (D - \alpha I)$. Using the causal solution, $G_\alpha(x) = H(x)e^{\alpha x}$, we define the first order exponential B-spline $\beta_\alpha : \mathbb{R} \to \mathbb{R}$ by

$$\beta_\alpha := G_\alpha - e^\alpha \, G_\alpha(\bullet - 1). \tag{2.78}$$

From this definition, it follows immediately that $\operatorname{supp} \beta_\alpha = [0, 1)$ and that $\beta_\alpha \geq 0$. (See Figure 2.12.)

Note that $\beta_0 = B_{1,\mathbb{Z}}$ on $[0, 1)$. For the general case, $k \in \mathbb{N}$, let $\alpha := \{\alpha_0, \alpha_1, \ldots, \alpha_k\}$ denote the set of real roots of the symbol $\mathcal{L}(s)$ of the k-th order linear differential operator $\mathcal{L}$, and define the general k-order exponential B-spline $\beta_\alpha : [0, k) \to \mathbb{R}^+$ by k-fold convolution

$$\beta_\alpha := \underset{i=1}{\overset{k}{*}} \beta_{\alpha_i}. \tag{2.79}$$

The length of support and the nonnegativity for β_α are an immediate consequence of the convolution product and the fact that $\operatorname{supp} \beta_\alpha = [0, 1)$ and $\beta_\alpha \geq 0$. Furthermore, since the convolution product increases regularity, we also have that $\beta_\alpha \in C^{k-2}[0, k)$. In case that $\alpha := \{0, 0, \ldots, 0\}$, i.e., $\mathcal{L} = D^k$, the above definitions reduce to the known formulae for k-order polynomial cardinal B-splines.

Figure 2.13 shows the graphs of a second and third order exponential B-spline.

For the following, we introduce the *exponential difference operator*.

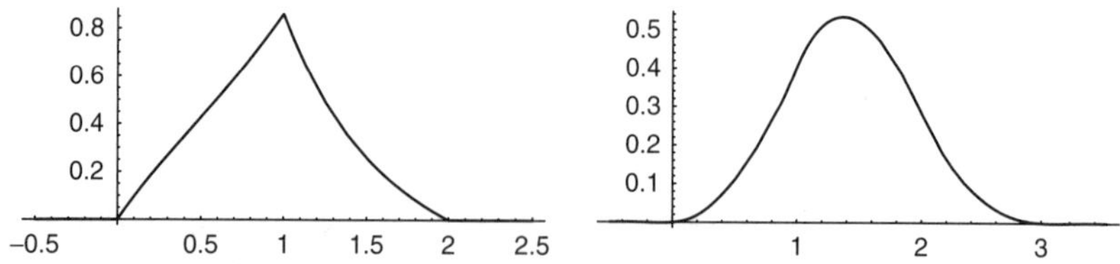

Figure 2.13 The graphs of the exponential B-splines $\beta_{\{-2,1\}}$ and $\beta_{\{-1,1,1\}}$.

Definition 2.91 (Exponential Difference Operator)   Let $k \in \mathbb{N}_0$ and let $\boldsymbol{\alpha} :=$ $\{\alpha_0, \alpha_1, \ldots, \alpha_k\} \subset \mathbb{R}$. Suppose that $f \in \mathrm{Map}(\mathbb{R}, \mathbb{R})$. Then the *exponential difference operator* $\nabla_{\boldsymbol{\alpha}} : \mathrm{Map}(\mathbb{R}, \mathbb{R}) \to \mathrm{Map}(\mathbb{R}, \mathbb{R})$ associated with $\boldsymbol{\alpha}$ is defined by

$$(\nabla_{\alpha_0} f)(t) := f(t) - e^{\alpha_0} f(t-1),$$

and

$$\nabla_{\{\alpha_0, \alpha_1, \ldots, \alpha_\varkappa\}} f := \nabla_{\alpha_\varkappa}(\nabla_{\{\alpha_0, \alpha_1, \ldots, \alpha_{\varkappa-1}\}} f), \quad \varkappa = 1, \ldots, k.$$

For simplicity, we set $\nabla_{\boldsymbol{\alpha}} := \nabla_{\{\alpha_0, \alpha_1, \ldots, \alpha_k\}}$.

Assuming that $f \in L^1(\mathbb{R}) \cap L^2(\mathbb{R}^2)$, we take the Fourier transform $\mathscr{F}$ of $\nabla_{\boldsymbol{\alpha}} f$ and obtain, using Property (2) of Proposition 2.26, the following equation:

$$\mathscr{F}(\nabla_{\boldsymbol{\alpha}} f)(\omega) = \prod_{\varkappa=0}^{k} (1 - e^{\alpha_\varkappa} e^{-i\omega})(\mathscr{F} f)(\omega) =: \mathcal{D}_{\boldsymbol{\alpha}}(e^{i\omega})(\mathscr{F} f)(\omega).$$

If we set $z := e^{i\omega}$, then the coefficient of the negative power $z^{-\varkappa}$ in $\mathcal{D}_{\boldsymbol{\alpha}}(z)$ corresponds to the coefficient of $f(\bullet - \varkappa)$, $\varkappa = 0, 1, \ldots, k$. In other words, the exponential difference operator $\nabla_{\boldsymbol{\alpha}}$ is completely determined by its *symbol* $\mathcal{D}_{\boldsymbol{\alpha}}(z)$.

Since $\beta_{\boldsymbol{\alpha}} = G_{\boldsymbol{\alpha}} - e^{\alpha} G_{\boldsymbol{\alpha}}(\bullet - 1) = \nabla_{\boldsymbol{\alpha}} G_{\boldsymbol{\alpha}}$, we obtain for the Fourier transform of the exponential B-spline (2.79)

$$\widehat{\beta}_{\boldsymbol{\alpha}}(\omega) = \mathcal{D}_{\boldsymbol{\alpha}}(e^{i\omega}) \widehat{G}_{\boldsymbol{\alpha}}(\omega) \tag{2.80}$$

and thus

$$\beta_{\boldsymbol{\alpha}} = \nabla_{\boldsymbol{\alpha}} G_{\boldsymbol{\alpha}},$$

i.e., the exponential B-spline $\beta_{\boldsymbol{\alpha}}$ is identical to the exponential difference operator applied to the Green's function $G_{\boldsymbol{\alpha}}$. This then implies that $\beta_{\boldsymbol{\alpha}}$ is a linear combination of integer-valued shifts of $G_{\boldsymbol{\alpha}}$. As a consequence of this fact, we can reproduce the Green's function from the exponential B-splines via Equation (2.80) as follows.

The reciprocal of the symbol $\mathcal{D}_{\boldsymbol{\alpha}}$ is given by

$$\mathcal{D}_{\boldsymbol{\alpha}}^{-1}(z) = \prod_{\varkappa=0}^{k} \frac{1}{1 - e^{\alpha_\varkappa} z^{-1}} = \prod_{\mu=1}^{l} \frac{1}{(1 - e^{\alpha_\mu} z^{-1})^{k_\mu}} = \sum_{\mu=1}^{l} \sum_{\varkappa=1}^{k_\mu} \frac{c_{\mu\varkappa}}{(1 - e^{\alpha_\mu} z^{-1})^\varkappa}, \tag{2.81}$$

where in the last equality we used a partial fraction expansion with coefficients $c_{\mu\varkappa}$. Taking the inverse transform of (2.81) yields (see also Equations (2.74) and (2.75))

$$G_{\boldsymbol{\alpha}} = \sum_{k \in \mathbb{N}_0} g_{\boldsymbol{\alpha}}(k) \beta_{\boldsymbol{\alpha}}(\bullet - k),$$

where the coefficients $g_\alpha(k)$ are given as

$$g_\alpha(k) := \sum_{\mu=1}^{l} \sum_{\varkappa=1}^{k_\mu} c_{\mu\varkappa} \frac{g_{\alpha_\mu}^{[\varkappa-1]}(k)}{(\varkappa-1)!} \tag{2.82}$$

with

$$g_\alpha^{[\varkappa]}(k) := \begin{cases} e^{\alpha k}, & \varkappa = 0; \\ (k+1,\varkappa)\, e^{\alpha k}, & \varkappa \in \mathbb{N}. \end{cases} \tag{2.83}$$

Here, $(k+1,\varkappa) := \dfrac{(k+\varkappa)!}{k!}$ is the *Pochhammer* [15] *symbol*.

Equation (2.80) has an interesting implication. Recall that the Fourier transform of $\widehat{G}$ is given by (2.73) and, therefore,

$$\widehat{\beta}_\alpha(\omega) = \frac{\mathcal{D}_\alpha(e^{i\omega})}{\mathcal{L}(i\omega)}.$$

This description has many important consequences yielding new types of splines such as splines of fractional and complex orders, but those would lead us far from the objectives of this text. We refer the interested reader to the original essays [62, 63, 64, 153] on this subject.

To establish that every exponential spline can be expressed as an $\mathbb{R}$-linear combination of exponential B-splines, we need to introduce the concept of a *Riesz* [16] *basis* of a Hilbert space H.

Definition 2.92 (Riesz Basis in a Hilbert Space)   Suppose that $(\mathsf{H}, \langle\,,\,\rangle)$ is a Hilbert space over the reals $\mathbb{R}$ and $\{h_i \mid i \in \mathbb{Z}\}$ is an orthonormal basis for H. Furthermore, assume that $U : \mathsf{H} \to \mathsf{H}$ is a bounded bijective operator. Then the system $\{Uh_i \mid i \in \mathbb{Z}\}$ is called a *Riesz basis* for H.

The next result provides an alternate description of a Riesz basis.

Theorem 2.93   Suppose that $(\mathsf{H}, \langle\,,\,\rangle)$ is a Hilbert space over the reals $\mathbb{R}$. A sequence of vectors $\{h_n \mid n \in \mathbb{Z}\} \subset \mathsf{H}$ is a Riesz basis for H, if

1. there exist constants $0 < A \leq B < \infty$ so that for all ℓ^2-sequences $\{c_n \mid n \in \mathbb{Z}\}$:

$$A\,\|c\|_{\ell^2(\mathbb{Z})} \leq \left\| \sum_{n\in\mathbb{Z}} c_n\, h_n \right\|^2 \leq B\,\|c\|_{\ell^2(\mathbb{Z})};$$

2. $\mathrm{clos}_\mathsf{H}\,\mathrm{span}\{h_n \mid n \in \mathbb{Z}\} = \mathsf{H}$.

---

15 Leo August Pochhammer, 25 August 1841–24 March 1920. German mathematician known mainly known for his contributions to the theory of special functions.

16 Marcel Riesz, 16 November 1886–4 September 1969. Hungarian mathematician who is known for his work in classical analysis, partial differential equations, Clifford algebras, and number theory. He was the younger brother of the Hungarian mathematician Frigyes Riesz.

Here, $\|\ \| : H \to \mathbb{R}$ denotes the norm induced by the inner product $\langle\ ,\ \rangle$ and clos_H the closure in H.

Proof See, for instance [37]. $\square$

The following results are needed for our further developments.

Proposition 2.94 Every basis in a finite dimensional Hilbert space $(H, \langle\ ,\ \rangle)$ is a Riesz basis.

Proof Exercise! $\square$

Theorem 2.95 Let $\varphi \in L^1(\mathbb{R}) \cap L^2(\mathbb{R})$ and set $\varphi_n := \varphi(\bullet - n)$, $n \in \mathbb{N}$. Assume that $0 < A \le B < \infty$ are constants. Then the following two statements are equivalent:

1. $\forall\, \{c_n \mid n \in \mathbb{Z}\} \in \ell^2(\mathbb{Z}):\ A\,\|c\|_{\ell^2(\mathbb{Z})} \le \left\|\sum_{n\in\mathbb{Z}} c_n\, \varphi_n\right\|^2 \le B\,\|c\|_{\ell^2(\mathbb{Z})};$

2. $A \le \sum_{n\in\mathbb{Z}} |\widehat{\varphi}(\omega + 2\pi n)|^2 \le B$, where $\widehat{\varphi}$ denotes the Fourier-Plancherel transform of φ.

Proof We will outline the proof, leaving the details to the reader. Let $c := \{c_n \mid n \in \mathbb{N}\} \in \ell^2(\mathbb{N})$ and denote by

$$C(\omega) := \sum_{n\in\mathbb{Z}} c_n\, e^{in\omega}$$

the *symbol* associated with the sequence c. Using the fact that $\{e^{in} \mid n \in \mathbb{Z}\}$ is an orthonormal basis of $L^2(0, 2\pi)$ and employing Parseval's Identity

$$\frac{1}{2\pi} \int_0^{2\pi} |f(x)|^2\, dx = \sum_{n\in\mathbb{Z}} |c_n|^2, \quad f \in L^2(0, 2\pi),$$

to the function $f := \sum_{n\in\mathbb{Z}} c_n\, \varphi_n$ yields

$$\left\|\sum_{n\in\mathbb{Z}} c_n\, \varphi_n\right\|_{L^2}^2 = \frac{1}{2\pi} \sum_{n\in\mathbb{Z}} \int_0^{2\pi} |C(x)\, \widehat{\varphi}(x + 2\pi n)|^2\, dx.$$

Here, we used the fact that the Fourier transform is 2π-periodic. Set

$$\Phi := \sum_{n\in\mathbb{Z}} \int_0^{2\pi} |\widehat{\varphi}(\bullet + 2\pi n)|^2$$

and define $g(x) := |C(x)|^2/\|C\|_{L^2}$. Another application of Parseval's Identity shows that Statement 1 in Theorem 2.95 is equivalent to

$$A \leq \frac{1}{2\pi} \int_0^{2\pi} g(x)\Phi(x)\,dx \leq B. \tag{2.84}$$

The second statement in Theorem 2.95 clearly implies (2.84). To show that (2.84) implies the second statement, we set $g := L_\nu$ and let $\nu \to \infty$. (Here, L_ν denotes the Landau kernel of order ν introduced in (1.23).) $\quad\square$

Now, let $\mathsf{V}^\alpha := \{\beta \in \mathrm{Map}(\mathbb{R}, \mathbb{R}) \mid \beta \text{ cardinal exponential spline}\} \cap L^2(\mathbb{R})$. Clearly, the pair $(\mathsf{V}^\alpha, \langle\,,\,\rangle_{L^2(\mathbb{R})})$ constitutes a Hilbert space. Theorem 2.95 implies the next result.

Theorem 2.96   The sequence $\{\beta_\alpha(\bullet - k) \mid k \in \mathbb{Z}\}$ forms a Riesz basis of V^α.

Proof   Exercise! $\square$

Associated with the above results, we state the reconstruction formulae for the regular exponential function $e^{\alpha x}$ and the functions $x^m e^{\alpha x}$, which are elements of the nullspace of the operator $\mathcal{L} := D - \alpha I$ and $\mathcal{L} := (D - \alpha I)^m$, respectively:

$$e^{\alpha x} = \sum_{k\in\mathbb{Z}} e^{\alpha k}\,\beta_\alpha(x - k) \tag{2.85}$$

and

$$x^m e^{\alpha x} = \sum_{k\in\mathbb{Z}} g_\alpha^{[m]}(k)\,\beta_{\{\alpha,\dots,\alpha\}}(x - k). \tag{2.86}$$

The first of these reconstruction formulae is easy to verify using the definition of cardinal exponential B-splines, and the second one follows from the considerations leading to (2.82) and (2.83).

Finally, we state without proof a result about the approximation properties of exponential splines. For this purpose, we need to construct a sequence of nested subspaces of V^α corresponding to a decreasing sequence of knot spacings. First, we show that the space V^α is shift-invariant.

Definition 2.97 (Shift-Invariance)   Let V be a linear space of functions. Denote by $\tau : \mathsf{V} \to \mathsf{V}$ the shift operator defined by $\tau f := f(\bullet - 1)$. A linear space V is called *shift-invariant* whenever $f \in \mathsf{V}$ implies that $\tau f \in \mathsf{V}$.

Proposition 2.98   The space V^α is shift-invariant.

Proof   It suffices to show that $(x - 1)^n e^{\alpha(x-1)} \in \mathsf{V}^\alpha$, for $n \in \mathbb{N}_0$. This, however, follows from

$$(x - 1)^n e^{\alpha(x-1)} = \sum_{\nu=0}^n \binom{n}{\nu} x^\nu e^{-\alpha}\, e^{\alpha x} =: \sum_{\nu=0}^n c_\nu\, x^\nu e^{\alpha x}. \quad\square$$

Now, let $\alpha \in \mathbb{R}$ and consider the cardinal exponential B-spline β_α. Then, using (2.78), it is straight-forward to show that

$$\beta_\alpha(x) = \beta_{\alpha/2}(2x) + e^{\alpha/2}\beta_{\alpha/2}(2x-1)$$

$$= \beta_{\alpha/4}(4x) + e^{\alpha/4}\beta_{\alpha/4}(4x-1) + e^{\alpha/2}\beta_{\alpha/4}(4x-2) + e^{3\alpha/4}\beta_{\alpha/4}(4x-3)$$

and, more generally,

$$\beta_\alpha(x-m) = \sum_{\mu=0}^{2^j-1} e^{2^{-j}\alpha x}\beta_{2^{-j}\alpha}(2^j(x-m)-\mu), \quad j \in \mathbb{N},\ m \in \mathbb{Z}.$$

Setting

$$\beta_{\alpha;j,m} := \beta_{2^{-j}\alpha}(2^j \bullet - m), \tag{2.87}$$

and

$$\mathsf{V}_j^{2^{-j}\alpha} := \mathrm{clos}_{L^2}\,\mathrm{span}\{\beta_{\alpha;j,m} \mid m \in \mathbb{Z}\},$$

one has

$$\mathsf{V}_j^{2^{-j}\alpha} \subset \mathsf{V}_{j+1}^{2^{-(j+1)}\alpha}, \quad j \in \mathbb{N}.$$

Thus, the spaces $\{\mathsf{V}_j^{2^{-j}\alpha} \mid j \in \mathbb{N}_0\}$ form a filtration of $L^2(\mathbb{R})$.

We now extend the definition of $\beta_{\alpha;j,m}$ given in (2.87) to $\boldsymbol{\alpha} = \{\alpha_0, \alpha_1, \ldots, \alpha_k\}$ by setting for $j \in \mathbb{N}$ and $m \in \mathbb{Z}$

$$\beta_{\boldsymbol{\alpha};j,m} := \beta_{2^{-j}\boldsymbol{\alpha}}(2^j \bullet - m),$$

where $2^{-j}\boldsymbol{\alpha} := \{2^{-j}\alpha_0, 2^{-j}\alpha_1, \ldots, 2^{-j}\alpha_k\}$. Moreover, for $j \in \mathbb{N}$, we define the *exponential spline approximation spaces* $\mathsf{V}_j^{2^{-j}\boldsymbol{\alpha}}$ by

$$\mathsf{V}_j^{2^{-j}\boldsymbol{\alpha}} := \mathrm{clos}_{L^2}\,\mathrm{span}\{\beta_{\boldsymbol{\alpha};j,m} \mid m \in \mathbb{Z}\}.$$

We leave it to the reader to establish that the collection $\{\beta_{\boldsymbol{\alpha};j,m} \mid m \in \mathbb{Z}\}$ forms a Riesz basis for $L^2(\mathbb{R})$ and that the spaces $\mathsf{V}_j^{2^{-j}\boldsymbol{\alpha}}$, $j \in \mathbb{N}$, are therefore closed in $L^2(\mathbb{R})$. (See also [152].) Similar to the exposition above, one shows that the spaces $\{\mathsf{V}_j^{2^{-j}\boldsymbol{\alpha}} \mid j \in \mathbb{N}_0\}$ form a filtration of $L^2(\mathbb{R})$.

The next result gives the exact asymptotic L^2-approximation error for functions in $H^k(\mathbb{R})$ when represented in terms of exponential B-splines.

Theorem 2.99 (Approximation Order of Exponential Splines)   Suppose that $f \in H^k(\mathbb{R})$, $k \in \mathbb{N}$. Denote by $\pi_j : L^2(\mathbb{R}) \to \mathsf{V}_j^{2^{-j}\boldsymbol{\alpha}}$ the projection onto $\mathsf{V}_j^{2^{-j}\boldsymbol{\alpha}}$. Then

$$\|f - \pi_j f\|_{L^2} = c_k \|\mathcal{L}_\alpha f\|_{L^2},$$

where $c_k = \sqrt{2\zeta(2k)}/(2\pi)^k$, where $\zeta(s) := \sum_{n=1}^{\infty} n^{-s}$ is the *Riemann zeta function*.

Proof   For the proof, we refer to [152]. $\square$

2.13 $\mathcal{L}$-Splines

Here, we very briefly mention an extension of the concepts and ideas introduced in Section 2.12 to more general differential operators $\mathcal{L}$ following the presentation provided in [142]. To this end, we consider linear differential operators of the form

$$\mathcal{L} := \sum_{j=0}^{k} a_j(x)\, D^j,$$

with coefficient functions $a_j \in C^j[a, b]$, $j = 0, 1, \ldots, k$, where $0 < \sup\{a_k(x)\,|\, x \in [a, b]\} \leq A$, for some $A \in \mathbb{R}^+$. Under these assumptions, it follows that the null space $\mathcal{N}(\mathcal{L}) := \{u \in C^k[a, b]\,|\,(\mathcal{L}u)(x) = 0,\ x \in [a, b]\}$ of the operator $\mathcal{L}$ has dimension $\dim \mathcal{N}(\mathcal{L}) = k$ and is spanned by the uniquely determined set of functions

$$\left\{u_j \in \mathcal{N}(\mathcal{L})\,|\,(D^\nu u_j)(a) = \delta_{\nu j},\ \forall j, \nu \in \{0, 1, \ldots, k\}\right\}. \qquad (2.88)$$

(As a reference, see for instance [158].)

Definition 2.100 ($\mathcal{L}$-Spline)   Let $k, N \in \mathbb{N}$ and let $\boldsymbol{\xi} := (\xi_0, \ldots, \xi_{N+1})$ be an ordered knot sequence. Furthermore, let $\boldsymbol{m} \in \mathbb{N}^N$ with $m_i \in \{1, \ldots, k-1\}$, $i = 1, \ldots, N$ be given. The linear space

$$S_{\mathcal{L}}^{m}(\boldsymbol{\xi}) := \left\{s \in \mathrm{Map}(\mathbb{R}, \mathbb{R})\,\Big|\, s\,|_{[\xi_\nu, \xi_{\nu+1}]} \in \mathcal{N}(\mathcal{L})\ \text{and}\ \mathrm{jump}_{\xi_\nu} D^\mu s = 0,\right.$$

$$\left. \forall \mu \in \{0, 1, \ldots, m_\nu\}\right\}$$

is called the *space of $\mathcal{L}$-splines* with (ordered) knot sequence $\boldsymbol{\xi}$ and smoothness conditions given by $\boldsymbol{m}$.

Example 2.101   If $\mathcal{L} := D^k$, then the space $S_{\mathcal{L}}^{m}(\boldsymbol{\xi})$ coincides with $\Pi_{\boldsymbol{\xi}}^k$. (See also Section 2.7.2.)

Analogously to the previous section, one can define a Green's function for the operator $\mathcal{L}$ defined in (2.88). For this purpose, recall the following result from the theory of differential equations. (Cf. for instance [158].) Suppose $f \in C^k[a, b]$ is a given function. Then there exists for each $x \in [a, b]$ exactly one element $u \in \mathcal{N}(\mathcal{L})$ with the property that

$$(D^\varkappa u)(t) = D^\varkappa f,\ \forall \varkappa = 0, 1, \ldots, k-1,$$

and a uniquely determined Green's function $G(x, t) \in C([a, b] \times [a, b])$ satisfying

$$f(x) = u(x) + \int_a^b G(x, t)(\mathcal{L}f)(t)\, dt. \qquad (2.89)$$

Example 2.102   Suppose that $\mathcal{L}$ is a linear differential operator with constant coefficients. Then $\mathcal{L}$ is a translation-invariant operator, i.e., $\mathcal{L}(\tau_y f) = \tau_y(\mathcal{L}f)$, where τ_y is the translation operator defined by $(\tau_y f)(x) := f(x - y)$. Then the Green's function $G(x, t)$ in (2.89) has to be of the form $G(x, t) = G(x - t)$ (reflecting the translation-invariance). Moreover, for $\mathcal{L} := D^k$, (2.89) yields the Taylor formula with integral remainder:

$$f(x) = \sum_{j=0}^{k-1} \frac{(x - a)^j}{j!} (D^j f)(a) + \frac{1}{(k-1)!} \int_a^b (x - t)_+^{k-1} (D^k f)(t)\, dt.$$

(Compare with the Peano Kernel for divided differences!) The above equality also shows that the Green's function $G(x, t)$ associated with the operator $\mathcal{L}$ given by (2.88) is a generalization of the truncated power function $(x - t)_+^{k-1}$. Conversely, every truncated power function can be regarded as a Green's function for the operator $\mathcal{L} = D^k$. We made implicit use of this fact already in the proof of Theorem 2.4 and also in Section 2.7.2.

Based on the above considerations, general theorems can be proven regarding the representation of $\mathcal{L}$-splines in terms of the Green's functions $G(x, t)$ and expansions in bases elements similar to polynomial cardinal B-splines and cardinal exponential splines. However, we will not pursue this issue any further and instead refer the interested reader to the literature of which [90, 93, 134, 138, 140, 141] is but a small excerpt.

In summary, we like to mention that the form of a linear differential operator $\mathcal{L}$ with *constant* coefficients gives rise to the following classes of splines:

1. **Exponential Splines:** $\mathcal{L} = (D - \alpha_1) \cdots (D - \alpha_k)$, for real $\alpha_\varkappa$, $\varkappa = 1, \ldots, k$;

2. **Hyperbolic Splines:** $\mathcal{L} = D^2 (D^2 - \alpha_2^2) \cdots (D^2 - \alpha_k^2)$, for real $\alpha_\varkappa$, $\varkappa = 2, \ldots, k$;

3. **Trigonometric Splines:** $\mathcal{L} = D^2 (D^2 + 1) \cdots (D^2 + k)$.

2.14 Wavelets and Splines

In this section, we show that a class of B-splines has multiscale properties, thus defining a wavelet basis for $L^2(\mathbb{R})$. To this end, we first introduce the concept of a *wavelet function* and define what is called a *multiresolution analysis of* $L^2(\mathbb{R})$. Next, we consider fundamental cardinal splines and show that they form wavelet bases. The exposition of this section is based on [39] and the reader is referred to it for more details and proofs.

Consider the (unit) translation operator

$$\mathcal{T} : L^2(\mathbb{R}) \to L^2(\mathbb{R}),$$

$$f \longmapsto f(\bullet - 1),$$

and the dyadic dilation operator

$$\mathcal{D} : L^2(\mathbb{R}) \to L^2(\mathbb{R}),$$

$$f \longmapsto \sqrt{2} f(2 \bullet).$$

It is straight-forward to show that $\mathcal{T}$ and $\mathcal{D}$ are unitary operators on $L^2(\mathbb{R})$ and that $\mathcal{D}\mathcal{T} = \mathcal{T}^2\mathcal{D}$. (The reader is encouraged to verify this statement.)

Definition 2.103 (Wavelet)   A function $\psi \in L^2(\mathbb{R})$ is called a *wavelet* if the affine system $\{\mathcal{D}^j\mathcal{T}^k\psi \mid j, k \in \mathbb{Z}\}$ forms an orthonormal basis of $L^2(\mathbb{R})$. It is customary to write $\mathcal{D}^j\mathcal{T}^k\psi$ as ψ_{jk}.

Definition 2.104 (Discrete Wavelet Transform)   Suppose that ψ is a wavelet. The representation of a function $f \in L^2(\mathbb{R})$ in terms of the affine system $\{\mathcal{D}^j\mathcal{T}^k\psi \mid j, k \in \mathbb{Z}\}$,

$$f \mapsto c(f) := \mathscr{W}f := \{c_{jk} := \langle f, \psi_{jk}\rangle_{L^2} \mid j, k \in \mathbb{Z}\}$$

is called a *discrete wavelet transform*. The *inverse discrete wavelet transform* is given by

$$c := \{c_{jk} \mid j, k \in \mathbb{Z}\} \mapsto \mathscr{W}^*(f) := \sum_{j,k \in \mathbb{Z}} c_{jk}\, \psi_{jk}.$$

Remark 2.105   Since $\{\psi_{jk} \mid j, k \in \mathbb{Z}\}$ is an *orthonormal* basis of $L^2(\mathbb{R})$, $\mathscr{W}^*\mathscr{W} = \mathrm{id}_{L^2(\mathbb{R})}$.

Example 2.106 (Haar Wavelet)   Let B_1 denote the first oder B-spline defined on the half-open interval $[0, 1)$, and let $\psi(x) := B_1(2x) - B_1(2x - 1) = (\nabla B_2)(2x) = (DB_2)(2x)$. (See Theorem 2.22, item 5!) The explicit expression for ψ is given by

$$\psi(x) = \begin{cases} +1, & x \in [0, 1/2); \\ -1, & x \in [1/2, 1). \end{cases}$$

It is easy to verify that

$$\int_{\mathbb{R}} \psi_{jk}\, \psi_{jl}\, dx = \delta_{jl}, \quad j, k, l \in \mathbb{Z}.$$

Thus, the system $\{\psi_{jk} \mid j, k \in \mathbb{Z}\}$ is orthonormal in $L^2(\mathbb{R})$. Alfréd Haar [17] introduced this system in 1909 and showed that it is an orthonormal basis of $L^2(\mathbb{R})$ [77].

The Haar wavelet is the simplest wavelet in the family of Daubechies' wavelets. The Daubechies family of wavelets consists of a system of compactly supported orthonormal wavelets of increasing regularity. We refer the

---

17 ALFRÉD HAAR, 11 October 1885–16 March 1933. Hungarian mathematician who worked in partial differential equations, Chebyshev approximations, and topological groups.

interested reader to [43] for the construction of this important family of wavelets.

Denoting by W_j the closure in $L^2(\mathbb{R})$ of $\mathrm{span}\{\psi_{jk} \mid k \in \mathbb{Z}\}$, we obtain from the definition of wavelet that

$$L^2(\mathbb{R}) = \bigoplus_{j \in \mathbb{Z}} W_j,$$

where $\bigoplus$ denotes the orthogonal sum decomposition. The spaces $W_j, j \in \mathbb{Z}$, are called *wavelet spaces*, and the spaces

$$V_j := \bigoplus_{\substack{l \in \mathbb{Z} \\ l < j}} W_l$$

are called *approximation spaces* of $L^2(\mathbb{R})$. Note that

$$V_j = V_{j-1} \oplus W_{j-1}, \quad \forall j \in \mathbb{Z},$$

with W_{j-1} being the orthogonal complement of V_{j-1} in V_j. Clearly, the system $\{V_j \mid j \in \mathbb{Z}\}$ forms a filtration of $L^2(\mathbb{R})$. Moreover,

$$\bigcap_{j \in \mathbb{Z}} V_j = \{0\} \quad \text{and} \quad \mathrm{clos}_{L^2} \bigcup_{j \in \mathbb{Z}} V_j = L^2(\mathbb{R}).$$

Since $V_j = V_{j-1} \oplus W_{j-1}, j \in \mathbb{Z}$, one can describe the wavelet spaces also as *detail spaces*: The elements of W_{j-1} add more detail to the approximation space V_{j-1} to obtain a better approximation to a function $f \in L^2(\mathbb{R})$ at level j.

These thoughts give rise to the following definition.

Definition 2.107 (Multiresolution Analysis)   A system of closed subspaces $\{V_j \mid j \in \mathbb{Z}\}$ of $L^2(\mathbb{R})$ forms a *multiresolution analysis* (MRA) of $L^2(\mathbb{R})$ provided that

1. $\{0\} = \bigcap_{j \in \mathbb{Z}} V_j \subset \cdots \subset V_j \subset V_{j+1} \subset \cdots \subset \mathrm{clos}_{L^2} \bigcup_{j \in \mathbb{Z}} V_j = L^2(\mathbb{R});$

2. $f \in V_j \iff \mathcal{D}f \in V_{j+1}, j \in \mathbb{Z};$

3. There exists a function $\varphi \in L^2(\mathbb{R})$ such that $\{\mathcal{T}^k \varphi \mid k \in \mathbb{Z}\}$ is a Riesz basis for V_0.

The function φ is called a *scaling function* associated with the MRA.

Property 2 in Definition 2.107 expresses the fact that all spaces V_j are dyadic dilates of a base space, for instance, V_0: $V_j = \mathcal{D}^j V_0$. Property 3 in Definition 2.107 shows that the space V_0, and thus all other spaces $V_j, j \in \mathbb{Z} \setminus \{0\}$, are *shift-invariant under* $\mathcal{T}$. These concepts are parts of a general definition given below.

Definition 2.108 (Dilation-Invariance)   Let V be a space of functions over $\mathbb{R}^n$, $n \in \mathbb{N}$, and let $\mathcal{D}$ be a dilation operator represented on $\mathbb{R}^n$ by a matrix all of whose eigenvalues have modulus greater than one. Then V is called *dilation-invariant* (with respect to $\mathcal{D}$) provided that

$$\forall f \in V : f \in V \Longrightarrow \mathcal{D}^{-1} f \in V,$$

or, equivalently, $\mathcal{D}^{-1} V \subseteq V$.

Definition 2.109 (Shift-Invariance)   Let V be a space of functions over $\mathbb{R}^n$, $n \in \mathbb{N}$, and let $\{\boldsymbol{\gamma}_\nu \in \mathbb{R}^n \,|\, \nu = 1, \ldots, n\}$ be a basis of $\mathbb{R}^n$. The set $\Gamma :=$
$\left\{ \sum\limits_{\nu=1}^{n} z_\nu \, \boldsymbol{\gamma}_\nu \,\middle|\, z_\nu \in \mathbb{Z} \right\}$ is called a *full-rank lattice* of $\mathbb{R}^n$. Let $\mathcal{T} := \mathcal{T}_\Gamma$ denote the translation operator along the lattice Γ: $(\mathcal{T}f)(\boldsymbol{x}) = f(\boldsymbol{x} + \boldsymbol{\gamma})$, $\boldsymbol{\gamma} \in \Gamma$. Then V is called *shift-invariant* with respect to $\mathcal{T}$ (and thus Γ) iff

$$f \in V \Longrightarrow \mathcal{T} f \in V.$$

Example 2.110   Let $f \in L^2(\mathbb{R}^n)$, $n \in \mathbb{N}$, be a function and let $\Gamma := \mathbb{Z}^n$. The set

$$S(f) := \mathrm{clos}_{L^2} \mathrm{span}\{f(\bullet + \boldsymbol{v}) \,|\, \boldsymbol{v} \in \mathbb{Z}^n\}$$

defines a closed shift-invariant subspace of $L^2(\mathbb{R}^n)$. Since $S(f)$ is generated by a single function, these spaces are also called *principal shift-invariant (PSI) spaces*.

In case a finite collection of functions $\mathcal{F} := \{f_1, \ldots, f_m\} \subset L^2(\mathbb{R}^n)$, $1 < m \in \mathbb{N}$, is given, the set

$$S(\mathcal{F}) := \mathrm{clos}_{L^2} \mathrm{span}\{f_\mu(\bullet + \boldsymbol{v}) \,|\, \mu = 1, \ldots, m; \ \boldsymbol{v} \in \mathbb{Z}^n\}$$

is termed a *finitely generated shift-invariant (FSI) space*. Furthermore, if $\mathrm{span}\{f_\mu(\bullet + \boldsymbol{v}) \,|\, \mu = 1, \ldots, m; \ \boldsymbol{v} \in \mathbb{Z}^n\}$ is an orthogonal (orthonormal) basis for $S(\mathcal{F})$, then the generator $\mathcal{F}$ is called *orthogonal (orthonormal)*.

Example 2.111   Recall that the cardinal spline space $S^k(\mathbb{Z})$ is the pointwise linear span of the integer-translates of the cardinal B-spline $B_{k,\mathbb{Z}}$ (cf. 2.19). Thus, in this restricted interpretation we have $S^k(\mathbb{Z}) = S(B_{k,\mathbb{Z}})$.

This last example can be extended to the $L^2(\mathbb{R})$-setting as follows. Recall that $S_0^k := S^k(\mathbb{Z})$ is the cardinal spline space for the knot set $\mathbb{Z}$. However, one can also consider cardinal spline spaces $S_j^k := S^k(2^{-j}\mathbb{Z})$ for the knot sets $2^{-j}\mathbb{Z}$, $j \in \mathbb{Z}$, and observe that any spline with knot set $2^{-j}\mathbb{Z}$ is also a spline with knot sequence $2^{-l}\mathbb{Z}$ with $j < l$. Thus, the cardinal spline spaces $\{S_j^k \,|\, j \in \mathbb{Z}\}$ are nested: $S_j^k \subset S_{j+1}^k$, $j \in \mathbb{Z}$. Defining V_j^k to be the L^2-closure of $S_j^k \cap L^2(\mathbb{R})$, we obtain a filtration of closed subspaces of $L^2(\mathbb{R})$ of the form $\cdots \subset V_j^k \subset V_{j+1}^k \subset \cdots$.

In addition, it is a consequence of the properties of splines that

$$\bigcap_{j\in\mathbb{Z}} V_j^k = \{0\} \quad \text{and} \quad \text{clos}_{L^2}\bigcup_{j\in\mathbb{Z}} V_j^k = L^2(\mathbb{R}). \tag{2.90}$$

The basis $\mathcal{B}_k$ of cardinal B-splines introduced in Section 2.4 can be shown to be a Riesz basis for the space V_0^k with Riesz bounds $A = \sum_{k\in\mathbb{Z}} |\widehat{B}_k(\pi + 2\pi k)|^2$ and $B = 1$. (Cf., for instance, [39].) Hence, we have the following result.

Theorem 2.112 (Cardinal Spline MRA) Suppose that S_j^k is the cardinal spline space with knot set $2^{-j}\mathbb{Z}$. Denote by V_j^k the L^2-closure of $S_j^k \cap L^2(\mathbb{R})$. Then the system $\{V_j^k \,|\, j \in \mathbb{Z}\}$ forms a multiresolution analysis of $L^2(\mathbb{R})$.

Returning now to the general case of an MRA, note that since the scaling function $\varphi \in V_0$, $\mathcal{D}\varphi \in V_1 \supset V_0$. Thus, there exist real numbers g_k, $k \in \mathbb{Z}$, so that

$$\varphi(x) = \sum_{k\in\mathbb{Z}} g_k\, \varphi(2x - k). \tag{2.91}$$

Similarly, since the wavelet space $W_0 \subset V_1$, there exist real numbers h_k, $k \in \mathbb{Z}$, so that

$$\psi(x) = \sum_{k\in\mathbb{Z}} h_k\, \varphi(2x - k). \tag{2.92}$$

Equations of the type (2.91) and (2.92) are called *refinement* or *dilation equations*. Sometimes, the term *two-scale equation*, referring to the dilation by "2" term, is also used.

Example 2.113 We now return to the cardinal spline MRA and derive the refinement equation for the scaling function $\varphi = B_k$. For this purpose, set

$$B_k(x) = \sum_{v\in\mathbb{Z}} g_v^k\, B_k(2x - v),$$

for some $g_v^k \in \mathbb{R}$. Note that the sequence $\{g_v^k \,|\, v \in \mathbb{Z}\}$ has to be in $\ell^2(\mathbb{Z})$. Taking the Fourier transform of both sides yields

$$\widehat{B}_k(\omega) = \frac{1}{2}\left(\sum_{v\in\mathbb{Z}} g_v^k e^{-iv\omega/2}\right) \widehat{B}_k(\omega/2). \tag{2.93}$$

The Fourier transform of $\widehat{B}_k$ is given by $\widehat{B}_k(\omega) = (2\pi)^{-1/2}(1 - e^{-i\omega})^k (i\omega)^{-k}$ and, therefore, (2.93) reads

$$\frac{1}{2\pi}\frac{(1 - e^{-i\omega})^k}{(i\omega)^k} = \frac{1}{2}\left(\sum_{v\in\mathbb{Z}} g_v^k e^{-iv\omega/2}\right)\left[\frac{1}{2\pi}\frac{(1 - e^{-i\omega/2})^k}{(i\omega/2)^k}\right].$$

Division by the term in brackets and simplification produces

$$\left(\frac{1+e^{-i\omega/2}}{2}\right)^{k} = \frac{1}{2}\left(\sum_{v\in\mathbb{Z}} g_{v}^{k} e^{-iv\omega/2}\right).$$

Applying the binomial formula to the left-hand side gives

$$\frac{1}{2^{k}}\sum_{\varkappa=0}^{k}\binom{k}{\varkappa} e^{-i\varkappa\omega/2} = \frac{1}{2}\left(\sum_{v\in\mathbb{Z}} g_{v}^{k} e^{-iv\omega/2}\right),$$

and, by a comparison of coefficients, one obtains for the g_{v}^{k} the formula

$$g_{v}^{k} = \begin{cases} \dfrac{1}{2^{k-1}}\dbinom{k}{v}, & \varkappa = 0, 1, \ldots, k \\ 0, & \text{otherwise.} \end{cases}$$

Thus,

$$B_{k}(x) = \sum_{\varkappa=0}^{k} \frac{1}{2^{k-1}}\binom{k}{\varkappa} B_{k}(2x-\varkappa).$$

Next, we construct the wavelets associated with the cardinal spline MRA defined above. For this purpose, we recall the Haar wavelet basis defined in Example 2.106. It is not difficult to verify that the scaling function φ associated with the Haar wavelet ψ is simply the characteristic function on the interval $[0, 1)$, i.e., the B-spline B_{1} on the interval $[0, 1)$. Moreover, one has the refinement equations

$$\varphi(x) = \varphi(2x) + \varphi(2x-1) \quad \text{and} \quad \psi = \varphi(2x) - \varphi(2x-1).$$

Since for each $k \in \mathbb{N}$, the B-splines B_{k} form an MRA and since for $k = 1$, $\varphi(x) = B_{1}(x)$ and $\psi(x) = (DB_{2})(2x) = (DL_{2})(2x - 1)$, it is suggestive to choose

$$\psi(x) := (D^{k}L_{2k})(2x - 1)$$

as a wavelet associated with $\varphi = B_{k}$. In order to justify this choice, we define

$$\mathsf{W}_{j} := \{\psi_{jl} \mid l \in \mathbb{Z}\}$$

and will show that

1. $\mathsf{W}_{0} \perp \mathsf{V}_{0}$, from which follows that $\mathsf{W}_{j} \perp \mathsf{V}_{j}, \forall j \in \mathbb{Z}$;

2. $\mathsf{W}_{0} \subset V_{1}$, form which follows that $\mathsf{W}_{j} \subset \mathsf{V}_{j+1}, \forall j \in \mathbb{Z}$;

3. and $V_{1} = V_{0} + \mathsf{W}_{0}$, which together with items 1 and 2, will show that the family $\{\mathsf{W}_{j} \mid j \in \mathbb{Z}\}$ are the wavelet spaces associated with the cardinal spline MRA.

To establish item 1, note that for $l, m, n \in \mathbb{Z}$ with $n := l - m$, and using the definition of fundamental cardinal B-spline (see (2.25)), one has

$$\int_{\mathbb{R}} \phi(x-l)\psi(x-m)\,dx = \int_{\mathbb{R}} \phi(x-n)\psi(x)\,dx = \int_{\mathbb{R}} (D^k L_{2k})(2x-1)B_k(x-n)\,dx$$

$$= \frac{(-1)^k}{2^k} \int_{\mathbb{R}} L_{2k}(2x-1)(D^k B_k)(x-n)\,dx$$

$$= \frac{1}{2^k} \sum_{\varkappa=0}^{k} (-1)^{k-\varkappa} \binom{k}{\varkappa} \int_{\mathbb{R}} L_{2k}(2x-1)\delta(x-n-\varkappa)\,dx$$

$$= \frac{1}{2^k} \sum_{\varkappa=0}^{k} (-1)^{k-\varkappa} \binom{k}{\varkappa} L_{2k}(2n+2\varkappa-1) = 0.$$

Here, we used integration by parts and formula (2.63) for $D^k B_k$. To obtain the statement in item 2, we make use of the following lemma.

Lemma 2.114 Let $C(\omega) := \sum_{v \in \mathbb{Z}} c_v^k e^{-iv\omega/2}$, where $\{c_v^k \mid v \in \mathbb{Z}\}$ are the coefficients of the representation of L_k in terms of centered cardinal B-splines. (See (2.25).) Then,

$$\widehat{\psi}(\omega) = \frac{1}{2} \frac{(1-z)^k}{z^{k-1}} C(\omega)\widehat{\varphi}(\omega/2).$$

Proof Taking the Fourier transform of both sides of (2.25) and using identity (2.94) in Exercise 8, namely

$$(D^k B_{2k})(x) = \sum_{\varkappa=0}^{k} (-1)^{\varkappa} \binom{k}{\varkappa} B_k(x - \varkappa),$$

yields the result. (The reader is encouraged to verify this statement.) $\qquad \square$

Item 3 is established when we can exhibit two $\ell^2(\mathbb{Z})$ sequences $\{a_v \mid v \in \mathbb{Z}\}$ and $\{b_v \mid v \in \mathbb{Z}\}$ so that

$$\varphi(2x - \varkappa) = \sum_{v \in \mathbb{Z}} a_{\varkappa-2v}\varphi(x - v) + \sum_{v \in \mathbb{Z}} b_{\varkappa-2v}\psi(x - v).$$

The existence of two such sequences is proven in [39] and we direct the reader to this reference for details.

Exercises

1. Prove formula (2.14).

2. Establish the validity of formula (2.15).

3. Suppose that $k, N \in \mathbb{N}$ and let $\mathcal{G}_N^k := \{(\bullet - v)_+^{k-1} \mid v = -N - k + 1, \ldots, N - 1\}$. Show that $\mathcal{G}_N^k$ is a basis for $S_N^k(\mathbb{Z})$.

4. Let $m \in \mathbb{N}$ and let ∇^m denote the mth-order backward difference operator. Show that the following identities hold.

   (a) $\nabla^m p = 0, \forall p \in \Pi^m$;

   (b) $(\nabla^m f)(x) = \displaystyle\sum_{v=0}^{m} (-1)^v \binom{m}{v} f(x - v), \; x \in \mathbb{R}$;

   (c) $(\Delta^m f)(x) = (\nabla^m f)(x + m), \; x \in \mathbb{R}$.

5. Prove Theorem 2.22.

6. Use the properties of cardinal B-splines summarized in Theorem 2.22 to obtain a closed formula for a cubic and quintic cardinal B-spline.

7. Given that

$$B_3(x) = p(x)B_1(x) + q(x)B_1(x - 1) + r(x)B_1(x - 2),$$

   find the polynomials $p, q, r \in \Pi^2$.

8. Prove the identity

$$D^k B_{2k} = \sum_{\varkappa=0}^{k} (-1)^\varkappa \binom{k}{\varkappa} B_k(x - \varkappa). \tag{2.94}$$

9. Prove the statement made in Remark 2.24.

10. Prove the properties of the Fourier transform listed in Proposition 2.26.

11. Prove Theorem 2.27.

12. Prove properties (2.28)–(2.31) of the spaces Π_ξ^k.

13. Show that the functionals $\Lambda_{v\mu}$ defined in (2.32) and the functions $\varphi_{v\mu}$ given in (2.33) are dual to each other, i.e., that

$$\Lambda_{v\mu}\varphi_{\lambda\varkappa} = \delta_{v\lambda}\delta_{\mu\varkappa}.$$

   holds for all $\lambda, v = 0, 1, \ldots, n$ and all $\mu, \varkappa = 1, \ldots, k$. Conclude from this, that the set of functions $\{\varphi_{v\mu} \mid v = 0, 1, \ldots, N; \; \mu = 1, \ldots, k\}$ forms a basis for Π_ξ^k.

14. Show that for all $k \in \mathbb{N}$ and all functions $f \in C^k(\mathbb{R})$,

$$k! \, [\xi_v, \ldots, \xi_{v+k}]f = (D^k f)(\xi_v),$$

   in case $\xi_v = \xi_{v+1} = \cdots = \xi_{v+k}$.

15. Prove that the set of functions $\{J_\nu, S_\nu \mid \nu = 1, \ldots, N\}$ forms a basis for $\Pi^4_{\mathbb{Z},2}$ and that each $f \in \Pi^4_{\mathbb{Z},2}$ allows the unique representation of the form

$$f = \sum_{\nu=1}^{N} \left[f(\xi_\nu) J_\nu + (Df)(\xi_\nu) S_\nu \right].$$

16. Show that for all polynomials $p \in \Pi^k$:

$$D(\lambda_{\nu k} p)(t) = 0,$$

where $\lambda_{\nu k}$ denotes the deBoor–Fix functional and D is the differential operator with respect to the variable t.

17. Show that the deBoor–Fix functionals $\lambda_{\nu k} : C^{k-1}[x_\nu, x_{\nu+k}] \to \mathbb{R}$ are linear and bounded.

18. Assume that $X := \{x_1, \ldots, x_{n+k}\}$ is a set of nondecreasing knots and that $\{B_{\nu k} \mid \nu = 1, \ldots, n\}$ is the set of B-splines defined on this knot set. Let $f \in \mathscr{S}^k(X)$ and let $x \in [x_\nu, x_{\nu+1}]$, $\nu = 1, \ldots, n$, be fixed. Show that $f(x)$ is a convex combination of the B-spline coefficients c_ν and deduce from this that the following inequalities hold:

$$\min\{c_{\nu+1-k}, \ldots, c_\nu\} \le f(x) \le \max\{c_{\nu+1-k}, \ldots, c_\nu\}.$$

19. Show that $\|\ \|_k$, $k \in \mathbb{N}$, defines a norm on $H^k[a, b]$.

20. Verify that the functional $|\ |_k : H^k(\Omega) \to \mathbb{R}$ given by

$$|f|_k := \sqrt{\sum_{|m|=k} \int_\Omega |(D^m f)(x)|^2 dx}.$$

is a semi-norm on $H^k(\Omega)$.

21. Show that the functional $\|\ \|_{H^\infty} : H^\infty(\Omega) \to \mathbb{R}$ defined by

$$\|f\|_{H^\infty} := \operatorname{ess\,sup}\{|f(x)| \mid x \in \Omega\}$$

is a norm on $H^\infty(\Omega)$.

22. Verify the statements made in Example 2.56.

23. Show that the functionals $\|\ \|_{k,p} : W^{k,p}(\Omega) \to \mathbb{R}$ are norms for all $k \in \mathbb{N}$ and $p \in [1, \infty]$.

24. Prove the two statements in Example 2.59.

25. Provide the details in the proof of Lemma 2.62.

26. Provide the details in the proof of Lemma 2.63.

27. Suppose that $k \in \mathbb{N}$ and that $m \in \{0, 1, \ldots, k\}$. Show that

$$\int_a^b (D^{2k-m} s)(t)(D^m f)(t)\,dt = 0,$$

for $s|_{[\xi_\nu, \xi_{\nu+1})} \in \Pi^{2k}$ and $f \in H^k[a, b]$ with $T_m f = 0$.

28. Compute $\dim \mathcal{N}^{2k}(\Xi)$.

29. Recall the definition of the Landau kernels (1.23). Show that if φ is a test function,

$$\lim_{\nu \to \infty} \int_{\mathbb{R}} \varphi(x) L_\nu(x)\,dx = \varphi(0) = \delta(\varphi).$$

Hence, one may define the Dirac delta distribution as the "limit" of Landau kernels.

30. Prove that there exists no locally integrable function $f \in \mathrm{Map}(\mathbb{R}, \mathbb{R})$ so that

$$\delta(\varphi) = \varphi(0) = \int_{\mathbb{R}} f(x)\varphi(x)\,dx.$$

(*Hint: Assume to the contrary that there exists such a function f. Use as test functions the class of functions φ_r introduced in (2.59), estimate the integral on the right-hand side, and show that is converges to 0 as $r \to 0$, giving a contradiction to the value on the left-hand side.*)

31. Using (2.61), derive a formula for $\delta^{(k)}$, $k \in \mathbb{N}$, and give an interpretation.

32. Establish Equation (2.63).

33. Show that $(x_+)' = H(x)$, and generalize this to higher-order powers and higher-order distributional derivatives.

34. Prove Theorem 2.79.

35. Show that $\widehat{\delta} = 1$.

36. Derive (2.74) from (2.75).

37. Apply the process described in Subsection 2.12.2 to the differential operator

$$\mathcal{L} := D^3 - 4D^2 + 5D - 2$$

to obtain the Green's function G.

38. Derive an explicit formula for the exponential B-spline $\beta_{\{-2,1\}}$.

39. Show that the second-order exponential B-spline $\beta_{\{1,1,1\}}$ is explicitly given by

$$\beta_{\{-1,1,1\}}(x) = \begin{cases} \frac{1}{4}e^{-x}(-1+e^{2x}-2x), & x \in [0,1); \\ \frac{1}{4}e^{-x-2}(-2e^{2x}+e^2[2x-3]+e^4[2x-1]), & x \in [1,2); \\ \frac{1}{4}e^{-x-4}(e^{2x}+e^6[-2x+5]), & x \in [2,3). \end{cases}$$

40. Suppose that β_α and β_β are two exponential B-splines of order k and l, respectively, where $\boldsymbol{\alpha} := \{\alpha_0, \alpha_1, \ldots, \alpha_k\}$ and $\boldsymbol{\beta} := \{\beta_0, \beta_1, \ldots, \beta_l\}$. Show that the convolution of these two exponential B-splines produces another exponential B-spline of order $k + l$ corresponding to $\{\alpha_0, \alpha_1, \ldots, \alpha_k, \beta_0, \beta_1, \ldots, \beta_l\}$.

41. Establish Equations (2.82) and (2.83).

42. Prove Proposition 2.94.

43. Fill in the details for the proof of Theorem 2.95.

44. Use the results from Theorem 2.95 to establish Theorem 2.96.

45. Derive Equations (2.85) and (2.86).

46. For $f \in \mathrm{Map}(\mathbb{R}, \mathbb{R})$ and $\alpha \in \mathbb{R}$, define the exponential difference operator of stepsize j, $j \in \mathbb{N}_0$, by

$$\nabla_{\alpha,j}f := \begin{cases} \nabla_\alpha f, & j = 0, \\ \dfrac{f - e^{2^{-j}\alpha}f(\bullet - 2^{-j})}{2^{-j}}, & j \in \mathbb{N}. \end{cases}$$

For $\boldsymbol{\alpha} = \{\alpha_0, \alpha_1, \ldots, \alpha_k\} \subset \mathbb{R}^{k+1}$, we set $\nabla_{\boldsymbol{\alpha},j} := \overset{k}{\underset{\varkappa=1}{*}} \nabla_{\alpha_\varkappa,j}$. Show that
$$\beta_{\boldsymbol{\alpha},j} = 2^{-j}\nabla_{\boldsymbol{\alpha},j}G_{\boldsymbol{\alpha}}.$$

47. Show that the Fourier transform of $\beta_{\boldsymbol{\alpha},j}$ is given by

$$2^{-j(k-1)}\prod_{\varkappa=1}^{k}\frac{1 - e^{2^{-j}(\alpha_\varkappa - i\omega)}}{i\omega - \alpha_\varkappa}.$$

48. Show that a linear differential operator with constant coefficients is translation-invariant.

49. Show that the unit translation operator $\mathcal{T}$ and the dyadic dilation operator $\mathcal{D}$ are unitary and satisfy $\mathcal{D}\mathcal{T} = \mathcal{T}^2\mathcal{D}$.

50. Every $f \in \mathsf{V}_j^k$ is of the form $f = \sum_{v \in \mathbb{Z}} c_v B_k(\bullet - v)$. As $f \in L^2(\mathbb{R})$, what can be said about the sequence of coefficients $\{c_v \mid v \in \mathbb{Z}\}$?

51. Show the validity of (2.90).

52. Fill in the details in the proof of Lemma 2.114. □

3

Interpolation in $\mathbb{R}^s$, $s > 1$

In this chapter, we briefly discuss interpolation in $\mathbb{R}^s$, where $s > 1$. The focus will be on polynomial interpolation, interpolation with box splines, and tensor product B-splines. We will prove uniqueness of polynomial interpolation for regions $G \subseteq \mathbb{R}^s$ and derive error estimates for rectangular regions. Next, we introduce box splines and discuss some of their properties. Tensor-product B-splines are presented and their shortcomings for interpolation and approximation purposes discussed. A short section on Kergin interpolation completes this chapter.

Since the topic of multivariate spline interpolation is multi-faceted, it is impossible to go into much detail regarding higher-dimensional interpolation and approximation, and we therefore refer the interested reader to the literature on multi-dimensional spline theory. The material presented in this section uses [38, 44] and [135] as sources and references.

3.1 Multivariate Polynomial Interpolation

In the following, we denote by G a domain in $\mathbb{R}^s$, where $s \in \mathbb{N}$ is always greater than 1, and by $\overline{G}$ its closure (with respect to the Euclidean topology on $\mathbb{R}^s$). We need to introduce the analogue of the polynomial spaces on $\mathbb{R}$. To this end, let $n \in \mathbb{N}$ and let $\boldsymbol{\alpha} := (\alpha_1, \ldots, \alpha_s) \in \mathbb{R}^s$ be a multi-index. Define the linear space of all real-valued multivariate polynomials of *total degree* $< n$ or, equivalently, of *total order* n by

$$\Pi_s^n := \left\{ p \in \mathrm{Map}(\mathbb{R}^s, \mathbb{R}) \,\middle|\, p(\boldsymbol{x}) = \sum_{0 \le |\boldsymbol{\alpha}| < n} a_{\boldsymbol{\alpha}}\, \boldsymbol{x}^{\boldsymbol{\alpha}}, \ a_{\boldsymbol{\alpha}} \in \mathbb{R} \right\},$$

and $\Pi_1^n := \Pi^n$. A straight-forward counting argument shows that the dimension of Π_s^n is given by

$$d_s := \dim \Pi_s^n = \binom{n-1+s}{s}.$$

(The reader is encouraged to verify this statement.) There exists another class of polynomial spaces, defined in the following way. Suppose that $\boldsymbol{m} := (m_1, \ldots, m_s) \in \mathbb{R}^s$ is a multi-index. We define the linear space of all polynomials of degree less than $\boldsymbol{m}$ by

$$\Pi_s^{\boldsymbol{m}} := \left\{ p \in \mathrm{Map}(\mathbb{R}^s, \mathbb{R}) \,\middle|\, p(\boldsymbol{x}) = \sum_{0 \leq \alpha < m} a_\alpha \, \boldsymbol{x}^\alpha, \ a_\alpha \in \mathbb{R} \right\}.$$

Note that for $s := 1$, we have $\Pi_1^{\boldsymbol{m}} = \Pi_1^n = \Pi^n$. The dimension of $\Pi_s^{\boldsymbol{m}}$ is readily seen to be

$$d_{\boldsymbol{m},s} := \dim \Pi_s^{\boldsymbol{m}} = \prod_{i=1}^{s} m_i.$$

To state the interpolation problem in $\mathbb{R}^s$, we introduce the set

$$X := \{\boldsymbol{x}_\nu \in \mathbb{R}^s \,|\, \boldsymbol{x}_\nu \text{ are pairwise distinct for } \nu = 0, 1, \ldots, d\},$$

where $d := d_{n,s}$ or $d := d_{\boldsymbol{m},s}$, and the set

$$Z := \{z_\nu \in \mathbb{R} \,|\, \nu = 0, 1, \ldots, d\}.$$

INTERPOLATION PROBLEM IN $\mathbb{R}^s$: Given an interpolating set $X \times Z \subseteq \mathbb{R}^s \times \mathbb{R}$ of cardinality d, where d is either $d_{n,s}$ or $d_{\boldsymbol{m},s}$, determine a polynomial $p \in \Pi_s^n$, respectively, $p \in \Pi_s^{\boldsymbol{m}}$, so that

$$p|_X = Z.$$

(Here, we abbreviated $p(x_\nu) = z_\nu$, $\nu = 0, 1, \ldots, d$ by $p|_X = Z$.)

For the reminder of this section, we will restrict ourselves to the case $s := 2$, as most of the important issues are already present in this special setting.

To this end, assume that G is a domain in $\mathbb{R}^2$ and $\overline{G}$ its closure. Moreover, assume that $f : \overline{G} \to \mathbb{R}$ is a function whose values are known on a knot set $X \subseteq \overline{G}$ of cardinality $d = d_{n,2} = \binom{n+1}{2}$.

Theorem 3.1 (Existence and Uniqueness of Bivariate Interpolating Polynomials) Let the set X be defined as above. Then, there exists exactly one polynomial $p \in \Pi_2^n$ so that $p|_X = f|_X$.

Proof Let $p \in \Pi_2^n$. We express p in the form

$$p(x, y) := \sum_{0 \leq i+j < n} a_{ij} x^i y^j = \sum_{k=0}^{n-1} q_k(x) y^{n-1-k},$$

with $q_k \in \Pi_1^k = \Pi^k$, $k = 1, \ldots, n$. Note that by the linearity of the interpolation problem, it suffices to show that $p|_X = 0$ has a unique solution.

Notice that for fixed x_0, $p(x_0, \cdot) \in \Pi^n$ and $p(x_0, y_\nu) = 0$, for all $\nu = 0, 1, \ldots, n - 1$. The existence and uniqueness of the one-dimensional interpolation problem now implies that $p(x_0, y) \equiv 0$ and, therefore, $q_k(x_0) = 0$, for all $k = 1, \ldots, n$. Since $q_0 \in \mathbb{R}$, we obtain $q_0 = 0$ and thus $p(x, \cdot) \in \Pi^{n-1}$.

Now consider $x = x_1$. Then, $p(x_1, \cdot) \in \Pi^{n-1}$ and $p(x_1, y_\nu) = 0$, for all $\nu = 1, \ldots, n - 1$. As above, we conclude that $p(x_1, y) \equiv 0$ and $q_k(x_1) = 0$, for all $k = 2, \ldots, n$. But then, $q_1 = 0$ implying that $p(x, \cdot) \in \Pi^{n-2}$.

Continuing in this fashion, we arrive at $p \equiv 0$. $\quad\square$

Now suppose that $\overline{G} := [a, b] \times [c, d]$, where $[a, b]$ and $[c, d]$ are nonempty intervals in $\mathbb{R}$. Let $X := \{(x, y) \in \mathbb{R}^2 \, | \, a := x_0 < x_1 < \cdots x_m =: b; \ c := y_0 < y_1 < \cdots < y_n =: d\}$. Note that $\operatorname{card} X = (m + 1) \times (n + 1) = \dim \Pi_2^{(m+1, n+1)}$. Next, we like to prove existence and uniqueness of interpolation for polynomials from $\Pi_2^{(m+1, n+1)}$ defined above the domain $\overline{G}$.

Theorem 3.2 (Existence and Uniqueness for Bivariate Polynomial Interpolation on Rectangular Regions)   Suppose that $f : [a, b] \times [c, d] \to \mathbb{R}$ is a function whose values are known on the set X defined above. Then there exists exactly one polynomial $p \in \Pi_2^{(m+1, n+1)}$ so that $p|_X = f|_X$.

Proof   Exercise! $\square$

Finally, we like to obtain error estimates for polynomial interpolation on rectangular regions. These estimates are given in the next theorem.

Theorem 3.3 (Error Estimates for Bivariate Interpolation on Rectangular Regions)   Let $\overline{G} := [a, b] \times [c, d]$ and assume that $f \in C^k(\overline{G})$, where $k := \min\{m + 1, n + 1\}$. Moreover, let $\|x\| := \max\{|x_{i+1} - x_i| \, | \, i = 0, 1, \ldots, m - 1\}$ and $\|y\| := \max\{|y_{j+1} - y_j| \, | \, j = 0, 1, \ldots, n - 1\}$. Then,

$$\|f - p\|_{\infty, \overline{G}} \le \frac{\|\partial_x^{m+1} f\|_{\infty, \overline{G}} \, \|x\|^{m+1}}{4(m + 1)} + \frac{\|\partial_y^{n+1} f\|_{\infty, \overline{G}} \, \|y\|^{n+1}}{4(n + 1)}$$

$$+ \frac{\|\partial_x^{m+1} f\|_{\infty, \overline{G}} \, \|\partial_y^{n+1} f\|_{\infty, \overline{G}} \, \|x\|^{m+1} \, \|y\|^{n+1}}{16(m + 1)(n + 1)},$$

where $\partial_x := \partial / \partial x$, $\partial_y := \partial / \partial y$, and where $p \in \Pi_2^{(m+1, n+1)}$ is the bivariate interpolation polynomial for f whose existence is guaranteed by Theorem 3.2.

Proof   To prove the theorem, one uses the results from one-dimensional interpolation. The details are left to the reader. $\square$

In the special case of an equal number of uniformly spaced knots on a square domain $\overline{G} := [a, b] \times [a, b]$, the estimate in Theorem 3.3 simplifies further. For this purpose, let $x_i := a + ih$ and $y_i := a + ih$, $i = 0, 1, \ldots, m$, where $h := (b - a)/m$.

Corollary 3.4 (Error Estimates for Bivariate Interpolation on Uniformly Spaced Knots on Square Regions)

$$\|f - p\|_{\infty,\overline{G}} \leq \frac{(\|\partial_x^{m+1} f\|_{\infty,\overline{G}} + \|\partial_y^{m+1} f\|_{\infty,\overline{G}})\, h^{m+1}}{4(m+1)}$$

$$+ \frac{\|\partial_x^{m+1} f\|_{\infty,\overline{G}}\, \|\partial_y^{m+1} f\|_{\infty,\overline{G}}\, h^{2m+2}}{16(m+1)^2}.$$

3.2 Spline Interpolation in $\mathbb{R}^s$

In this section, we briefly discuss multivariate spline interpolation and present some of the most important properties. Since there are several generalizations of a univariate B-spline to the multivariate setting, we only consider so-called *box splines* and refer the interested reader to the multivariate spline literature for additional information.

In the following, we will enumerate vectors in $\mathbb{R}^s$ by superscripts and components of such vectors by subscripts. For example, $\{x^1, x^2, x^3\}$ is a collection of vectors in $\mathbb{R}^s$ and x_i^1 denotes the i-th component of vector x^1.

Let $n \in \mathbb{N}$ and let $X_n := (x^1, \ldots, x^n)$ be a matrix whose columns are vectors $x^1, \ldots, x^n \in \mathbb{Z}^s \setminus \{0\}$ with the property that span $\{x^1, \ldots, x^n\} = \mathbb{R}^s$, i.e., rank $X_n = s$. The collection of columns in the matrix X_n is called a *direction set*. Although it is an abuse of language, we also refer to X_n as the direction set. With this set of directions X_n one associates the geometric set

$$[X_n] := [x^1, \ldots, x^n] := \left\{ \sum_{\nu=1}^{n} t_\nu x^\nu \,\middle|\, 0 \leq t_\nu \leq 1,\ \nu = 1, \ldots, n \right\}.$$

Since the column space of X_n spans $\mathbb{R}^s$, the s-dimensional volume $\mathrm{vol}_s[X_n]$ of $[X_n]$ is positive. We assume, by possibly reindexing the vectors in X_n, that $\mathrm{vol}_s[x^1, \ldots, x^s] > 0$.

Definition 3.5 (Box Spline)   The *box spline* $B(\bullet \,|\, [X_n]) : \mathbb{R}^s \to \mathbb{R}$ with direction set X_n is recursively defined by

$$B(x \,|\, [x^1, \ldots, x^s]) := \begin{cases} \dfrac{1}{\mathrm{vol}_s[x^1, \ldots, x^s]}, & x \in [x^1, \ldots, x^s]; \\ 0, & \text{otherwise.} \end{cases}$$

For $m = s + 1, \ldots, n$, define

$$B(x \,|\, [x^1, \ldots, x^m]) := \int_0^1 B(x - t x^m \,|\, [x^1, \ldots, x^{m-1}])\, dt$$

and

$$B(x \,|\, [x^1, \ldots, x^n]) := B(x \,|\, [X_n]).$$

Example 3.6 Let $s := 1$ and set $x^1 = \cdots = x^n := 1$. Then $[X_n] = [0, 1]$ and $B(x \,|\, x^1) = \chi_{[0,1]}(x)$. Moreover, for $m = 2, \ldots, n$,

$$B(x \,|\, [x^1, \ldots, x^m]) = \int_0^1 B(x - t \,|\, [x^1, \ldots, x^{m-1}]) dt = \left(\mathop{\ast}_{i=1}^{m} \chi_{[0,1]} \right)(x).$$

(See item 8 in Theorem 2.22.) Thus, $B(x \,|\, [x^1, \ldots, x^n]) = B_{0,n,\mathbb{Z}}(x)$, the n-th order cardinal B-spline.

Just like B-splines, box splines satisfy a Hermite–Genocchi formula. The precise statement will be given next.

Theorem 3.7 For all functions $f \in C(\mathbb{R}^s)$, we have that

$$\int_{\mathbb{R}^s} B(x \,|\, [X_n]) f(x) \, dx = \int_{W^n} f\left(\sum_{\nu=1}^{n} t_\nu x^\nu \right) dt_1 \ldots dt_n,$$

with $W^n := [0, 1]^n$.

Proof Induction over $m = s, \ldots, n$. Exercise! □

Example 3.8 Let $e^1 := (1, 0)^\top$, $e^2 := (0, 1)^\top \in \mathbb{R}^2$ and set

$$X_n := (\underbrace{e^1, \ldots, e^1}_{a}, \underbrace{e^2, \ldots, e^2}_{b}, \underbrace{e^1 + e^2, \ldots, e^1 + e^2}_{c}, \underbrace{e^1 - e^2, \ldots, e^1 - e^2}_{d})$$

$$=: X_{a+b+c+d},$$

where $a + b + c + d = n$. Moreover, we set $B_{a,b,c,d}(x, y) := B(x, y \,|\, X_{a+b+c+d})$. For $a := b := 1$, it is easy to see that $B_{1,1,0,0}(x, y) = B(x, y \,|\, [e^1, e^2]) = (\chi_{[0,1] \times [0,1]}(x, y))^{-1} = (\chi_{[0,1]}(x))^{-1}(\chi_{[0,1]}(y))^{-1} = B_{0,1,\mathbb{Z}}(x) B_{0,1,\mathbb{Z}}(y)$. For general a and b in $\mathbb{N}$, it then follows from Example 3.6 that $B_{a,b,0,0}(x, y) = B_{0,a,\mathbb{Z}}(x) B_{0,b,\mathbb{Z}}(y)$. Thus, the support of the box spline $B_{a,b,0,0}$ is the rectangle $[0, a] \times [0, b]$.

Proposition 3.9 $\displaystyle\int_{\mathbb{R}^s} B(x \,|\, X_n) \, dx = 1.$

Proof Note that for $s \le m \le n$, we have

$$\int_{\mathbb{R}^s} B(x \,|\, [x^1, \ldots, x^m]) \, dx = \int_{\mathbb{R}^s} \int_0^1 B(x - tx^m \,|\, [x^1, \ldots, x^{m-1}]) \, dt \, dx$$

$$= \int_0^1 \int_{\mathbb{R}^s} B(x - tx^m \,|\, [x^1, \ldots, x^{m-1}]) \, dx \, dt.$$

The statement now follows by induction over $m = s, \ldots, n$, together with the fact that

$$\int_{\mathbb{R}^s} B(x \,|\, [x^1, \ldots, x^s]) \, dx = 1. \quad □$$

Next, we summarize some of the properties of box splines. We will state these without proofs and refer to, for instance, [38, 47] for more details.

Proposition 3.10 The box spline $B(\bullet \,|\, [X_n]) : \mathbb{R}^s \to \mathbb{R}$ with direction set X_n has the following properties.

1. $B(\bullet \,|\, [X_n])$ does not depend on the order of the directions $x^1, \ldots, x^n$.

2. $B(\bullet \,|\, [X_n]) > 0$ in the interior of $[X_n]$.

3. $\operatorname{supp} B(\bullet \,|\, [X_n]) = [X_n]$.

4. $B(\bullet \,|\, [X_n])$ is symmetric about the center of its support.

5. $B(\bullet \,|\, [X_n])$ is a piecewise polynomial function of total degree $n - s$.

6. $B(\bullet \,|\, [X_n]) \in C^r(\mathbb{R}^s)$, where

$$r := r(X_n) := \min\{\operatorname{card} Y \,|\, Y \subset X_n, \ \operatorname{span} X_n \setminus Y \neq \mathbb{R}^s\} - 2.$$

As an example, we consider the box splines $B_{1,1,0,0}$ and $B_{1,1,1,0}$. Then $\operatorname{supp} B_{1,1,0,0} = [0, 1] \times [0, 1]$ and $B_{1,1,0,0} \in C^{-1}(\mathbb{R}^s)$. The support of $B_{1,1,1,0}$ is shown in Figure 3.1. Moreover, it is easily seen that $B_{1,1,1,0} \in C^0(\mathbb{R}^s)$.

The next result shows that the translates of box splines along the lattice $\mathbb{Z}^s$ are linearly independent provided a certain condition on the direction set X_n holds.

Theorem 3.11 The set $\{B(\bullet - k \,|\, [X_n]) \,|\, k \in \mathbb{Z}^s\}$ is linearly independent iff

$$\det(x^{i_1}, \ldots, x^{i_s}) = \pm 1,$$

for all $\{x^{i_1}, \ldots, x^{i_s}\} \subseteq X_n$ with $\operatorname{span}\{x^{i_1}, \ldots, x^{i_s}\} = \mathbb{R}^s$.

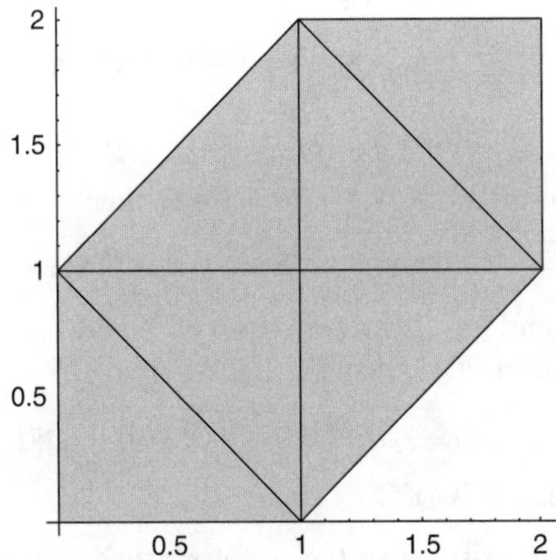

Figure 3.1 The support of the box spline $B_{1,1,1,0}$.

Remark 3.12   A direction set $X_n := (x^1, \ldots, x^n)$ satisfying $|\det(x^{i_1}, \ldots, x^{i_s})| = 1$, for all $\{x^{i_1}, \ldots, x^{i_s}\} \subseteq X_n$ with $\mathrm{span}\{x^{i_1}, \ldots, x^{i_s}\} = \mathbb{R}^s$ is called *unimodular.*

Now suppose that X_n is unimodular. The linear space

$$S(X_n) := \mathrm{span}\{B(\bullet - k \,|\, [X_n]) \,|\, k \in \mathbb{Z}^s\}$$

is called the *box spline space with unimodular direction set X_n.*

Theorem 3.13 $\Pi_s^{r(X_n)+1} \subseteq S(X_n)$ and the L^2-order of approximation of $S(X_n)$ equals $r(X_n) + 2$.

3.3 Tensor Products of B-Splines

In this section, we introduce the tensor product of spline functions defined over domains of $\mathbb{R}^s$, $1 < s \in \mathbb{N}$. This, of course, is not the most general setting but the construction reflects some of the difficulties and shortcomings when extending splines to higher dimensions via the tensor product.

Before commencing with the construction, the definition of *tensor product* is needed. The definition that will be given is not the most general one, but it is more than adequate for our purposes.

Definition 3.14 (Tensor Product)   Let X and Y be vector spaces over $\mathbb{R}$ of dimension n and m, respectively. With the spaces X and Y one associates an $n \cdot m$-dimensional vector space $X \otimes Y$ in the following way. For $x \in X$ and $y \in Y$, let $\theta : X \times Y \to X \otimes Y$ be the mapping that associates with the pair (x, y) an element of $X \otimes Y$, denoted by $x \otimes y$. The mapping θ is to satisfy three conditions:

1. Distributive Law: If $x, x_1, x_2 \in X$ and $y, y_1, y_2 \in Y$, then

$$x \otimes (y_1 + y_2) = x \otimes y_1 + x \otimes y_2,$$

$$(x_1 + x_2) \otimes y = x_1 \otimes y + x_2 \otimes y.$$

2. Associative Law: If $x \in X$, $y \in Y$, and $\alpha \in \mathbb{R}$, then

$$\alpha x \otimes y = x \otimes \alpha y = \alpha (x \otimes y).$$

3. If $\{x_1, \ldots, x_n\}$ and $\{y_1, \ldots, y_m\}$ are bases of X and Y, respectively, then the collection of $n \cdot m$ elements

$$\{x_i \otimes y_j \,|\, i = 1, \ldots, n, \ j = 1, \ldots, m\}$$

is to form a basis of $X \otimes Y$.

If such a mapping θ exists, then the vector space $X \otimes Y$ is called the *tensor product of X with Y* and the element $x \otimes y$ the tensor product of the vectors x and y.

Example 3.15   An example of a tensor product is provided by the linear spaces Π^k. Suppose that $k := 2$. Then $\{e_1 := 1, e_2 := \mathrm{id}_{\mathbb{R}}\}$ is a basis for the vector space Π^2. The tensor product $\Pi^2 \otimes \Pi^2$ is then isomorphic to the vector space $\Pi_2^{(2,2)}$ of all bivariate real polynomials p of the form

$$p(x, y) = \sum_{i=0}^{1} \sum_{j=0}^{1} a_{ij} x^i y^j,$$

where $x^i y^j := e_i \otimes e_j(x, y)$ and the a_{ij}, $i, j = 0, 1$, are real numbers.

3.3.1 The Construction of Tensor-Product B-Splines

A *tensor product surface* is the graph of the tensor product of s functions $f_1, \ldots, f_s : J \to \mathbb{R}$ and thus is given by

$$\left(\bigotimes_{i=1}^{s} f_i \right)(x_1, \ldots, x_s) = f_1(x_1) \cdot \ldots \cdot f_s(x_s), \quad \forall x_1, \ldots, x_s \in J,$$

where J is a nonempty compact interval in $\mathbb{R}$. Note that it is no loss of generality to choose the same interval for all functions $f_1, \ldots, f_s$.

Let us briefly consider the case $s := 2$. Then, the graph $f_1 \otimes f_2$ is a surface in $\mathbb{R}^3$ above the square $J \times J$. Suppose that $f_1, f_2 \in \Pi^k_{\xi, m}$, that $J := [x_{1,k}, x_{1,n+1}]$, and that we use the same knot sequence for both intervals. (See Theorem 2.40 for notation and terminology.) For a fixed value $\bar{y}$ of y, we can write $f_1 \otimes f_2$ in the form (2.40) as

$$(f_1 \otimes f_2)(x_1, \bar{x}_2) = \sum_{\mu=1}^{n} c_\mu(\bar{y}) B_{\mu k}(x_1) \tag{3.1}$$

on the interval J. The coefficient functions $c_v : J \to \mathbb{R}$ can now themselves be written in the form

$$c_v(x_2) = \sum_{v=1}^{n} c_{\mu v} B_{vk}(x_2), \tag{3.2}$$

for real numbers $c_{\mu v}$, $\mu, v = 1, \ldots, n$. Combining (3.1) and (3.2) yields

$$(f_1 \otimes f_2)(x_1, x_2) = \sum_{\mu=1}^{n} \sum_{v=1}^{n} c_{\mu v} B_{\mu k}(x_1) B_{vk}(x_2) \tag{3.3}$$

on $J \times J$.

From this short description, it should be clear how to extend univariate B-splines to multivariate B-spline surfaces with different knot sequences and different orders along the s directions, provided that the Curry–Schoenberg

Theorem applies. We only state the generalization and leave it to the reader to provide the details.

Definition 3.16 (Tensor Product B-Spline Surface)   Suppose $s \in \mathbb{N}$ and $k_i \in \mathbb{N}$, $i = 1, \ldots, s$. Let $\{X_i \mid i = 1, \ldots, s\}$ be a sequence of knot sequences $X_i := \{x_{i,1}, \ldots, x_{i,n_i+k_i}\}$ whose tensor product is defined by

$$X := \bigotimes_{i=1}^{s} X_i := \{ \boldsymbol{x}_l = (x_{1,l_1}, \ldots, x_{s,l_s}) \mid 1 \leq l \leq \boldsymbol{n} + \boldsymbol{k} \}.$$

Define the *tensor product B-spline* $B_{\boldsymbol{k}}(\bullet \mid X) : J^s \to \mathbb{R}$ (for the notation, see Remark 2.16) via

$$B_{\boldsymbol{k}}(\boldsymbol{x} \mid X) := \prod_{i=1}^{s} B_{k_i}(x_i \mid X_i).$$

In Figure 3.2, the graph of the tensor product B-spline $B_3 \otimes B_4$ with knot sequence $\{0, 1, 2, 3\}$, respectively, $\{0, 1, 2, 3, 4\}$ is displayed.

The following result is a direct consequence of the univariate case.

Proposition 3.17   The tensor product B-splines $B_{\boldsymbol{k}}(\bullet \mid X)$ form a nonnegative partition of unity.

Proof   Induction over s. Exercise! $\square$

Now, define the *vector-valued multiplicity* $\mathbf{mult} := \mathbf{mult}_{\boldsymbol{x}_l} : \mathbb{R} \to (\mathbb{N}^\infty)^s$ of a knot $\boldsymbol{x}_l$ by

$$\mathbf{mult}\,\boldsymbol{x}_l := (\mathrm{mult}\,x_{i,l_i} \mid i = 1, \ldots, s)^{\top},$$

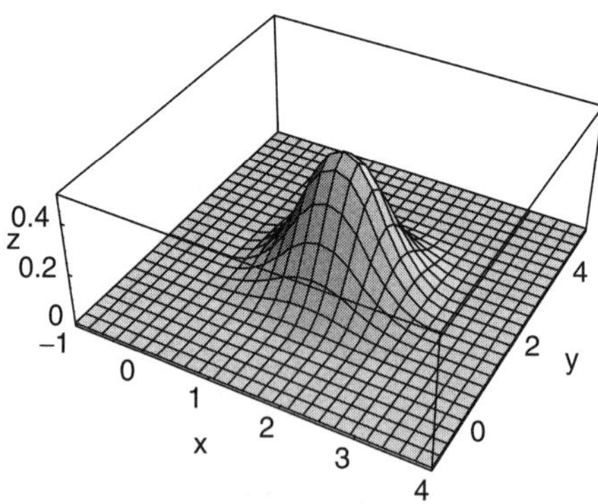

Figure 3.2   The tensor product B-spline $B_3 \otimes B_4$.

where $1 \le l \le n + k$. This then gives rise to the tensor product spline space $S^k(X)$ defined by

$$
S^k(X) := \left\{ p \in \Pi_s^k \; \middle| \; \frac{\partial^{|l|} p}{\partial x^l}(x_i) \text{ exists and is continuous } \forall\, l \le k - \text{mult}\, x_i, \right.
$$
$$
\left. i = 1, \ldots, s \right\}.
$$

Note that in the case $s := 1$, the above definition is a slight reformulation of the characterization of the spaces $\Pi_{\xi m}^k$ given in Section 2.7.

Consequently, we have the Curry–Schoenberg Theorem for tensor product B-splines.

Theorem 3.18 (Basis Property of Tensor Product B-Splines) The set $\{B_k(\bullet \,|\, X)\}$ of tensor product B-splines forms a basis of $S^k(X)$.

Proof The statement follows from the Curry–Schoenberg Theorem 2.40. Exercise! □

In similar fashion, we can derive the Schoenberg–Whitney Theorem for tensor product splines. For this purpose, suppose that we are given the projections Ξ_i, $i = 1, \ldots, s$, of a set of interpolation points in $\mathbb{R}^s$ onto the coordinate axes. Set

$$
\Xi := \mathop{\otimes}_{i=1}^{s} \Xi_i = \{ \xi_l := (\xi_{1,l_1}, \ldots, \xi_{s,l_s})^\top \,|\, 1 \le l \le n \}.
$$

Then, we have the following multivariate version of the Schoenberg–Whitney Theorem.

Theorem 3.19 (Interpolation Conditions for Tensor Product Splines) The interpolation problem in $S^k(X)$ is uniquely solvable iff $x_l \in J_l^k$ for all $1 \le l \le n$, where $J_l^k := \bigotimes_{i=1}^{s} [x_{i,l_i}, x_{i,l_i+k_i+1}]$ is the support parallelepiped of the s-dimensional B-spline $B_k(\bullet \,|\, X)\}$.

Proof Use the discussion in Section 2.9 and the one-dimensional Schoenberg–Whitney conditions. Exercise! □

3.3.2 Shortcomings of the Tensor Product Approach

The tensor product approach seems to be a straight-forward extension from the univariate to the multivariate case. However, there are also some serious disadvantages of this approach. Here, we briefly summarize them. (See also [135].)

- **Homogeneity:** All directions are treated equally and this induces an apparent homogeneity in the tensor product B-spline basis. If these functions are to be used to model or represent data that do not have homogeneous features, then information will be in general discarded. Essentially, we are dealing with s one-dimensional problems.

- **Domain of Definition:** The domain of definition has to be an s-dimensional parallelepiped. In applications, however, one may have to choose a different geometry consisting of triangles or more general compact sets.

- **Exponential Complexity:** The number of coefficients in an s-dimensional representation of the form (3.3), i.e., the dimension of $S^k(X)$, is $\prod_{i=1}^{s} n_i$. In the case where we have the same number n of knots in all s directions, this complexity is about n^s. Solving interpolation problems gives a complexity of about n^{3s} (n^{2s} in the matrix and n^s "fast" operations to solve the system), which for many realistic high-parameter problems is too large to solve.

- **Lattice Structure:** In order to use tensor product B-splines for modeling purposes, the underlying geometry and all geometric structures of the model must have tensor character. Compare this with the assumption for Theorem 3.19, where the interpolation points had to form a lattice. Many real-world data sets, for instance tomography or MRI data, do not have lattice structure.

3.4 Kergin Interpolation

In this short section, we briefly introduce Kergin interpolation, which is a generalization of the Newton univariate polynomial interpolation method to $\mathbb{R}^s$. We present the definition and basic properties of Kergin interpolation and refer to the original paper [96] and, in particular, to [119] for more details and proofs.

To this end, recall the form of the Newton interpolating polynomial $N(X; f)$ (1.14) for a knot set X of cardinality $n+1$ in $\mathbb{R}$ and the Hermite–Genocchi formula (1.34) from Exercise 33 in Chapter 1. Suppose that $x_0, x_1, \ldots, x_n$, $n \in \mathbb{N}$, are – not necessarily distinct – points in $\mathbb{R}$. Let J be an interval containing the convex hull of $X := \{x_0, x_1, \ldots, x_n\}$ an let $f \in C(J)$. Define

$$\int_{[X]} f := \int_{\mathrm{conv}\,\{x_0,\ldots,x_n\}} f := \begin{cases} \int_{\Sigma^n} f(u_0 x_0 + \cdots + u_n x_n)\, du, & 1 < n \in \mathbb{N}; \\[2ex] f(x_0), & n = 1. \end{cases} \tag{3.4}$$

Note that, by definition of an integral over Σ^n (see (1.33)), the result does not depend on the particular ordering of the knots x_v, $v = 0, 1, \ldots, n$, and it is therefore admissible to write $\int_{[X]} f$. The continuous linear mapping

$$\sigma : C(J) \to \mathbb{R}, \tag{3.5}$$

$$f \longmapsto \int_{[X]} f \tag{3.6}$$

is called the *simplex functional*. Formula (1.15) has an analogue in the current setting.

Proposition 3.20 Let $X := \{x_0, x_1, \ldots, x_n\} \subset \mathbb{R}$ and let J be an interval containing conv X. Suppose that $f \in C^1(J)$. Then

$$(x_\mu - x_\nu) \int_{[X]} Df = \int_{[X \setminus \{x_\nu\}]} f - \int_{[X \setminus \{x_\mu\}]} f,$$

for any $x_\mu, x_\nu \in X$.

Proof If $x_\mu = x_\nu$, then there is nothing to prove. Suppose then that $x_\mu \neq x_\nu$. Since the simplex functional σ is a symmetric function of the knots, it suffices to prove the statement for $\mu := n$ and $\nu := n - 1$. First, consider the case $n := 1$. Then,

$$(x_1 - x_0) \int_{[x_0, x_1]} Df = \int_0^1 (x_1 - x_0)(Df)(x_0 + t(x_1 - x_0)) \, dt = f(x_1) - f(x_0)$$

$$= \int_{[x_1]} f - \int_{[x_0]} f.$$

For the case $n > 1$, notice that

$$\int_{[X]} Df = \int_0^1 \int_0^{u_1} \cdots \int_0^{u_{n-1}} (Df)(x_0 + \sum_{\nu=0}^n u_\nu(x_\nu - x_{\nu-1})) \, du_1 \cdots du_n. \tag{3.7}$$

(The reader is encouraged to verify this statement.) Moreover,

$$(x_n - x_{n-1}) \int_0^{u_{n-1}} (Df)\Big(x_0 + \sum_{\nu=0}^{n-2} u_\nu(x_\nu - x_{\nu-1}) + u_{n-1}(x_{n-1} - x_{n-2})$$

$$+ u_n(x_n - x_{n-1})\Big) du_n$$

$$= f\Big(x_0 + \sum_{\nu=0}^{n-2} u_\nu(x_\nu - x_{\nu-1}) + u_{n-1}(x_n - x_{n-2})\Big)$$

$$- f\Big(x_0 + \sum_{\nu=0}^{n-2} u_\nu(x_\nu - x_{\nu-1}) + u_{n-1}(x_{n-1} - x_{n-2})\Big).$$

Substitution into (3.7) and elimination of the variable x_n yields the result. The details are left to the reader. $\square$

Using the simplex functional we can restate the Hermite–Genocchi formula as follows.

Theorem 3.21 Suppose that $X := \{x_0, \ldots, x_n\} \subset \mathbb{R}$ and that J is an interval containing $\operatorname{conv} X$. Then, for any $f \in C^n(J)$,

$$[X]f = \int_{[X]} D^n f. \tag{3.8}$$

Proof The proof is by induction on n. The case $n := 0$ is trivial, so suppose then that the statement is true for $n - 1$. First, we consider the situation when all knots coalesce. By (2.38) and Exercise 36 in Chapter 1, we obtain

$$\int_{[X]} D^n f = (D^n f)(x_0)\operatorname{vol}(\Sigma^n) = \frac{(Df)(x_0)}{n!} = [X]f.$$

Now, assume that $\exists \mu, \nu \in \{0, 1, \ldots, n\}$ with the property that $x_\mu \neq x_\nu$. The induction hypothesis, Proposition 3.20, and (1.15) yield

$$\int_{[X]} D^n f = \frac{\int_{[X \setminus \{x_\mu\}]} D^{n-1} f - \int_{[X \setminus \{x_\nu\}]} D^{n-1} f}{x_\nu - x_\mu}$$

$$= \frac{[X \setminus \{x_\mu\}]f - [X \setminus \{x_\nu\}]f}{x_\nu - x_\mu} = [X]f. \qquad \square$$

After these preliminaries, we reconsider the interpolation error for the Newton interpolating polynomial, Theorem (1.29),

$$f(x) - \mathbf{N}(x_0, \ldots, x_n; f)(x) = [x_0, \ldots, x_n, x]f \prod_{\nu=0}^{n} (x - x_\nu),$$

and relate it to the integral representation of the divided difference operator in terms of the B-spline $M_{\nu k, X}$, Equation (2.7), adapted to the present setting:

$$[x_0, \ldots, x_n, x]f = \frac{1}{(n+1)!} \int_{\mathbb{R}} M_{\nu, n+1, X}(t)(D^{n+1} f)(t)\, dt,$$

where $X := \{x_0, \ldots, x_n, x\}$. Note that it is admissible to take the integral over $\mathbb{R}$ since the B-spline $M_{\nu, n+1, X}$ has compact support. With Theorem 3.21, we thus obtain the following result.

Proposition 3.22 Assume that $\{x_0, \ldots, x_n\}$ is a knot set consisting of different knots and that $f \in C^{n+1}(\mathbb{R})$. Then

$$f(x) - \mathbf{N}(x_0, \ldots, x_n; f)(x) = \frac{\prod_{\nu=0}^{n}(x - x_\nu)}{(n+1)!} \int_{\mathbb{R}} M_{\nu, n+1, X}(t)(D^{n+1} f)(t)\, dt$$

$$= \int_{[X]} D_{x-x_0} \cdots D_{x-x_n} f.$$

Proof   Only the last equality needs to be established; it, however, follows immediately from the definition of directional derivative. $\square$

Using the Hermite–Genocchi formula (1.34) and (3.8), the Newton interpolating polynomial $\mathbf{N}(X; f)(x)$ can also be rewritten in the form

$$\mathbf{N}(X; f)(x) = \sum_{\nu=0}^{n} [x_0, \ldots, x_\nu] f \prod_{\mu=0}^{\nu-1} (x - x_\mu) = \sum_{\nu=0}^{n} \prod_{\mu=0}^{\nu-1} (x - x_\mu) \int_{[x_0,\ldots,x_\nu]} D^\nu f$$

$$= \sum_{\nu=0}^{n} \int_{[x_0,\ldots,x_\nu]} D_{x-x_0} \cdots D_{x-x_{\nu-1}} f. \tag{3.9}$$

The generalization of (3.9) to $\mathbb{R}^s$ leads to the *Kergin interpolant* $\mathbf{K}(X; f)$ of $f \in C^n(\mathbb{R}^s)$.

Definition 3.23 (Kergin Interpolant)   For a knot set $X := \{x_0, \ldots, x_n\} \subset \mathbb{R}^s$ consisting of distinct knots, and for an $f \in C^n(\mathbb{R}^s)$, the *Kergin interpolant* $\mathbf{K}(X; f)$ of f is defined as

$$\mathbf{K}(X; f)(x) := \sum_{\nu=0}^{n} \int_{[x_0,\ldots,x_\nu]} D_{x-x_0} \cdots D_{x-x_{\nu-1}} f, \quad x \in \mathbb{R}.$$

Here, $\int_{[x_0,\ldots,x_\nu]}$ is the extension of (3.4) to the s-dimensional setting and $D_u f$ is the directional derivative of f in the direction of the vector u: $D_u f := \langle \operatorname{grad} f, u \rangle$.

Remark 3.24   In case the knots $x_0, \ldots, x_n$ are in general position, the Kergin interpolant can be continuously extended to $C^{s-1}(\mathbb{R}^s)$, thus weakening the strong requirement that $f \in C^n(\mathbb{R}^s)$. For details, see [96, 119].

To establish the existence and uniqueness of a Kergin interpolation operator $\mathbf{K}(X; \bullet)$, we need the following notation. To a polynomial $q \in \Pi_s^n$, $n \in \mathbb{N}_0$,

$$q(x) := \sum_{0 \le |\alpha| < n} q_\alpha x^\alpha,$$

we associate the partial differential operator $q(D)$ with constant coefficients

$$q(D) := q\left(\frac{\partial}{\partial x_1}, \ldots, \frac{\partial}{\partial x_s} \right) := \sum_{0 \le |\alpha| < n} q_\alpha \frac{\partial^{|\alpha|}}{\partial x^\alpha}.$$

Example 3.25   Let $s := 3$. To the polynomial

$$q(x_1, x_2, x_3) := 1 + 2x_3 + 4x_2^2 + 3x_1^2 x_2 - x_1 x_2^2 x_3^3 \in \Pi_3^7$$

one associates the differential operator

$$q(D) = 1 + 2\frac{\partial}{\partial x_3} + 4\frac{\partial^2}{\partial x_2^2} + 3\frac{\partial^3}{\partial x_1^2 \partial x_2} - \frac{\partial^6}{\partial x_1 \partial x_2^2 \partial x_3^3}.$$

Theorem 3.26 (Kergin Interpolation Theorem)   Let $s \in \mathbb{N}$ and $n \in \mathbb{N}_0$. Suppose that $\{x_0, \ldots, x_n\} \subset \mathbb{R}^s$ is a collection of not necessarily distinct vectors. Let $X := \{x_0, \ldots, x_n\}$. Then there exists a unique Kergin interpolant $\mathbf{K}(X; \bullet) : C^n(\mathbb{R}^s) \to \Pi_s^n$ having the following properties.

1. $\mathbf{K}(X; \bullet)$ is linear;

2. For every $f \in C^n(\mathbb{R}^s)$ and every homogeneous differential operator $q(D)$ with constant coefficients, $q \in \Pi_s^n$, and every $\Upsilon \subseteq \{0, 1, \ldots, n\}$ with card $\Upsilon = \deg q + 1$, there exists an $x \in \mathrm{conv}\,\{x_j \,|\, j \in \Upsilon\}$ so that

$$(q(D)\mathbf{K}(X; f))(x) = (q(D)f)(x).$$

Proof   The proof is quite elaborate and is therefore omitted. We refer to [96] for details. $\square$

Remark 3.27   It can be shown that conditions 1. and 2. in Theorem 3.26 imply that the Kergin interpolant $\mathbf{K}(X; \bullet) : C^n(\mathbb{R}^s) \to \Pi_s^n$ is continuous. For the proof, we again refer to the original paper [96].

Example 3.28   Suppose that $s := 3$, and $n := 4$. Let $x_0 := (0, 0, 0)^\top$, $x_1 := (1, 0, 0)^\top$, $x_2 := (0, 1, 0)^\top$, and $x_3 := (0, 0, 1)^\top$, and set $X := \{x_0, x_1, x_2, x_3\}$. Further assume that

$$q(D) := \frac{\partial^2}{\partial x_2 \partial x_3}.$$

Then $q \in \Pi_3^3$ and, therefore, $\deg q = 2$. Let $\Upsilon := \{0, 2, 3\}$. Hence, there exists an $x := (\xi_1, \xi_2, \xi_3) \in \mathrm{conv}\,\{x_0, x_2, x_3\}$ and a polynomial $p := \mathbf{K}(X; f)$ of total degree less than three so that

$$\frac{\partial^2 p}{\partial x_2 \partial x_3}(\xi_1, \xi_2, \xi_3) = \frac{\partial^2 f}{\partial x_2 \partial x_3}(\xi_1, \xi_2, \xi_3).$$

As a final result in this brief section, we will prove an error estimate for Kergin interpolation.

Theorem 3.29 (Error Estimate for Kergin Interpolation)   Suppose that $f \in C^{n+1}(\mathbb{R}^s)$ and that $x_0, \ldots, x_n \in \mathbb{R}^s$. Let $n \in \mathbb{N}_0$ and let $X := \{x_0, \ldots, x_n\}$. Then

$$(f - \mathbf{K}(X; f))(x) = \int_{[x_0, \ldots, x_n, x]} D_{x - x_0} \cdots D_{x - x_n} f, \quad x \in \mathbb{R}^s,$$

and, for any compact $K \subset \mathbb{R}^s$ with $x_\nu \in K$, $\nu = 0, 1, \ldots, n$,

$$\|f - \mathbf{K}(\mathbf{X}; f)\|_{K,\infty} \leq \frac{(s|K|_\infty)^{n+1}}{(n+1)!} \max\{\|D^\alpha f\|_{K,\infty} \mid \alpha \in \Gamma_{n+1}\},$$

where $\Gamma_{n+1} := \{\alpha \in \mathbb{N}_0^s \mid |\alpha| = n+1\}$ and $|K|_\infty := \max\limits_{x,y \in K} \max\limits_{j=1,\dots,s} |x_j - y_j|$.

Proof The proof is by induction on n. We write $\mathbf{X}_k := \{\mathbf{x}_0, \dots, \mathbf{x}_k\}$, $k = 0, 1, \dots, n$. Proposition 3.20 yields

$$f(\mathbf{x}) - \mathbf{K}(\mathbf{X}_n; f)(\mathbf{x}) = f(\mathbf{x}) - \mathbf{K}(\mathbf{X}_{n-1}; f)(\mathbf{x}) + \mathbf{K}(\mathbf{X}_{n-1}; f)(\mathbf{x}) - \mathbf{K}(\mathbf{X}_n; f)(\mathbf{x})$$

$$= \int_{[\mathbf{x}_0,\dots,\mathbf{x}_{n-1},\mathbf{x}]} D_{\mathbf{x}-\mathbf{x}_0} \cdots D_{\mathbf{x}-\mathbf{x}_{n-1}} f - \int_{[\mathbf{x}_0,\dots,\mathbf{x}_n]} D_{\mathbf{x}-\mathbf{x}_0} \cdots D_{\mathbf{x}-\mathbf{x}_{n-1}} f$$

$$= \int_{[\mathbf{x}_0,\dots,\mathbf{x}_n,\mathbf{x}]} D_{\mathbf{x}-\mathbf{x}_0} \cdots D_{\mathbf{x}-\mathbf{x}_n} f.$$

Furthermore, for $\mathbf{x} \in K$,

$$|(D_{\mathbf{x}-\mathbf{x}_0} \cdots D_{\mathbf{x}-\mathbf{x}_n} f)(\mathbf{x})| \leq \sum_{j_0=1}^{s} \cdots \sum_{j_n=1}^{s} \prod_{v=0}^{n} \left| (\mathbf{x} - \mathbf{x}_v)_{j_v} \frac{\partial^{n+1} f}{\partial x_{j_0} \cdots \partial x_{j_n}}(\mathbf{x}) \right|$$

$$\leq s^{n+1} |K|_\infty^{n+1} \max\{\|D^\alpha\|_{K,\infty} \mid \alpha \in \Gamma_{n+1}\}.$$

Noting that $\displaystyle\int_{[\mathbf{x}_0,\dots,\mathbf{x}_n,\mathbf{x}]} 1 = \frac{1}{(n+1)!}$, gives the final result. $\square$

Exercises

1. Prove the dimension formulae for the polynomial spaces Π_s^m and Π_s^n.

2. Use the existence and uniqueness of the one-dimensional interpolation problem to prove Theorem 3.2.

3. Complete the proof of Theorem 3.7.

4. Given the direction set $\mathbf{X}_3 := \begin{pmatrix} 1 & -1 & -1 \\ 0 & 1 & -2 \end{pmatrix}$, choose $\mathbf{x}^1 := \begin{pmatrix} 1 \\ 0 \end{pmatrix}$ and $\mathbf{x}^2 := \begin{pmatrix} -1 \\ 1 \end{pmatrix}$, and derive an explicit formula for the box spline $B(\bullet \mid \mathbf{x}^1, \mathbf{x}^2)$. Then compute $B(\bullet \mid [\mathbf{X}_3])$. For each of these box splines, indicate its degree of smoothness..

5. Explicitly compute the box splines $B_{1,1,1,0}$ and $B_{1,1,1,1}$.

6. Provide the details on which Definition 3.16 is based.

7. Prove Proposition 3.17.

8. Show that the support of the tensor product B-spline $B_k(\bullet \,|\, X)$ is the s-dimensional parallelepiped

$$J_n^k := \bigotimes_{i=1}^{s} [x_{i,n_i}, x_{i,n_i+k_i+1}].$$

9. Provide a proof of Theorem 3.18.

10. Establish Theorem 3.19.

11. Show that the simplex functional (3.5) is linear and continuous.

12. Fill in the details for the proof of Proposition 3.20.

13. Compute the Kergin interpolating polynomial for Example 3.28. $\quad\square$

4

Fractals

The term "fractal," which is derived from the Latin word *fractus* meaning "broken," was first introduced by Benoît Mandelbrot in 1975, and is used to describe or characterize a class of sets that exhibit a high degree of scale invariance or self-similarity. This, for instance, occurs when a set is made up of a number of smaller copies of itself. Geometric objects with this property distinguish themselves in essential aspects from ordinary smooth figures. The study of fractals is the subject of fractal geometry and fractal analysis and both areas have connections to complex analysis (Fatou[1] and Julia[2] sets), computability theory, and dynamical systems. Fractal sets provide a natural means to model complex structures such as natural objects or images that cannot be easily represented in terms of classical Euclidean or metric geometry. These interpolation and approximation properties are the primary subject of this chapter.

There exist several mathematical constructions of fractals, which may be grouped into algebraic, analytic, and stochastic methods. All of these methods describe and represent slightly different aspects of the complex structure of a fractal. In this text, we will exclusively use one of the methods of fractal construction, namely the analytic generation of fractals by *iterated function systems*. The reader who is interested in other types of approaches or in variations of the themes discussed in this chapter is referred to the following, albeit incomplete, list of references [5, 6, 48, 51, 56, 57, 58, 76, 113].

1 PIERRE JOSEPH LOUIS FATOU, 28 February 1878–10 August 1929. French mathematician who worked in astronomy and complex dynamical systems.
2 GASTON MAURICE JULIA, 3 February 1893–19 March 1978. French mathematician who is known for his work on the iteration of complex rational functions and dynamical systems.

The structure of this chapter is as follows. The concept of topological and metric dimension is introduced, discussed, and set into perspective with fractals and their properties. Iterated function systems are defined and several of their properties considered. Next, the Collage Theorem, which allows for approximation of compact sets by fractal sets, is proven. Formulae for the Hausdorff and box dimension for a class of iterated function systems are presented. It is also shown how the box dimension of fractals can be computed numerically. The important concept of code space associated with an iterated function system is introduced next. Fractal transformations are briefly considered and conditions under which two fractals are homeomorphic are given. The final section deals with iterated function systems with probabilities and their properties. In this connection, the concept of fractal or IFS-invariant measure is introduced. A list of references for the material presented in this chapter is [8, 9, 57, 58] and [113], as well as the original papers [7, 11, 59, 60, 87, 127, 150].

4.1 Definitions

A *fractal (set)* $\mathfrak{F}$ is a geometric object whose basic structure repeats at all scales or at all levels of magnification. Such a set has in general the following characteristics.

1. $\mathfrak{F}$ is (approximately) a finite union of images of itself in the sense that there exist a finite set of non-surjective homeomorphisms $f_i : \mathfrak{F} \to \mathfrak{F}$, $i = 1, \ldots, N$, such that

$$\mathfrak{F} = \bigcup_{i=1}^{N} f_i(\mathfrak{F}); \tag{4.1}$$

An equation of the type (4.1) is called a *self-referential equation.*

2. $\mathfrak{F}$ exhibits a complex fine structure, which does not simplify upon magnification.

3. $\mathfrak{F}$ can in general not be easily described by traditional Euclidean geometry.

4. $\mathfrak{F}$ has a Hausdorff[3] dimension that is in general greater than its topological dimension.

5. $\mathfrak{F}$ is recursively defined.

---

3 FELIX HAUSDORFF, 8 November 1868–26 January 1942. German mathematician who was one of the founders of modern topology. He also made significant contributions to set theory, measure theory, and real analysis.

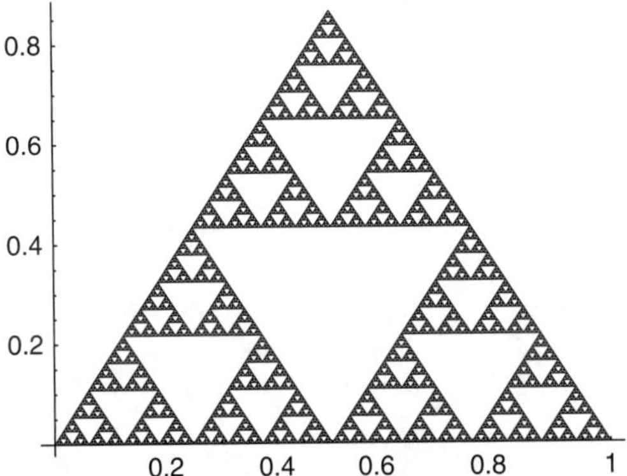

Figure 4.1 The Sierpiński triangle $\mathfrak{S}$.

Example 4.1 The classical ternary Cantor set $\mathfrak{C}$ is an example of a fractal. Here $\mathfrak{C}$ exhibits exact self-similarity under the two non-surjective homeomorphisms $f_1 : \mathfrak{C} \to \mathfrak{C}, x \mapsto \frac{1}{3}x$ and $f_2 : \mathfrak{C} \to \mathfrak{C}, x \mapsto \frac{1}{3}x + \frac{2}{3}$.

Example 4.2 Another example of a well-known mathematical object that is a fractal is provided by the Sierpiński[4] triangle $\mathfrak{S}$. (See Figure 4.1.) Again, $\mathfrak{S}$ has exact self-similarity under the non-surjective homeomorphisms $f_i : \mathfrak{S} \to \mathfrak{S}$, $i = 1, 2, 3$, where

$$f_1(x,y) = \left(\frac{x}{2}, \frac{y}{2}\right), \quad f_2(x,y) = \left(\frac{x}{2} + \frac{1}{2}, \frac{y}{2}\right), \quad f_3(x,y) = \left(\frac{x}{2} + \frac{1}{4}, \frac{y}{2} + \frac{\sqrt{3}}{4}\right).$$

Before proceeding with the theory of fractals, we need to introduce the concept of topological dimension in $\mathbb{R}^n$.

4.1.1 Topological Dimension in $\mathbb{R}^n$

Suppose that M is a set in $\mathbb{R}^n$. The *topological dimension of M*, written as $\dim M$, is inductively defined as follows.

1. $\dim \emptyset := -1$.

2. The topological dimension of M at a point $p \in M$ is $\leq n$, written $\dim_p M \leq n$, if there exist arbitrarily small neighborhoods of p whose boundaries have topological dimension at most $n - 1$.

4 WACŁAW FRANCISZEK SIERPIŃSKI, 14 March 1882–21 October 1969. Polish mathematician who made contributions to set theory, point set topology, and real analysis.

3. M has topological dimension at most n if it has topological dimension at most n at each of its points p:

$$\dim M \le n \quad \Longleftrightarrow \quad \forall p \in M : \dim_p M \le n.$$

In addition, $\dim_p M := \infty$, if Property 2 above does not hold for any $n \in \mathbb{N}$, and $\dim M := \infty$ in case Property 3 does not hold for any $n \in \mathbb{N}$.

Example 4.3   The sets $\mathbb{N}$, $\mathbb{Z}$, and $\mathbb{Q}$ all have topological dimension 0.

Example 4.4   Let $p \in \mathbb{R}^p$. Then $\dim_p \mathbb{R}^n \le n$ and $\dim \mathbb{R}^n \le n$. Moreover, one can show that $\dim \mathbb{R}^n = n$.

Example 4.5   The Hilbert cube $W := [0,1]^{\mathbb{N}} = \mathrm{Map}(\mathbb{N}, [0,1])$ has the property that $\dim W = \infty$.

4.1.2 The Hausdorff and Box Dimension in $\mathbb{R}^n$

The topological dimension of a set $M \subseteq \mathbb{R}^n$ makes use of the topological properties of $\mathbb{R}^n$ but ignores the metric properties. The concept of *Hausdorff dimension* employs these metric properties to estimate the size of M.

To this end, let M be a bounded subset of $\mathbb{R}^n$ and $x_0 \in M$. Define

$$B_r(\boldsymbol{x}_0) := \{\boldsymbol{x} \in M \mid \|\boldsymbol{x} - \boldsymbol{x}_0\| < r\}$$

to be the ball of radius $r > 0$ with center at $\boldsymbol{x}_0$. Here $\| \ \|$ denotes the usual Euclidean norm on $\mathbb{R}^n$. Choose $\varepsilon > 0$ and cover M by balls of radius $0 < r_i \le \varepsilon$ centered at points $\boldsymbol{x}_i \in M$:

$$M \subseteq \bigcup_{i \in \mathbb{N}} B_{r_i}(\boldsymbol{x}_i).$$

For $s \ge 0$, consider the quantity

$$\mathscr{H}_\varepsilon^s(M) := \inf \left\{ \sum_{i=1}^{\infty} |r_i|^s \ \bigg| \ M \subseteq \bigcup_{i=1}^{\infty} B_{r_i}(\boldsymbol{x}_i); \ 0 < r_i \le \varepsilon \right\}$$

and define

$$\mathscr{H}^s(M) := \limsup_{\varepsilon \to 0+} \mathscr{H}_\varepsilon^s(M). \tag{4.2}$$

Since $\mathscr{H}_{\varepsilon_1}^s(M) \ge \mathscr{H}_{\varepsilon_2}^s(M)$, for $\varepsilon_2 > \varepsilon_1 > 0$, the above limit superior exists and is an element of $[0, \infty]$. It is easy to verify that the function $s \mapsto \mathscr{H}_\varepsilon^s(M)$ is nondecreasing for $0 < \varepsilon < 1$.

Now assume that $0 < \varepsilon < 1$ and that $t > s \ge 0$. Then,

$$\sum_{i=1}^{\infty} |r_i|^t \le \varepsilon^{t-s} \sum_{i=1}^{\infty} |r_i|^s$$

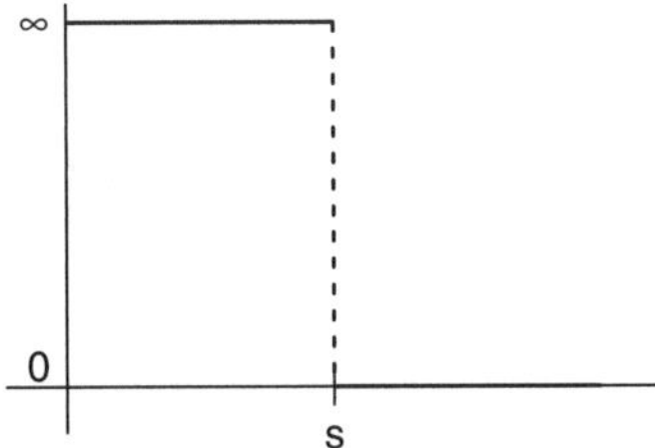

Figure 4.2 The 0–∞ behavior of $\mathcal{H}^s(M)$.

and, therefore,

$$\mathcal{H}_\varepsilon^t(M) \le \varepsilon^{t-s}\,\mathcal{H}_\varepsilon^s(M).$$

If $\mathcal{H}^s(M) < \infty$, then

$$0 \le \limsup_{\varepsilon \to 0+}\mathcal{H}_\varepsilon^t(M) \le \left(\lim_{\varepsilon \to 0+}\varepsilon^{t-s}\right)\left(\limsup_{\varepsilon \to 0+}\mathcal{H}_\varepsilon^s(M)\right) = 0.$$

On the other hand, if $\mathcal{H}^t(M) < \infty$, then

$$\limsup_{\varepsilon \to 0+}\mathcal{H}_\varepsilon^s(M) \ge \left(\limsup_{\varepsilon \to 0+}\varepsilon^{s-t}\right)\left(\limsup_{\varepsilon \to 0+}\mathcal{H}_\varepsilon^t(M)\right) = \infty.$$

The quantity $\mathcal{H}^s(M)$ thus exhibits a 0–∞ behavior that is depicted in Figure 4.2. Based on these observations, one defines the *Hausdorff–Besicovitch* [5] *dimension of M* by

$$\dim_H M := \inf\{s \ge 0 \mid \mathcal{H}^s(M) = 0\} = \sup\{s \ge 0 \mid \mathcal{H}^s(M) = \infty\},$$

i.e., as that value of s at which $\mathcal{H}^s(M)$ jumps from 0 to ∞.

We note that the numerical value of $\mathcal{H}^s(M)$ for $s = \dim_H M$, can be any number between 0 and ∞, inclusive. Sets for which $\mathcal{H}^{\dim_H M}(M) < \infty$ are termed *s-sets*.

Next, we consider a simple example of how one computes the Hausdorff–Besicovitch dimension of a set.

Example 4.6 Let $M := [0, 1] \subset \mathbb{R}$ and let $B_r(x_i) := [x_i - r, x_i + r] \subset M, r > 0$, $x_i \in M$, $i \in \mathbb{N}$. For a given $\varepsilon > 0$, choose $n(\varepsilon) > 1/\varepsilon$ and let $x_i := (i + \frac{1}{2})\varepsilon$ and $r_i := 1/(2n(\varepsilon))$, $i = 1, \ldots, n(\varepsilon)$. Then,

$$[0, 1] \subseteq \bigcup_{i=1}^{n(\varepsilon)} B_{r_i}(x_i) = \bigcup_{i=1}^{n(\varepsilon)} [x_i - r_i, x_i + r_i].$$

5 ABRAM SAMOILOVITCH BESICOVITCH, 24 January 1891–2 November 1970. Russian mathematician who primarily worked on combinatorial questions and in real analysis.

For $s > 1$, we have

$$\mathcal{H}_\varepsilon^s([0, 1]) \leq \sum_{i=1}^{n(\varepsilon)} |2r_i|^s = n(\varepsilon) \left(\frac{1}{n(\varepsilon)}\right)^s = \left(\frac{1}{n(\varepsilon)}\right)^{s-1} < \varepsilon^{s-1},$$

and, thus,

$$\limsup_{\varepsilon \to 0+} \mathcal{H}_\varepsilon^s([0, 1]) = 0.$$

Now suppose $s < 1$. The collection $\{B_{r_i}(x_i) \mid i = 1, \ldots, n(\varepsilon)\}$ covers the interval $[0, 1]$ and therefore, $\sum_{i=1}^{n(\varepsilon)} |2r_i| \geq 1$. Then,

$$\sum_{i=1}^{n(\varepsilon)} |2r_i|^s = \sum_{i=1}^{n(\varepsilon)} |2r_i|^{s-1} |2r_i| \geq \left(\frac{1}{\varepsilon}\right)^{s-1} \sum_{i=1}^{n(\varepsilon)} |2r_i| \geq \left(\frac{1}{\varepsilon}\right)^{s-1},$$

implying that $\limsup\limits_{\varepsilon \to 0+} \mathcal{H}_\varepsilon^s([0, 1]) = \infty$. Combining these two results gives

$$\inf\{s \geq 0 \mid \mathcal{H}^s([0, 1]) = 0\} = 1 = \sup\{s \geq 0 \mid \mathcal{H}^s([0, 1]) = \infty\},$$

establishing that $\dim_H[0, 1] = 1$, as expected.

Example 4.7 Let $\emptyset \neq M$ be an arbitrary open set in $\mathbb{R}^n$, where n is the smallest integer such that $M \subset \mathbb{R}^n$. Then $\dim_H M = n$, for M contains an n-dimensional ball.

Next, we investigate how the Hausdorff–Besicovitch dimension of a set M behaves under Lipschitz mappings.

Theorem 4.8 Let $M \subset \mathbb{R}^n$ and let $f \in \text{Lip}^\alpha(M, \mathbb{R}^m)$, where $m, n \in \mathbb{N}$. Then

$$\dim_H f(M) \leq \frac{1}{\alpha} \dim_H M.$$

Proof Let $\varepsilon > 0$ and let $\mathcal{H}_\varepsilon^s$ be defined as above. Suppose that $M \subseteq \bigcup_{i=1}^{\infty} B_{r_i}(x_i)$. Note that since f is a Lipschitz function, $f(B_r(x)) \subseteq B_{cr^\alpha}(f(x))$, where c is the Lipschitz constant of f. Therefore, if

$$\bigcup_{i=1}^{\infty} B_{r_i}(x_i) \supseteq M, \quad \text{where } 0 < r_i \leq \varepsilon,$$

then

$$\bigcup_{i=1}^{\infty} B_{\rho_i}(f(x_i)) \supseteq f(M), \quad \text{where } 0 < \rho_i := cr_i^\alpha \leq \eta := c\varepsilon^\alpha.$$

Hence,

$$\sum_{i=1}^{\infty} |f(B_{r_i}(\boldsymbol{x}_i) \cap M)|^{s/\alpha} \le \sum_{i=1}^{\infty} [\mathrm{diam}(B_{r_i}(f(\boldsymbol{x}_i)))]^{s/\alpha} = c^{s/\alpha} \sum_{i=1}^{\infty} r_i^s,$$

and, thus,

$$\mathscr{H}_{\eta}^{s/\alpha}(f(M)) \le c^{s/\alpha} \, \mathscr{H}_{\varepsilon}^s(M).$$

Now, since $\varepsilon \to 0+$ implies $\eta \to 0+$, $\mathscr{H}^{s/\alpha}(f(M)) \le c^{s/\alpha} \, \mathscr{H}^s(M)$, and one has for $s > \dim_H M$ that $\mathscr{H}^{s/\alpha}(f(M)) \le c^{s/\alpha} \, \mathscr{H}^s(M) = 0$. Therefore,

$$\dim_H f(M) \le \frac{s}{\alpha}, \quad \forall \, s > \dim_H M,$$

proving the claim. $\square$

Corollary 4.9 Assume that $M \subseteq \mathbb{R}^n$ and $f \in \mathrm{Lip}^1(M, \mathbb{R}^m)$, $m, n \in \mathbb{N}$.

(a) Then, $\dim_H f(M) \le \dim_H M$.

(b) In addition, suppose that $m = n$ and $N := f(M)$. Moreover, suppose that $f^{-1} \in \mathrm{Lip}^1(N, \mathbb{R}^n)$. Then $\dim_H f(M) = \dim_H M$.

Proof The first claim follows from setting $\alpha := 1$ in Theorem 4.8. To establish the second, observe first that $f^{-1} \in \mathrm{Lip}^1(N, \mathbb{R}^n)$ is equivalent to the existence of a constant $c' > 0$ so that for all $\boldsymbol{x}', \boldsymbol{y}' \in N$,

$$\|f^{-1}(\boldsymbol{x}') - f^{-1}(\boldsymbol{y}')\| \le c' \, \|\boldsymbol{x}' - \boldsymbol{y}'\|.$$

However, $\exists \, \boldsymbol{x}, \boldsymbol{y} \in M : \boldsymbol{x}' = f(\boldsymbol{x})$ and $\boldsymbol{y}' = f(\boldsymbol{y})$. Combining this with the fact that $f \in \mathrm{Lip}^1(M, \mathbb{R})$, one obtains

$$\frac{1}{c'} \, \|\boldsymbol{x} - \boldsymbol{y}\| \le \|f(\boldsymbol{x}) - f(\boldsymbol{y})\| \le c \, \|\boldsymbol{x} - \boldsymbol{y}\|,$$

where $c > 0$ is the Lipschitz constant of f. The statement now follows from (a) applied to f^{-1} and $N := f(M)$. $\square$

Example 4.10 Let M be an arbitrary nonempty subset of the interval $[0, 10]$ and suppose that $f : [0, 10] \to \mathbb{R}, \, x \mapsto \sqrt{x+1}$. Then, $\dim_H f(M) = \dim_H M$.

Remark 4.11 Instead of using balls to cover the set M, one can also use cubes or any other connected and convex set. The actual value of $\mathscr{H}^s(M)$ changes, but the location of the $0 - \infty$ jump does not.

The explicit computation of the Hausdorff–Besicovitch dimension of a set $M \subset \mathbb{R}^n$ is rather difficult since it involves taking the infimum over covers consisting of balls of radius *less than or equal to* a given $\varepsilon > 0$. A slight simplification is obtained by considering only covers by balls of radius *equal* to ε. This gives rise to the concept of *fractal* or *box dimension*.

To this end, let M be a bounded subset of $\mathbb{R}^n$, $n \in \mathbb{N}$, and let $\varepsilon > 0$ be given. For $\boldsymbol{x}_0 \in M$, denote by

$$W_\varepsilon(\boldsymbol{x}_0) := \{x \in \mathbb{R}^n \mid \|\boldsymbol{x} - \boldsymbol{x}_0\|_\infty = \varepsilon/2\}$$

the n-dimensional cube with side length $\varepsilon > 0$ and center $\boldsymbol{x}_0$. Denote by $\mathcal{N}_\varepsilon(M)$ the minimum number of such cubes required to cover the set M, i.e.,

$$\mathcal{N}_\varepsilon(M) := \inf \left\{ n \in \mathbb{N} \;\middle|\; \bigcup_{i=1}^n W_\varepsilon(\boldsymbol{x}_i) \supseteq M; \; \boldsymbol{x}_i \in M \right\}.$$

Definition 4.12 (Box Dimension) The *box dimension* of a bounded set $M \subset \mathbb{R}^n$, $n \in \mathbb{N}$, is defined by

$$\dim_B M := \lim_{\varepsilon \to 0+} \frac{\log \mathcal{N}_\varepsilon(M)}{\log \varepsilon^{-1}}, \quad \text{provided the limit exists.}$$

A direct consequence of this particular choice of covers is that, in general, $\dim_H M \le \dim_B M$. There exist, however, circumstances under which one has equality between these two metric dimensions. For instance, smooth curves and surfaces in $\mathbb{R}^n$ are such that the two dimensions agree. Below, we give conditions under which fractal sets share this property, too.

Next, we consider a simple example of how to compute the box dimension of a set.

Example 4.13 Let $M := [0, 1] \times [0, 1] \subset \mathbb{R}^2$ and let $\varepsilon > 0$ be given. Choose $n \in \mathbb{N}$ so that $n > 1/\varepsilon$. Then,

$$[0, 1] \times [0, 1] \subseteq \bigcup_{i,j=1}^n [(i-1)\varepsilon, i\varepsilon] \times [(j-1)\varepsilon, j\varepsilon)].$$

If we denote by $\lceil \; \rceil : \mathbb{R} \to \mathbb{Z}$, $x \mapsto \min\{n \in \mathbb{Z} \mid n \ge x\}$, the ceiling function, then the smallest number of such ε-cubes to cover M is given by $\mathcal{N}_\varepsilon(M) = \lceil \varepsilon^{-1} \rceil \cdot \lceil \varepsilon^{-1} \rceil$. Thus,

$$\dim_B M = \lim_{\varepsilon \to 0+} \frac{\log \lceil \varepsilon^{-1} \rceil^2}{\log \varepsilon^{-1}} = 2,$$

since $\varepsilon^{-1} \le \lceil \varepsilon^{-1} \rceil < \varepsilon^{-1} + 1$.

Remark 4.14 Instead of obtaining an exact expression for the minimum number of ε-cubes to cover a set M, it suffices to work with functional inequalities of the form

$$c_1 \varepsilon^{-d} \le \mathcal{N}_\varepsilon(M) \le c_2 \varepsilon^{-d},$$

for constants $0 < c_1 < c_2$ and some $d \in \mathbb{R}^+$. Then, $\dim_B M = d$. (The reader is encouraged to verify this statement.)

Remark 4.15   Suppose that M is a bounded subset of $\mathbb{R}^n$ and let $\varepsilon > 0$ be given. Assume that $a > 1$. Then there exists an $m \in \mathbb{N}$ so that $a^{-(m+1)} < \varepsilon \leq a^{-m}$. Set $W_m := W_{a^m}$ and $\mathcal{N}_m(M) := \min\{n \in \mathbb{N} \mid \bigcup_{i=1}^n W_m(\boldsymbol{x}_i) \supseteq M; \ \boldsymbol{x}_i \in M\}$. Then

$$\dim_B M = \lim_{m \to \infty} \frac{\log \mathcal{N}_m(M)}{\log a^m}.$$

Proof   Set $d := \lim_{\varepsilon \to 0+} \dfrac{\log \mathcal{N}_\varepsilon(M)}{\log \varepsilon^{-1}}$. The inclusions $W_{m+1} \subset W_\varepsilon \subset W_m$, imply that $\mathcal{N}_{m+1}(M) \geq \mathcal{N}_\varepsilon(M) \geq N_m(M)$. Together with the inequalites $\log a^m \leq \log \varepsilon^{-1} \leq a^{m+1}$, one obtains

$$\frac{\log a^m}{\log a^{m+1}} \frac{\log \mathcal{N}_m(M)}{\log a^m} \leq \frac{\log \mathcal{N}_\varepsilon(M)}{\log \varepsilon^{-1}} \leq \frac{\log \mathcal{N}_{m+1}(M)}{\log a^{m+1}} \frac{\log a^{m+1}}{\log a^m}.$$

As $m \to \infty$ implies $\varepsilon \to 0+$, the statement is proved. $\square$

Remark 4.16   There exist several other notions of topological dimension such as the *small* and *large inductive dimension* and the *covering dimension*. What we called topological dimension in this section is actually, more precisely, the small inductive dimension. However, for the type of spaces we are interested in the three types of dimension coincide. In case the set M allows a metric structure, there are also several other metric dimensions that are employed to describe the "size" of the set M. Examples of such metric dimensions are the *(upper and lower) Minkowski dimension* and the *(upper and lower) packing dimension*. In addition, there exist a number of probabilistic dimensions such as the *Billingsley dimension*. For a discussion of these dimensions in the context of fractals, we refer the interested reader to [113] and the references to the literature therein.

4.2 Iterated Function Systems

We now present a method of construction of fractal sets. It is one of several different approaches to fractals and was introduced in [11] and in slightly different form also in [87]. We refer the interested reader to [113] and the references given therein for a summary of other fractal construction methods.

Definition 4.17 (Iterated Function System)   Let (X, d) be a complete metric space and let $1 < N \in \mathbb{N}$. The collection of functions $\mathcal{F}_N := \{f_i \in \mathrm{Lip}^1(\mathsf{X}, \mathsf{X}) \mid i = 1, \ldots, N\}$ is called an *iterated function system (IFS)* on X provided that $\lambda := \max\{|\lambda_i| \mid i = 1, \ldots, N\} < 1$, where $\lambda_i := \mathrm{Lip}(f_i)$.

Remark 4.18   What we defined as an IFS, some authors call a *hyperbolic* iterated function system. As we deal exclusively with contractive mappings in this textbook, this more restrictive definition suffices for our purposes. For a more general discussion of IFSs, we refer the interested reader to [92, 97, 99].

We denote an IFS by $(\mathbf{X}; \mathcal{F}_N)$ or, if the number N of maps is understood, simply by $(\mathbf{X}; \mathcal{F})$.

Remark 4.19 A function $f \in \mathrm{Lip}^1(\mathbf{X}, \mathbf{X})$ whose Lipschitz constant $\mathrm{Lip}(f) < 1$ is also called a *contraction* on $\mathbf{X}$.

In order to construct fractals via IFSs, we require the notion of convergence of compact subsets of $\mathbf{X}$. For this purpose, we introduce the following class of sets.

Definition 4.20 Suppose that $(\mathbf{X}, d)$ is a complete metric space. The collection of all nonempty compact subsets of $\mathbf{X}$ is called the *hyperspace of compact subsets* (of $\mathbf{X}$) and is denoted by $\mathcal{H}(\mathbf{X})$.

Clearly, $\mathcal{H}(\mathbf{X}) \subset \mathscr{P}(\mathbf{X})$. On the set $\mathcal{H}(\mathbf{X})$ we define a function $d_{\mathcal{H}} : \mathcal{H}(\mathbf{X}) \times \mathcal{H}(\mathbf{X}) \to \mathbb{R}$ by

$$d_{\mathcal{H}}(A, B) := \max \left\{ \max_{a \in A} \min_{b \in B} d(a, b), \max_{b \in B} \min_{a \in A} d(b, a) \right\}.$$

Proposition 4.21 The function $d_{\mathcal{H}}$ is a metric on $\mathcal{H}(\mathbf{X})$ and the resulting metric space $(\mathcal{H}(\mathbf{X}), d_{\mathcal{H}})$ is complete.

Proof Exercise! $\square$

The function $d_{\mathcal{H}}$ is referred to as the *Hausdorff metric* or *Hausdorff distance* for $\mathcal{H}(\mathbf{X})$. There exists another, equivalent, definition of Hausdorff metric using the ε-parallel body of a nonempty compact set. We refer the interested reader to the exercise section at the end of this chapter for more details.

Next, we investigate some of the properties of the Hausdorff metric and show how the Hausdorff metric $d_{\mathcal{H}}$ changes under the application of functions from $\mathrm{Lip}^1(\mathbf{X}, \mathbf{X})$.

Lemma 4.22 Let $f \in \mathrm{Lip}^1(\mathbf{X}, \mathbf{X})$ with Lipschitz constant $\lambda := \mathrm{Lip}(f) < 1$. The numerical function f induces a set-valued mapping $\overline{f} : \mathcal{H}(\mathbf{X}) \to \mathcal{H}(\mathbf{X})$ with the property that

$$d_{\mathcal{H}}(\overline{f}(A), \overline{f}(B)) \le \lambda \, d_{\mathcal{H}}(A, B),$$

i.e., $\overline{f} \in \mathrm{Lip}^1(\mathcal{H}(\mathbf{X}), \mathcal{H}(\mathbf{X}))$ with $\mathrm{Lip}(\overline{f}) = \lambda < 1$.

Proof As $f \in \text{Lip}^1(\mathbf{X}, \mathbf{X})$, f is continuous and maps compact sets to compact set. Thus, we can define a set-valued mapping $\bar{f} : \mathcal{H}(\mathbf{X}) \to \mathcal{H}(\mathbf{X})$ by

$$\bar{f}(A) := \{f(a) \mid a \in A\}.$$

Now, let $A, B \in \mathcal{H}(\mathbf{X})$. Then

$$d_{\mathcal{H}}(\bar{f}(A), \bar{f}(B)) = \max \left\{ \max_{f(a) \in \bar{f}(A)} \min_{f(b) \in \bar{f}(B)} d(f(a), f(b)), \right.$$

$$\left. \max_{f(b) \in \bar{f}(B)} \min_{f(a) \in \bar{f}(A)} d(f(b), f(a)) \right\},$$

and since

$$\max_{f(a) \in \bar{f}(A)} \min_{f(b) \in \bar{f}(B)} d(f(a), f(b)) \le \max_{a \in A} \min_{b \in B} \lambda \cdot d(a, b)$$

and

$$\max_{f(b) \in \bar{f}(B)} \min_{f(a) \in \bar{f}(A)} d(f(b), f(a)) \le \max_{b \in B} \min_{a \in A} \lambda \cdot d(a, b),$$

one obtains that $d_{\mathcal{H}}(\bar{f}(A), \bar{f}(B)) \le \lambda \cdot d_{\mathcal{H}}(A, B)$. $\square$

In order to present the definition of a fractal generated by an IFS we need two lemmata.

Lemma 4.23 Let $A, B, C \in \mathcal{H}(\mathbf{X})$. Then

$$d_{\mathcal{H}}(A \cup B, C) = \max\{d_{\mathcal{H}}(A, C), d_{\mathcal{H}}(B, C)\}.$$

Proof Note that

$$d(A \cup B, C) = \max_{\alpha \in A \cup B} d(\alpha, C) = \max\{\max_{\alpha \in A} d(\alpha, C), \max_{\alpha \in B} d(\alpha, C)\}$$

$$= \max\{d(A, C), d(B, C)\},$$

which implies that claim. (The reader is encouraged to verify this.) $\square$

Now, each IFS $(\mathbf{X}; \mathcal{F}_N)$ induces an IFS $(\mathcal{H}(\mathbf{X}); \mathcal{H}(\mathcal{F}_N))$ by setting

$$\mathcal{H}(\mathcal{F}_N) := \{\bar{f}_i \mid i = 1, \ldots, N\}.$$

For, Lemma 4.22 immediately implies that $\bar{f}_i \in \text{Lip}^1(\mathcal{H}(\mathbf{X}), \mathcal{H}(\mathbf{X}))$ with $\text{Lip}(\bar{f}_i) = \text{Lip}(f_i)$, $i = 1, \ldots, N$.

In particular, there exists a mapping $F : \mathcal{H}(\mathbf{X}) \to \mathcal{H}(\mathbf{X})$ given by

$$F(A) := \bigcup_{i=1}^{N} \bar{f}_i(A).$$

The next result shows that F is a contraction on the complete metric space $(\mathcal{H}(\mathbf{X}), d_{\mathcal{H}})$.

Lemma 4.24 The mapping F as defined above is an element of $\text{Lip}^1(\mathcal{H}(\mathbf{X}), \mathcal{H}(\mathbf{X}))$ and $\text{Lip}(F) = \lambda := \max_{i=1,\dots,N} |\text{Lip}(f_i)|$.

Proof The proof is by induction on N. We show the case $N := 2$ and leave the remainder of the proof to the reader.

$$
\begin{aligned}
d_{\mathcal{H}}(F(A), F(B)) &= d_{\mathcal{H}}(\bar{f}_1(A) \cup \bar{f}_2(A), \bar{f}_1(B) \cup \bar{f}_2(B)) \\
&\leq \max\{d_{\mathcal{H}}(\bar{f}_1(A), \bar{f}_1(B)), d_{\mathcal{H}}(\bar{f}_2(A), \bar{f}_2(B))\} \\
&\leq \max\{\lambda_1 \, d_{\mathcal{H}}(A, B), \lambda_2 \, d_{\mathcal{H}}(A, B)\} \leq \lambda \cdot d_{\mathcal{H}}(A, B),
\end{aligned}
$$

where the first inequality is a consequence of Lemma 4.23. $\square$

Now we are ready to state the main result in this section.

Theorem 4.25 Let $(\mathbf{X}, d)$ be a complete metric space and $(\mathbf{X}; \mathcal{F}_N)$ an IFS over $\mathbf{X}$. Then there exists a mapping $F : \mathcal{H}(\mathbf{X}) \to \mathcal{H}(\mathbf{X})$ defined by

$$
F(A) := \bigcup_{i=1}^{N} \bar{f}_i(A), \tag{4.3}
$$

which is a contraction on $(\mathcal{H}(\mathbf{X}), d_{\mathcal{H}})$ with Lipschitz constant $\lambda = \max\{|\text{Lip}(f_i)| \mid i = 1, \dots, N\} < 1$. The uniquely determined fixed point $\mathfrak{A} \in \mathcal{H}(\mathbf{X})$ of F is called the *attractor of the IFS* $(\mathbf{X}; \mathcal{F}_N)$ or the *fractal (set) generated by the IFS* $(\mathbf{X}; \mathcal{F}_N)$.

Proof The results follow readily from the above lemmata and the Banach Fixed Point Theorem 1.35. $\square$

The unique fixed point $\mathfrak{A}$ of F satisfies the fixed point or self-referential equation

$$
\mathfrak{A} = \bigcup_{i=1}^{N} \bar{f}_i(\mathfrak{A}), \tag{4.4}
$$

which expresses the fact that the fractal $\mathfrak{A}$ is a finite union of the images under the mappings $\bar{f}_i$ of itself.

The Banach Fixed Point Theorem 1.35 also provides a means of obtaining the fractal $\mathfrak{A}$ as the limit of a sequence of sets from $\mathcal{H}(\mathbf{X})$. To this end, let $A_0 \in \mathcal{H}(\mathbf{X})$ and define

$$
A_n := F(A_{n-1}), \quad n \in \mathbb{N}. \tag{4.5}
$$

Then, $\mathfrak{A} = \lim_{n \to \infty} A_n$, where the limit is taken in the Hausdorff metric $d_{\mathcal{H}}$:

$$
\forall \varepsilon > 0 \, \exists n_0 \in \mathbb{N} \, \forall n \geq n_0 : d_{\mathcal{H}}(\mathfrak{A}, A_n) < \varepsilon.
$$

In particular, if $A_0 := \{a_0\}$, $a_0 \in \mathbf{X}$ arbitrary, then

$$
\mathfrak{A} = \text{clos}_{d_{\mathcal{H}}} \{F^n(\{a_0\}) \mid n \in \mathbb{N}\},
$$

where $F^n := \underbrace{F \circ \cdots \circ F}_{n \text{ times}}$ and where the closure is taken with respect to the Hausdorff metric $d_{\mathcal{H}}$. Note that Equation (4.5) gives a *deterministic* algorithm for the construction of fractals. For the generation of fractals whose IFS consists of many mappings, this deterministic algorithm may be computationally prohibitive, both in regard to memory allocation and to computing speed. Later in this chapter, in Theorem 4.92, we will provide a stochastic algorithm for the generation of fractals that overcomes these impediments.

Definition 4.26 (Prefractal)   Let $(X, \mathcal{F}_N)$ be an IFS and let $A_0 \in \mathcal{H}(X)$. Then any set A_n of the form (4.5) is called a *prefractal (for $\mathfrak{A}$) of rank n*.

Remark 4.27   The graphical display of a fractal $\mathfrak{A}$ actually shows a prefractal A_n of high enough rank n.

Example 4.28   Let $X := [0, 1]$ endowed with the Euclidean metric, and let $f_1 : [0, 1] \to [0, 1]$, $x \mapsto \frac{1}{3}x$ and $f_2 : [0, 1] \to [0, 1]$, $x \mapsto \frac{1}{3}x + \frac{2}{3}$. Then $([0, 1]; \{f_1, f_2\})$ is an IFS with $\mathrm{Lip}(f_1) = \mathrm{Lip}(f_2) = \frac{1}{3} < 1$. The fractal set generated by this IFS is the classical Cantor set $\mathfrak{C}$.

Example 4.29   Let $X := [0, 1] \times [0, 1]$ endowed with the Euclidean metric, and define functions $f_i : [0, 1] \times [0, 1] \to [0, 1] \times [0, 1]$, $i = 1, 2, 3$, by

$$f_1(x, y) = \left(\frac{x}{2}, \frac{y}{2} \right),$$

$$f_2(x, y) = \left(\frac{x}{2} + \frac{1}{2}, \frac{y}{2} \right),$$

$$f_3(x, y) = \left(\frac{x}{2} + \frac{1}{4}, \frac{y}{2} + \frac{\sqrt{3}}{4} \right).$$

Then $([0, 1] \times [0, 1]; \{f_1, f_2, f_3\})$ is an IFS with $\mathrm{Lip}(f_1) = \mathrm{Lip}(f_2) = \mathrm{Lip}(f_3) = \frac{1}{2} < 1$. The fractal set generated by this IFS is the Sierpiński triangle $\mathfrak{S}$.

Given an IFS $(X, \mathcal{F}_N)$, it is sometimes useful to work with a compact subset K of X instead of using X itself. The following proposition shows that this can be easily achieved.

Proposition 4.30   Suppose that $K \subseteq X$ is compact. Then there exists a $\widehat{K} \in \mathcal{H}(X)$ such that $\widehat{K} \supseteq K$ and $F(\widehat{K}) \subseteq \widehat{K}$.

Proof   Set $\widehat{K} := \mathrm{clos}_{d_{\mathcal{H}}} \bigcup_{n=0}^{\infty} F^n(K)$. Then $\widehat{K}$ is compact and, by its definition, satisfies $F(\widehat{K}) \subseteq \widehat{K}$. $\square$

Definition 4.31 (Overlapping IFS)   Let $(\mathbf{X}, \mathcal{F}_N)$ be an IFS with $\mathcal{F}_N :=$ $\{f_1, \ldots, f_N\}$ and attractor $\mathfrak{A}$. Then $(\mathbf{X}, \mathcal{F}_N)$ is called *overlapping* if the set

$$\{\overline{f}_i(\mathfrak{A}) \cap \overline{f}_j(\mathfrak{A}) \mid i, j = 1, \ldots, N; \ i \neq j\}$$

contains a nonempty open ball.

4.3 Approximation and the Collage Theorem

Next, we consider the approximation of complex and intricate geometric objects by fractals. It is in general extremely difficult to approximate or model such sets using classical, non-fractal, geometries. For a large class of objects with complicated multi-scale structures, *fractal models* provide a reasonable approximation or description.

To this end, we prove a theorem due to Barnsley [8].

Theorem 4.32 (Collage Theorem)   Let $(\mathbf{X}, d)$ be a complete metric space. Suppose that $\varepsilon > 0$ and $A \in \mathcal{H}(\mathbf{X})$ are given. Further assume that there exists an IFS $(\mathbf{X}; \mathcal{F}_N)$ so that $d_{\mathcal{H}}(A, F(A)) < \varepsilon$. If $\mathfrak{A} \in \mathcal{H}(\mathbf{X})$ denotes the fractal generated by the IFS $(\mathbf{X}; \mathcal{F}_N)$, then

$$d_{\mathcal{H}}(A, \mathfrak{A}) < \frac{\varepsilon}{1 - \lambda},$$

where $\lambda := \mathrm{Lip}(F)$.

Proof   The statement follows readily from the properties of the Hausdorff metric and its continuity. For,

$$d_{\mathcal{H}}(A, \mathfrak{A}) = d_{\mathcal{H}}(A, \lim_{n \to \infty} F^n(A)) = \lim_{n \to \infty} d_{\mathcal{H}}(A, F^n(A))$$

$$\leq \lim_{n \to \infty} \sum_{\nu=1}^{n} d_{\mathcal{H}}(F^{\nu-1}(A), F^{\nu}(A)) \leq \lim_{n \to \infty} \sum_{\nu=1}^{n} \lambda^{\nu-1} d_{\mathcal{H}}(A, F(A))$$

$$= \lim_{n \to \infty} \frac{1 - \lambda^n}{1 - \lambda} d_{\mathcal{H}}(A, F(A)) < \frac{\varepsilon}{1 - \lambda}. \qquad \square$$

Applications of the Collage Theorem include the modeling of natural objects such as leaves, images, clouds, mountain ranges, etc. by means of IFSs. For more examples along these lines, we refer the reader to [8, 9].

Example 4.33   As an illustrative example, we consider the well-known Barnsley fern $\mathfrak{F}$, which is a fractal model of a fern belonging to the black

Figure 4.3 The Barnszely fern $\mathfrak{F}$.

spleenwort variety. $\mathfrak{F}$ is generated by the IFS $([0,1] \times [0,1]; \{f_1, f_2, f_3, f_4\})$, where

$$f_1 \begin{pmatrix} x \\ y \end{pmatrix} := \frac{4}{5} \begin{pmatrix} \cos \frac{\pi}{60} & \sin \frac{\pi}{60} \\ -\sin \frac{\pi}{60} & \cos \frac{\pi}{60} \end{pmatrix} \begin{pmatrix} x \\ y \end{pmatrix} + \begin{pmatrix} 0 \\ \frac{3}{2} \end{pmatrix},$$

$$f_2 \begin{pmatrix} x \\ y \end{pmatrix} := \frac{1}{3} \begin{pmatrix} \cos \frac{5\pi}{18} & -\sin \frac{5\pi}{18} \\ \sin \frac{5\pi}{18} & \cos \frac{5\pi}{18} \end{pmatrix} \begin{pmatrix} x \\ y \end{pmatrix} + \begin{pmatrix} 0 \\ \frac{3}{2} \end{pmatrix},$$

$$f_3 \begin{pmatrix} x \\ y \end{pmatrix} := \begin{pmatrix} \frac{1}{3} \cos \frac{2\pi}{3} & \frac{9}{25} \sin \frac{5\pi}{18} \\ \frac{1}{3} \sin \frac{2\pi}{3} & -\frac{9}{25} \cos \frac{5\pi}{18} \end{pmatrix} \begin{pmatrix} x \\ y \end{pmatrix} + \begin{pmatrix} 0 \\ \frac{2}{5} \end{pmatrix},$$

$$f_4 \begin{pmatrix} x \\ y \end{pmatrix} := \begin{pmatrix} 0 & 0 \\ 0 & \frac{11}{50} \end{pmatrix} \begin{pmatrix} x \\ y \end{pmatrix}.$$

The angles and scaling factors in the matrices above are found applying a protractor and a ruler to a picture of a black spleenwort fern. The reader is encouraged to find IFSs for a number of naturally occuring leaves and to compare the modeled image to the original.

The main difficulty in applying the Collage Theorem consists in finding the IFS. There are many approaches to obtain a solution to the *Fractal Inverse Problem*:

Given an $\varepsilon > 0$ and an $A \in \mathcal{H}(\mathbf{X})$, find an $N \in \mathbb{N}$ and a collection of mappings $\mathcal{F}_N$, so that $d_{\mathcal{H}}(A, F(A)) < \varepsilon$.

Most of these approaches are very dependent on the structure of the underlying natural object, although some general statements can be made. An albeit incomplete list of papers that provide solutions to the fractal inverse problem and which are based on the concepts introduced here is given by [34, 66, 67, 68], and [69].

4.4 Dimensions of Fractal Sets

In this section, we investigate the Hausdorff–Besicovitch and box dimension of fractals generated by IFSs. For this purpose, we need to restrict the class of mappings that defines an IFS.

Definition 4.34 (Similitude)   A function $S : \mathbb{R}^n \to \mathbb{R}^n$ is called a *similitude* or a *similarity mapping* if there exists an $s > 0$ so that

$$\|S(x) - S(y)\| = s\,\|x - y\|, \quad \forall x, y \in \mathbb{R}^n.$$

The constant s is called the *similarity constant* of S. In the case where $0 \leq s < 1$, the similitude S is termed a *contracting similitude.*

Similitudes are closely related to the Euclidean group $E(n)$ of $\mathbb{R}^n$. The elements of this group are isometries associated with the Eulidean metric and are commonly called *Euclidean motions*. They are classified as follows.

Definition 4.35 (Euclidean Motions)

1. A function $H_s : \mathbb{R}^n \to \mathbb{R}^n$, $x \mapsto sx$, $s \in \mathbb{R}_0^+$, is called a *homothety* or *homothetic transformation.*

2. A function $O : \mathbb{R}^n \to \mathbb{R}^n$ is called an *orthogonal transformation* if $OO^\top = O^\top O = I$. The mappings O define the *orthogonal group* $O(n)$ of $\mathbb{R}^n$.

3. A function $\tau_v : \mathbb{R}^n \to \mathbb{R}^n$, $x \mapsto x + v$, $v \in \mathbb{R}^n \setminus \{0\}$ fixed, is called a *translation.*

The next proposition completely characterizes similitudes.

Proposition 4.36   The function $S : \mathbb{R}^n \to \mathbb{R}^n$ is a similitude iff $S = H_s \circ O \circ \tau_v$.

Proof Exercise! □

Now suppose that we are given an IFS all of whose mappings are (contracting) similitudes $S_i : \mathbf{X} \to \mathbf{X}$, $i = 1, \ldots, N$. We denote such a collection of similitudes by $\mathcal{S}_N$ and the associated IFS by $(\mathbf{X}; \mathcal{S}_N)$.

Example 4.37 The Cantor set $\mathfrak{C}$ and the Sierpiński triangle $\mathfrak{S}$ are generated by IFSs whose mappings are similitudes.

4.4.1 The Theoretical Computation of Dimension

To present a dimension result for fractals generated by a special class of IFSs, we require one more definition.

Definition 4.38 (Open Set Condition) Suppose that $\mathfrak{A}$ is a fractal generated by the IFS $(\mathbf{X}; \mathcal{S}_N)$ where $\mathbf{X}$ is a closed subset of $\mathbb{R}^n$. The set $\mathcal{S}_N$ satisfies the *open set condition (OSC)* iff there exists a nonempty bounded open set $G \subset \mathbb{R}^n$ so that

$$\dot{\bigcup_{i=1}^{N}} \overline{f}_i(G) \subseteq G,$$

where $\dot{\bigcup}$ denotes the disjoint union of sets.

Remark 4.39 Suppose that $(\mathbf{X}; \mathcal{S}_N)$ is an IFS satisfying the OSC. Then, a simple volume argument shows that

$$\sum_{i=1}^{N} s_i^n < 1,$$

where $s_i := \mathrm{Lip}(S_i)$, $i = 1, \ldots, N$. (The reader is encouraged to verify this statement.)

Theorem 4.40 Suppose that $\mathfrak{A}$ is a fractal generated by an IFS $(\mathbf{X}; \mathcal{S}_N)$ satisfying the OSC. If $s_i := \mathrm{Lip}(S_i)$, $i = 1, \ldots, N$, then

$$\dim_H \mathfrak{A} = \dim_B \mathfrak{A} = d,$$

where d is the unique positive solution of

$$\sum_{i=1}^{N} s_i^d = 1.$$

Proof We refer to [8, 56, 57, 87] or [113] for the proof. □

Example 4.41 For the Canter set $\mathfrak{C}$ and the Sierpiński triangle $\mathfrak{S}$ one obtains with $s_1 = s_2 = \frac{1}{3}$, respectively, $s_1 = s_2 = s_3 = \frac{1}{2}$, for the dimensions

$$\dim_H \mathfrak{C} = \dim_B \mathfrak{C} = \frac{\log 2}{\log 3}, \quad \text{respectively,} \quad \dim_H \mathfrak{S} = \dim_B \mathfrak{S} = \frac{\log 3}{\log 2}.$$

In connection to Theorem 4.40, the question arises under what conditions the Hausdorff–Besicovitch dimension and the box dimension agree for a given fractal set $\mathfrak{A}$. A partial answer is provided by the following result due to Falconer [59].

Theorem 4.42 Let $\mathbf{X}$ be a closed subset of $\mathbb{R}^n$ and let $\mathfrak{A}$ be a fractal generated by an IFS $(\mathbf{X}; \mathcal{F}_N)$. Suppose that all the $f_i \in \mathcal{F}_N$ have the form $f_i(x) = A_i x + b_i$, where $A_i \in \mathbb{R}^{n \times n}$ is a contractive matrix and $b_i \in \mathbb{R}^n$. Then, for almost all $(v_1, \ldots, v_N)^\top \in \mathbb{R}^{nN}$,

$$\dim_H \mathfrak{A} = \dim_B \mathfrak{A}.$$

4.4.2 The Numerical Computation of Dimension

The computation of the Hausdorff–Besicovitch dimension, or even the box dimension of a fractal set, is of considerable difficulty. There are only a few dimension results available, most of them for IFSs consisting of similitudes or affine mappings. For general fractals, there are no known estimates nor exact results for the Hausdorff–Besicovitch or the box dimension available. However, there exists a numerical approach to the computation of the box dimension. We will discuss this approach briefly.

To the end, let M be a nonempty compact subset of $\mathbb{R}^n$. By Remark 4.15, we can compute the box dimension of M via the formula

$$\dim_B M = \lim_{m \to \infty} \frac{\log \mathcal{N}_m(M)}{\log 2^m}.$$

This suggests the following procedure. For an increasing sequence of m, say $\{m_1, \ldots, m_\nu, \ldots\} \subset \mathbb{N}$, determine experimentally the numbers $\mathcal{N}_{m_\nu}(M)$ and graph the results in a log-log plot. If the number of the points $(\log 2^{m_\nu}, \log \mathcal{N}_{m_\nu}(M))$ plotted is statistically relevant, then a regression line can be computed. The slope of the regression line gives an approximate value for the box dimension of the set M. This procedure is demonstrated by the following simple example.

Example 4.43 We consider the well-known twin fragon fractal $\mathfrak{D}$, which is generated by the IFS $([-1, 1] \times [-1, 1], \{f_1, f_2\})$, with

$$f_1 \begin{pmatrix} x \\ y \end{pmatrix} := \begin{pmatrix} 0.5 & -0.5 \\ 0.5 & 0.5 \end{pmatrix} \begin{pmatrix} x \\ y \end{pmatrix},$$

$$f_2 \begin{pmatrix} x \\ y \end{pmatrix} := \begin{pmatrix} 0.5 & -0.5 \\ 0.5 & 0.5 \end{pmatrix} \begin{pmatrix} x \\ y \end{pmatrix} + \begin{pmatrix} 0.5 \\ 0.5 \end{pmatrix}.$$

We are interested in estimating the box dimension of the twin dragon's fractal boundary. (See Figure 4.4.)

To experimentally determine the box dimension of the bounday of $\mathfrak{D}$, we cover the fractal set by a grid of size 2^{-m}, $m \in \mathbb{N}$, as depicted on the right-hand

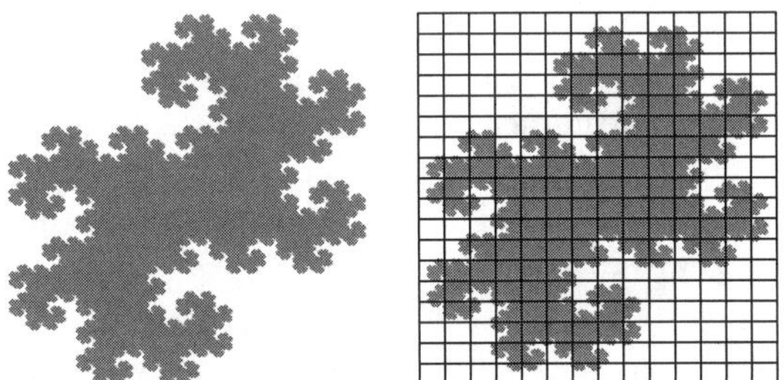

Figure 4.4 The twin dragon fractal $\mathfrak{D}$.

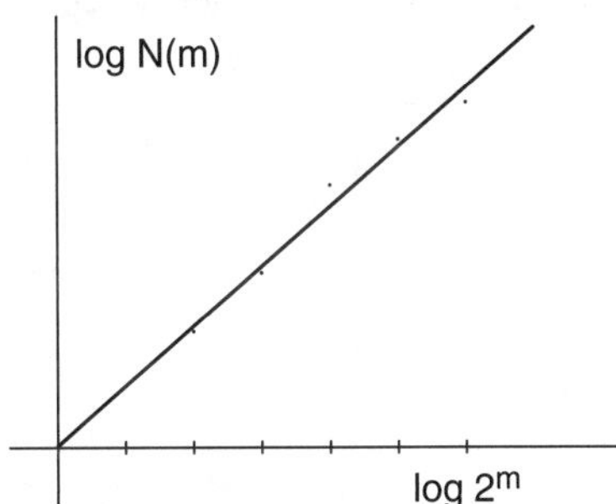

Figure 4.5 The regression line whose slope approximates $\dim_B \partial\mathfrak{D}$.

side of Figure 4.4, and count the number $\mathcal{N}(m)$ of $2^{-m} \times 2^{-m}$-squares that intersect the boundary $\partial\mathfrak{D}$:

$$\mathcal{N}(m) := \operatorname{card}\{g \in G(m) \,|\, g \cap \partial\mathfrak{D} \neq \emptyset\},$$

where $G(m)$ denotes the grid consisting of $2^{-m} \times 2^{-m}$-squares with sides parallel to the coordinate axes. For an increasing sequence of integers $m_1 < m_2 < \cdots < m_n$, $n \in \mathbb{N}$, we obtain numbers $\mathcal{N}(m_\nu)$, $\nu = 1, \ldots, n$, and graph the set of points $\{(m_\nu, \mathcal{N}(m_\nu)) \,|\, \nu = 1, \ldots, n)\}$ in a log–log coordinate system, creating a plot like the one depicted in Figure 4.5. Using regression analysis, we may fit a line through the data points $\{(\log m_\nu, \log \mathcal{N}(m_\nu)) \,|\, \nu = 1, \ldots, n)\}$ that minimizes the mean-square error. The slope of this line is then an approximate value for the box dimension of $\partial\mathfrak{D}$. In the given case, one obtains $\dim_B \partial\mathfrak{D} \approx 1.5234$, whereas the exact value is

$$\dim_B \partial\mathfrak{D} = \frac{2\log\frac{1}{3}\left[1 + (28 - 3\sqrt{87})^{1/3} + (28 + 3\sqrt{87})^{1/3}\right]}{\log 2} \approx 1.5236.$$

For a derivation of the above dimension formula, we refer the reader to [75].

In a similar fashion, the box dimension of several naturally occuring objects has been computed. For instance, for the surface of the human brain and the human lung, the values of $\dim_{B,\text{human brain}} \approx 2.79$ and $\dim_{B,\text{human lung}} \approx 2.97$ have been obtained. For vegetables such as cauliflower or broccoli the box dimension was computed exactly in the former case, since each cauliflower branch carries about 13 branches three times smaller giving $\dim_{B,\text{cauliflower}} = \log 13/\log 3 \approx 2.33$, whereas in the former case an experimental value of $\dim_{B,\text{broccoli}} \approx 2.66$ was computed.

4.5 The Code Space of an IFS

With every IFS $(\mathbf{X}; \mathcal{F}_N)$ one can associate a *code space* with N symbols. To this end, recall that if $\mathfrak{A}$ denotes the fractal generated by $(\mathbf{X}; \mathcal{F}_N)$, then

$$\mathfrak{A} = \lim_{n\to\infty} F^n(A),$$

where $A \in \mathcal{H}(\mathbf{X})$ is arbitrary. Note that

$$F^2(A) = (F \circ F)(A) = (\overline{f}_1 \circ \overline{f}_1)(A) \cup (\overline{f}_1 \circ \overline{f}_2)(A) \cup \cdots \cup (\overline{f}_N \circ \overline{f}_N)(A)$$

$$= \bigcup_{i_1,i_2=1}^{N} (\overline{f}_{i_1} \circ \overline{f}_{i_2})(A),$$

and, in general,

$$F^n(A) = \bigcup_{i_1,\ldots,i_n=1}^{N} (\overline{f}_{i_1} \circ \cdots \circ \overline{f}_{i_n})(A). \tag{4.6}$$

Now, set $\Sigma_N := \{1,\ldots,N\}^{\mathbb{N}} = \text{Map}(\mathbb{N}, \{1,\ldots,N\})$ and denote the elements of Σ_N by σ, τ, etc. The set Σ_N is called the *code space with N symbols* and $\sigma \in \Sigma_N$ a *code*. One refers to the collection $\{1,\ldots,N\}$ of symbols as the *alphabet* (of the code space). Codes of the form $\sigma = (\sigma_1,\ldots,\sigma_n,0,0,\ldots)$, $n \in \mathbb{N}$, are called *finite codes* and are denoted by $\sigma(n)$. The length of a finite code is denoted by $|\sigma|$. A *periodic code of period k* is a code of the form $\overline{\sigma_1\ldots\sigma_k} := (\sigma_1,\ldots,\sigma_k,\sigma_1,\ldots,\sigma_k,\ldots)$.

We define a product, called *concatenation*, on the set Σ_N^f of finite codes in Σ_N in the following way. Let $m, n \in \mathbb{N}$ and let $\sigma(m) := (\sigma_1,\sigma_2,\ldots,\sigma_m)$ and $\tau(n) := (\tau_1,\tau_2,\ldots,\tau_n)$ be finite codes. Define

$$c : \Sigma_N^f \times \Sigma_N^f \to \Sigma_N^f$$

$$c(\sigma(m),\tau(n)) := (\sigma_1,\ldots,\sigma_m,\tau_1,\ldots,\tau_n).$$

Note that c is a non-commutative operation on Σ_N^f. Instead of writing $c(\sigma(m),\tau(n))$ for the concatenation, we use the shorter notation $\sigma(m)\tau(n)$. Concatenation can be extended to an operation from $\Sigma_N^f \times \Sigma_N \to \Sigma_N$, by setting

$$\sigma(m)\tau := (\sigma_1,\ldots,\sigma_m,\tau_1,\tau_2,\ldots),$$

for all $m \in \mathbb{N}$ and all codes $\tau \in \Sigma_N$.

Defining a function $d_F : \Sigma_N \times \Sigma_N \to \mathbb{R}$ by

$$d_F(\sigma, \tau) := \sum_{n=1}^{\infty} \frac{|\sigma_n - \tau_n|}{(N+1)^n},$$

one readily shows that d_F is a metric on Σ_N and that with this metric (Σ_N, d_F) becomes a compact metric space. (The reader is encouraged to verify this statement.) The function d_F is called the *Fréchet*[6] *metric* on Σ_N.

Setting $\overline{f}_{\sigma(n)} := \overline{f}_{\sigma_1 \cdots \sigma_n} := \overline{f}_{\sigma_1} \circ \cdots \circ \overline{f}_{\sigma_n}$, one can rewrite (4.6) in the form

$$F^n(A) = \bigcup_{\sigma(n) \in \Sigma_N} \overline{f}_{\sigma(n)}(A),$$

and, thus,

$$\mathfrak{A} = \lim_{n \to \infty} \bigcup_{\sigma(n) \in \Sigma_N} \overline{f}_{\sigma(n)}(A).$$

Theorem 4.44 Suppose that $\mathfrak{A}$ is a fractal generated by an IFS $(\mathbf{X}; \mathcal{F}_N)$. Then there exists a surjective mapping $\gamma_{\mathcal{F}} := \gamma_{\mathcal{F}_N} : \Sigma_N \to \mathfrak{A}$, given by

$$\gamma_{\mathcal{F}}(\sigma) := \lim_{n \to \infty} f_{\sigma(n)}(x), \tag{4.7}$$

where the limit is independent of $x \in \mathbf{X}$.

Proof By Proposition 4.30, we may assume without loss of generality that $\mathbf{X}$ is compact. Let $x, y \in \mathbf{X}$ and let $\lambda := \max\{|\mathrm{Lip}(f_i)| \,|\, i = 1, \ldots, N\} < 1$. Then, for $m, n \in \mathbb{N}$,

$$|f_{\sigma(n)}(x) - f_{\sigma(m)}(y)| < \lambda^{\min\{m,n\}} \mathrm{diam}(\mathbf{X}), \quad \forall x, y \in \mathbf{X},$$

showing that the limit exists and is independent of $x \in \mathbf{X}$.

Now, choose an $x \in \mathbf{X}$ and set $\mathfrak{A}(x) := \lim_{n \to \infty} F^n(\{x\})$. Then, $\gamma_{\mathcal{F}}(\sigma) \in \mathfrak{A}(x)$. To show that $\gamma_{\mathcal{F}}$ is continuous, suppose that $\varepsilon > 0$ is given. Then, there exists an $n \in \mathbb{N}$ so that $\lambda^n \mathrm{diam}(\mathbf{X}) < \varepsilon$. Choose $\sigma, \tau \in \Sigma_N$ satisfying $\sigma(n) = \tau(n)$, but $\sigma(n+1) \neq \tau(n+1)$. Thus,

$$d_F(\sigma, \tau) = \sum_{j=n+1}^{\infty} \frac{|\sigma_j - \tau_j|}{(N+1)^j} < \sum_{j=n+1}^{\infty} \frac{N}{(N+1)^j} = \frac{1}{(N+1)^n},$$

and, therefore,

$$|\gamma_{\mathcal{F}}(\sigma) - \gamma_{\mathcal{F}}(\tau)| = \lim_{n \to \infty} |f_{\sigma(n)}(x) - f_{\tau(n)}(x)| < \lambda^n \mathrm{diam}(\mathbf{X}),$$

proving continuity.

To establish the surjectivity of $\gamma_{\mathcal{F}}$, we assume that $y \in \mathfrak{A}(x)$. Then, since $\mathcal{H}(\mathbf{X})$ is a complete metric space, there exists a sequence $\{\sigma_\nu(n) \,|\, \nu \in \mathbb{N}\} \subset$

---

6 Maurice René Fréchet, 2 September 1878–4 June, 1973. French mathematician who made fundamental contributions to topology and functional analysis.

Σ_N such that $\lim\limits_{n\to\infty} f_{\sigma_{\nu(n)}}(x) = y$. By the compactness of Σ_N there exists a subsequence $\{\sigma_{\nu_\mu} \,|\, \mu \in \mathbb{N}\}$ of $\{\sigma_\nu(n) \,|\, \nu \in \mathbb{N}\}$ with the property that $\sigma_{\nu_\mu} \to \tau$, as $\mu \to \infty$. But then, $d_F(\sigma_{\nu_\mu}, \tau) \to 0$, as $\mu \to \infty$.

Let $c(\mu) := \mathrm{card}\{\varkappa \in \mathbb{N} \,|\, (\sigma_{\nu_\mu})_j = \tau_j, \ j = 1, \ldots, \varkappa\}$. Then, by the above arguments, $c(\mu) \to \infty$ as $\mu \to \infty$. Hence,

$$|f_{\sigma_{\nu_\mu}(n)}(x) - f_{\tau(n)}(x)| < \lambda^{c(\mu)} \,\mathrm{diam}(\mathbf{X}),$$

which establishes that $y - \gamma_{\mathcal{F}}(\tau) = 0$ as $\mu \to \infty$ and $n \to \infty$. $\qquad\square$

Remark 4.45 If for all $i, j = 1, \ldots, N$, $i \neq j$, $\overline{f}_i(A) \cap \overline{f}_j(A) = \emptyset$, then $\gamma_{\mathcal{F}}$ is also injective. In this case, $\gamma_{\mathcal{F}} : \Sigma_N \to \mathfrak{A}$ is a bijection.

Remark 4.46 Suppose that $\sigma \in \Sigma_N$. Let $A_{\sigma(n)} := \overline{f}_{\sigma(n)}(A)$, for $A \in \mathcal{H}(\mathbf{X})$. Then Theorem 4.44 states that

$$\gamma_{\mathcal{F}}(\sigma) = \bigcap_{n \in \mathbb{N}} A_{\sigma(n)}.$$

The fact that to each point $\mathfrak{a} \in \mathfrak{A}$ there corresponds a code $\sigma \in \Sigma_N$ gives rise to the following definitions.

Definition 4.47 (Address Set and Address Function)   Let $\mathfrak{a} \in \mathfrak{A}$. The *address set* of $\mathfrak{a}$ is given by the set

$$\gamma_{\mathcal{F}}^{-1}(\mathfrak{a}) = \{\sigma \in \Sigma_N \,|\, \gamma_{\mathcal{F}}(\sigma) = \mathfrak{a}\}.$$

Each element of $\gamma_{\mathcal{F}}^{-1}(\mathfrak{a})$ is called an *address* of $\mathfrak{a}$. In this context, the mapping $\gamma_{\mathcal{F}} : \Sigma_N \to \mathfrak{A}$ is also called the *address function* (of the IFS $(\mathbf{X}; \mathcal{F}_N)$).

Definition 4.48 (Totally Disconnected IFS)   An IFS is called *totally disconnected* if $\gamma_{\mathcal{F}}^{-1}(\mathfrak{a})$ is one-elementary for all $\mathfrak{a} \in \mathfrak{A}$, i.e., if $\gamma_{\mathcal{F}}$ is injective.

Example 4.49 The IFS that generates the classical Cantor set $\mathfrak{C}$ is totally disconnected.

In the case that the IFS is totally disconnected, each point $\mathfrak{a} \in \mathfrak{A}$ has a unique representation in terms of the N symbols of the associated code space Σ_N. Such representations of $\mathfrak{a}$ are then given by N-adic expansions.

The code space Σ_N can be endowed with a strict linear order in the following way. Suppose that $\sigma, \omega \in \Sigma_N$. Then $\prec \subset \Sigma_N \times \Sigma_N$ is defined by

$$\sigma \prec \omega \quad \Longleftrightarrow \quad \omega_i < \sigma_i, \tag{4.8}$$

where i is the first index for which $\sigma_i \neq \omega_i$, $i \in \mathbb{N}$. The order relation $\prec$ is strictly linear in the sense that

    (i) $\forall \sigma \in \Sigma_N : \sigma \not\prec \sigma$ (irreflexivity)

    (ii) $\forall \rho, \sigma \omega \in \Sigma_N : \rho \prec \sigma \wedge \sigma \prec \omega \Longrightarrow \rho \prec \omega$ (transitivity) holds. (The reader is encouraged to verify this statement.) Let $\iota \in \Sigma_N$ be defined by $\iota := \overline{1} = 111\ldots$ and $\varpi \in \Sigma_N$ by $\varpi := \overline{N} = NNN\ldots$. Then, for all $\iota \neq \sigma \neq \varpi$, we have $\sigma \prec \iota$ and $\varpi \prec \sigma$. With this definition of strict linear order, it is easy to show that the set $\gamma_{\mathcal{F}}^{-1}(\mathfrak{a})$ has a unique maximal element. For, let σ_1 be the smallest first component of all elements in $\gamma_{\mathcal{F}}^{-1}(\mathfrak{a})$ and inductively define σ_i be the smallest i-th component of all elements in $\gamma_{\mathcal{F}}^{-1}(\mathfrak{a})$ whose first $(i-1)$-components are $\sigma_1 \ldots \sigma_{i-1}$, $1 < i \in \mathbb{N}$. Then $(\sigma_1 \ldots \sigma_i \ldots)$ is the unique maximal element of $\gamma_{\mathcal{F}}^{-1}(\mathfrak{a})$. Moreover, the continuity of $\gamma_{\mathcal{F}}$ implies that $(\sigma_1 \ldots \sigma_i \ldots)$ is an element of $\gamma_{\mathcal{F}}^{-1}(\mathfrak{a})$, i.e., an address of $\mathfrak{a}$.

These observations give rise to the following important definition [9].

Definition 4.50 (Tops Function and Tops Code Space)   Let $(\mathsf{X}; \mathcal{F}_N)$ be an IFS with attractor $\mathfrak{A}$. Suppose that $\gamma_{\mathcal{F}} : \Sigma_N \to \mathfrak{A}$ is the address function. For $\mathfrak{a} \in \mathfrak{A}$, let

$$\tau_{\mathcal{F}}(\mathfrak{a}) := \tau_{(\mathsf{X}; \mathcal{F}_N)}(\mathfrak{a}) := \max\{\sigma \in \Sigma \mid \gamma_{\mathcal{F}}(\sigma) = \mathfrak{a}\} = \max \gamma_{\mathcal{F}}^{-1}(\{\mathfrak{a}\}),$$

where the maximum is taken with respect to $\prec$. The set

$$\Omega_{\mathcal{F}} := \{\tau_{\mathcal{F}}(\mathfrak{a}) \mid \mathfrak{a} \in \mathfrak{A}\} \subset \Sigma_N$$

is called the *tops code space* and $\tau_{\mathcal{F}} : \mathfrak{A} \to \Omega_{\mathcal{F}}$ the *tops function* for the IFS $(\mathsf{X}; \mathcal{F}_N)$.

Remark 4.51   The tops function assigns a *unique* code to each point $\mathfrak{a} \in \mathfrak{A}$ even if the IFS is overlapping, i.e., each point $\mathfrak{a} \in \mathfrak{A}$ has a *unique* address under $\tau_{\mathcal{F}}$.

Hence, the tops function $\tau_{\mathcal{F}}$ is a bijective mapping from $\mathfrak{A}$ to $\Omega_{\mathcal{F}}$ and provides a right-inverse to the address function $\gamma_{\mathcal{F}}$: $\gamma_{\mathcal{F}} \circ \tau_{\mathcal{F}} = \mathrm{id}_{\mathfrak{A}}$. The inverse function

$$\tau_{\mathcal{F}}^{-1} : \Omega_{\mathcal{F}} \to \mathfrak{A}$$

is clearly bijective and, since it is the restriction of $\gamma_{\mathcal{F}}$ to $\Omega_{\mathcal{F}}$, also continuous. The tops function $\tau_{\mathcal{F}}$, however, is in general not continuous. (The reader is encouraged to verify this statement.)

Example 4.52   Consider the Sierpiński triangle $\mathfrak{S}$ and the IFS that generates it. (See Example 4.29!) Let the points $\mathfrak{a}, \mathfrak{b}, \mathfrak{c} \in \mathfrak{S}$ be given by $\mathfrak{a} := (0, 0)$, $\mathfrak{b} := (\frac{1}{2}, 0)$, and $\mathfrak{c} := (\frac{1}{4}, \frac{\sqrt{3}}{8})$. Then

$$\tau_{\mathcal{F}}(\mathfrak{a}) = \overline{1}, \quad \tau_{\mathcal{F}}(\mathfrak{b}) = \max\{1\overline{2}, 2\overline{1}\} = 1\overline{2}, \quad \tau_{\mathcal{F}}(\mathfrak{c}) = \max\{11\overline{3}, 13\overline{1}\} = 11\overline{3}.$$

Let $\overline{\Omega}_{\mathcal{F}}$ denote the closure of the tops code space $\Omega_{\mathcal{F}}$ (with respect to the Fréchet metric on Σ_N). Then, it is possible to continuously extend the mapping $\tau_{\mathcal{F}}^{-1}$ to $\overline{\Omega}_{\mathcal{F}}$. Thus obtaining the extension

$$\overline{\tau_{\mathcal{F}}^{-1}} : \overline{\Omega}_{\mathcal{F}} \to \mathfrak{A},$$

is continuous and surjective. Note that one can also interpret $\overline{\tau_{\mathcal{F}}^{-1}}$ as the restriction of the continuous mapping $\gamma_{\mathcal{F}}$ to $\overline{\Omega}_{\mathcal{F}} \subseteq \Sigma_N$. Finally, we remark that, since the attractor $\mathfrak{A}$ is compact and thus closed,

$$\operatorname{rg}\overline{\tau_{\mathcal{F}}^{-1}} = \operatorname{rg}\tau_{\mathcal{F}}^{-1} = \mathfrak{A}.$$

The concepts introduced above will be considered again in Chapter 7, where they play an important role in the construction of superfractals.

4.6 Fractal Transformations

The surjective mapping $\gamma_{\mathcal{F}}$ and the tops function $\tau_{\mathcal{F}}$ can be used to define *fractal transformations* between two fractals $\mathfrak{A}$ and $\mathfrak{B}$. The question under what condition(s) fractal transformations are continuous mappings between fractal sets $\mathfrak{A}$ and $\mathfrak{B}$ has a surprisingly simple answer in terms of the underlying code space structure.

To this end, let $(\mathbf{X}, d_{\mathbf{X}})$ and $(\mathbf{Y}, d_{\mathbf{Y}})$ denote two compact metric spaces and let $(\mathbf{X}; \mathcal{F})$ and $(\mathbf{Y}; \mathcal{G})$ represent IFSs on $\mathbf{X}$ and $\mathbf{Y}$, respectively. We assume that both IFSs have the *same* number, N, of contractions. Then, $\Sigma_N = \{1,\ldots,N\}^{\mathbb{N}}$ is the same code space for both $(\mathbf{X}; \mathcal{F})$ and $(\mathbf{Y}; \mathcal{G})$. The attractors of these IFSs are denoted by $\mathfrak{A}_{\mathcal{F}}$ and $\mathfrak{A}_{\mathcal{G}}$, respectively.

Definition 4.53 (Fractal Transformation) Suppose that $(\mathbf{X}; \mathcal{F})$ and $(\mathbf{Y}; \mathcal{G})$ are IFS on compact metric spaces $(\mathbf{X}, d_{\mathbf{X}})$ and $(\mathbf{Y}, d_{\mathbf{Y}})$ of the aforementioned type. The associated *fractal transformation* $\mathfrak{T}_{\mathcal{F}\mathcal{G}} : \mathfrak{A}_{\mathcal{F}} \to \mathfrak{A}_{\mathcal{G}}$ is defined by

$$\mathfrak{T}_{\mathcal{F}\mathcal{G}} := \gamma_{\mathcal{G}} \circ \tau_{\mathcal{F}}.$$

The graph of the fractal transformation $\mathfrak{T}_{\mathcal{F}\mathcal{G}}$ can be described as follows. Consider the IFS

$$(\mathbf{X} \times \mathbf{Y} \times \Sigma_N; \mathcal{K}),$$

where $\mathcal{K} := \{k_i \mid i = 1,\ldots,N\}$ with $k_i(x, y, \sigma) := (f_i(x), g_i(y), c(i,\sigma))$. Here, $f_i \in \mathcal{F}$, $g_i \in \mathcal{G}$, and $c(i,\sigma) := i\sigma$ denotes the concatenation of the symbol i with the code σ. Endowing the set $\mathbf{X} \times \mathbf{Y} \times \Sigma_N$ with the metric

$$d((x, y, \sigma), (x', y', \sigma')) := \max\{d_{\mathbf{X}}(x, x'), d_{\mathbf{Y}}(y, y'), d_F(\sigma, \sigma')\},$$

generates the compact metric space $(\mathbf{X} \times \mathbf{Y} \times \Sigma_N, d)$. The mappings k_i, $i = 1,\ldots,N$, are contractive in the metric d with maximum contractivity constant $\lambda_{\mathcal{K}} := \max\{\lambda_{\mathcal{F}}, \lambda_{\mathcal{G}}, \frac{1}{2}\}$, where $\lambda_{\mathcal{F}}$ and $\lambda_{\mathcal{G}}$ denotes the maximum contractivity constant of the IFS $(\mathbf{X}; \mathcal{F})$ and $(\mathbf{Y}; \mathcal{G})$, respectively. The unique attractor of the IFS $(\mathbf{X} \times \mathbf{Y} \times \Sigma_N; \mathcal{K})$ is the fractal set

$$\mathfrak{A}_{\mathcal{K}} = \{(x, y, \sigma) \mid \gamma_{\mathcal{F}}(\sigma) = x, \ \gamma_{\mathcal{G}}(y) = y, \ \sigma \in \Sigma_N\}. \tag{4.9}$$

Therefore, the graph of the fractal transformation $\mathfrak{T}_{\mathcal{F}\mathcal{G}}$ is identical to the set

$$\{(x, y) \mid (x, y, \sigma) \in \mathfrak{A}_{\mathcal{K}}, \ \tau \prec \sigma \text{ for all } (x, y', \tau) \in \mathfrak{A}_{\mathcal{K}}\}. \tag{4.10}$$

For applications of fractal transformations to images, we refer the interested reader to [9, 10].

Before addressing the question of the continuity of a fractal transformation, a few new concepts and ideas need to be introduced. For this purpose, let $(\mathsf{X}; \mathcal{F})$ be an IFS on a compact metric space and $\Omega_{\mathcal{F}}$ its tops code space.

Definition 4.54 (Address Structure) The *address structure* of the IFS $(\mathsf{X}; \mathcal{F})$ is the collection of sets

$$\mathcal{C}_{\mathcal{F}} := \left\{ \gamma_{\mathcal{F}}^{-1}(\{\mathfrak{a}\}) \cap \overline{\Omega}_{\mathcal{F}} \mid \mathfrak{a} \in \mathfrak{A}_{\mathcal{F}} \right\}.$$

Remark 4.55 The address structure of an IFS is sometimes also called its *code structure.*

Remark 4.56 Notice that a fractal $\mathfrak{A}$ may have several different address structures, as $\mathfrak{A}$ may be the attractor of several different IFS.

Example 4.57 Consider the unit interval $I := [0, 1]$ and the contractions $f_i : I \to I$, $x \mapsto \frac{1}{2}(x + i - 1)$, $i = 1, 2$. The IFS $(I; \mathcal{F} := \{f_1, f_2\})$ has I as its unique attractor. The tops code space $\Omega_{\mathcal{F}}$ satisfies $\overline{\Omega}_{\mathcal{F}} = \Sigma_2 = \{1, 2\}^{\mathbb{N}}$. Every nonempty set in the address structure of this IFS is either a singleton or a set of the form $\{\sigma 1\overline{0}, \sigma 0\overline{1}\}$, where σ is a finite code.

Suppose we are given two IFSs $(\mathsf{X}; \mathcal{F})$ and $(\mathsf{Y}; \mathcal{G})$ of the type considered above. Denote the address structure of the IFS $(\mathsf{Y}; \mathcal{G})$ by $\mathcal{C}_{\mathcal{G}}$. We introduce a binary relation $\preceq$ on address structures as follows.

$$\mathcal{C}_{\mathcal{F}} \preceq \mathcal{C}_{\mathcal{G}} \quad \Longleftrightarrow \quad \forall A \in \mathcal{C}_{\mathcal{F}} \; \exists B \in \mathcal{C}_{\mathcal{G}} : A \subseteq B.$$

We define $\mathcal{C}_{\mathcal{F}} = \mathcal{C}_{\mathcal{G}}$ to mean that $\mathcal{C}_{\mathcal{F}} \preceq \mathcal{C}_{\mathcal{G}}$ and $\mathcal{C}_{\mathcal{G}} \preceq \mathcal{C}_{\mathcal{F}}$. Note that $\mathcal{C}_{\mathcal{F}} = \mathcal{C}_{\mathcal{G}}$ implies that $\Omega_{\mathcal{F}} = \Omega_{\mathcal{G}}$.

Example 4.58 The Sierpiński triangle $\mathfrak{S}$ (left-hand side of Figure 4.6) and the modified Sierpiński triangle $\mathfrak{S}(\frac{3}{4}, \frac{1}{2}, \frac{1}{4})$ (right-hand side of Figure 4.6) are examples of two fractals with the same address structure.

Figure 4.6 Two fractal sets with the same address structure.

The next theorem gives a condition under which the fractal transformation $\mathfrak{T}_{\mathcal{F}\mathcal{G}}$ between two IFSs is continuous.

Theorem 4.59   Assume that $(\mathbf{X}; \mathcal{F})$ and $(\mathbf{Y}; \mathcal{G})$ are two IFSs of the type considered above with attractors $\mathfrak{A}_{\mathcal{F}}$ and $\mathfrak{A}_{\mathcal{G}}$, respectively. Let $\mathcal{C}_{\mathcal{F}}$ and $\mathcal{C}_{\mathcal{G}}$ be their respective address structures. If $\mathcal{C}_{\mathcal{F}} \preceq \mathcal{C}_{\mathcal{G}}$, then the fractal transformation $\mathfrak{T}_{\mathcal{F}\mathcal{G}} : \mathfrak{A}_{\mathcal{F}} \to \mathfrak{A}_{\mathcal{G}}$, $\mathfrak{T}_{\mathcal{F}\mathcal{G}} = \gamma_{\mathcal{G}} \circ \tau_{\mathcal{F}}$, is continuous. If $\mathcal{C}_{\mathcal{F}} = \mathcal{C}_{\mathcal{G}}$, then $\mathfrak{T}_{\mathcal{F}\mathcal{G}}$ is a homeomorphism.

Proof   First, we observe that the closure $\overline{\Omega}_{\mathcal{F}}$ of the tops code space $\Omega_{\mathcal{F}}$ is compact and that the mapping $\tau_{\mathcal{F}}^{-1}$ is continuous. Consider the mapping

$$\mathfrak{T}_{\mathcal{F}\mathcal{G}} \circ \overline{\tau_{\mathcal{F}}^{-1}} = \gamma_{\mathcal{G}} \circ \tau_{\mathcal{F}} \circ \overline{\tau_{\mathcal{F}}^{-1}} : \overline{\Omega}_{\mathcal{F}} \to \mathfrak{A}_{\mathcal{G}}.$$

Given any code $\sigma \in \overline{\Omega}_{\mathcal{F}}$, we have that σ and $(\tau_{\mathcal{F}} \circ \overline{\tau_{\mathcal{F}}^{-1}})(\sigma)$ are both elements of the same set in $\mathcal{C}_{\mathcal{F}}$. As $\mathcal{C}_{\mathcal{F}} \preceq \mathcal{C}_{\mathcal{G}}$, they both also belong to the same set in $\mathcal{C}_{\mathcal{G}}$. Thus,

$$(\gamma_{\mathcal{G}} \circ \tau_{\mathcal{F}} \circ \overline{\tau_{\mathcal{F}}^{-1}})(\sigma) = \gamma_{\mathcal{G}}(\sigma), \quad \forall \sigma \in \overline{\Omega}_{\mathcal{F}}.$$

The mapping $\gamma_{\mathcal{G}} : \Sigma_N \to \mathfrak{A}_{\mathcal{G}}$ is continuous and, therefore, the mapping $\mathfrak{T}_{\mathcal{F}\mathcal{G}} \circ \overline{\tau_{\mathcal{F}}^{-1}} : \overline{\Omega}_{\mathcal{F}} \to \mathfrak{A}_{\mathcal{G}}$ is also continuous. Proposition 1.34 now gives the continuity of $\mathfrak{T}_{\mathcal{F}\mathcal{G}} : \mathfrak{A}_{\mathcal{F}} \to \mathfrak{A}_{\mathcal{G}}$.

Now suppose that $\mathcal{C}_{\mathcal{F}} = \mathcal{C}_{\mathcal{G}}$. Then the mapping $\gamma_{\mathcal{G}} \circ \tau_{\mathcal{F}} : \mathfrak{A}_{\mathcal{F}} \to \mathfrak{A}_{\mathcal{G}}$ is a bijection with inverse $\gamma_{\mathcal{F}} \circ \tau_{\mathcal{G}}$. (The reader is encouraged to verify this statement.) The equality of the two address structures $\mathcal{C}_{\mathcal{F}}$ and $\mathcal{C}_{\mathcal{G}}$ implies in particular that $\mathcal{C}_{\mathcal{G}} \preceq \mathcal{C}_{\mathcal{F}}$, and by the first statement in the theorem we know that $\gamma_{\mathcal{F}} \circ \tau_{\mathcal{G}}$ is continuous. Hence, $\mathfrak{T}_{\mathcal{F}\mathcal{G}}$ is a homeomorphism. $\square$

The above result can be weakened a little bit by requiring that the address structures are homeomorphic instead of equal. For this purpose, we need to following definition.

Definition 4.60 (Homeomorphic Address Structures)   Assume that $(\mathbf{X}; \mathcal{F})$ and $(\mathbf{Y}; \mathcal{G})$ are two IFSs of the type considered above. The address structures $\mathcal{C}_{\mathcal{F}}$ and $\mathcal{C}_{\mathcal{G}}$ are called *homeomorphic*, in symbol $\mathcal{C}_{\mathcal{F}} \cong \mathcal{C}_{\mathcal{G}}$, iff there exists a homeomorphism $\eta : \overline{\Omega}_{\mathcal{F}} \to \overline{\Omega}_{\mathcal{G}}$ that respects the address structures, i.e., η maps each $A \in \mathcal{C}_{\mathcal{F}}$ onto a $B \in \mathcal{C}_{\mathcal{G}}$.

Example 4.61   A very simple example of homeomorphic address structures is given by a totally disconnected IFS $\mathcal{F}$, such as the one generating the Cantor set $\mathfrak{C}$, and an IFS $\mathcal{G}$ whose contractions are a nontrivial permutation of the contractions in $\mathcal{F}$.

A slightly weaker version of Theorem 4.59 is the following.

Theorem 4.62 Suppose that the address structures $\mathcal{C}_{\mathcal{F}}$ and $\mathcal{C}_{\mathcal{G}}$ of the IFSs $(\mathsf{X}; \mathcal{F})$ and $(\mathsf{Y}; \mathcal{G})$ are homeomorphic. Then the attractors $\mathfrak{A}_{\mathcal{F}}$ and $\mathfrak{A}_{\mathcal{G}}$ generated by these two IFSs are also homeomorphic.

Proof A slight modification of the arguments used to prove Theorem 4.59 yields the statement. It is straight-forward to show that the mapping $\gamma_{\mathcal{G}} \circ \eta \circ \tau_{\mathcal{F}} : \mathfrak{A}_{\mathcal{F}} \to \mathfrak{A}_{\mathcal{G}}$ is the desired homeomorphism. (The reader is encouraged to verify this.) $\square$

Example 4.63 Denote by $(\square; \mathcal{F})$, where $\square := I \times I$, the IFS that generates the Sierpiński triangle $\mathfrak{S}$ and let $(I; \mathcal{G})$ be the IFS that has the unit interval $I := [0, 1]$ as its attractor, Suppose that the contractions $g \in \mathcal{G}$ are given by $g_i(x) := \frac{1}{3}(x + i - 1), i = 1, 2, 3$. Since I is not homeomorphic to $\mathfrak{S}$, the address structures $\mathcal{C}_{\mathcal{F}}$ and $\mathcal{C}_{\mathcal{G}}$ cannot be homeomorphic.

Many more aspects and applications of fractal transformations and fractal homeomorphisms can be found in [9] and, in particular, [10]. These two references also provide additional examples for the concepts introduced in this section.

4.7 Fractal Measures

In this section, we introduce the concept of *measure* in a metric space and show that with each IFS one can associate a measure in a natural way, whose support is the attractor or fractal set generated by the IFS. To this end, we need some definitions and new terminology.

4.7.1 Measures

This short subsection provides a brief introduction to measure spaces and measures. Only those concepts and properties of the theory are given that are important and necessary for later developments. The reader interested in more details and a fuller account of measure theory is referred to the literature on this vast and diverse subject.

Definition 4.64 (Algebra and σ-Algebra) Suppose $\emptyset \neq \mathsf{X}$ is a set. A collection of subsets $\mathcal{A} \subseteq \mathscr{P}(\mathsf{X})$ is called an *algebra over* X provided the following conditions hold.

    (A1) $\emptyset \in \mathcal{A}$;

    (A2) $A \in \mathcal{A} \implies \complement A \in \mathcal{A}$;

    (A3) $A, B \in \mathcal{A} \implies A \cup B \in \mathcal{A}$.

$\mathcal{A}$ is called a *σ-algebra over* $\mathbf{X}$, if in addition to items **(A1)** and **(A2)**,

$$(\text{A4}) \quad \forall k \in \mathbb{N} : A_k \in \mathcal{A} \Longrightarrow \bigcup_{k \in \mathbb{N}} A_k \in \mathcal{A}.$$

holds.

Examples 4.65 The collections of sets $\mathcal{A}_1 := \{\emptyset, \mathbf{X}\}$, $\mathcal{A}_2 := \{\emptyset, A, \complement A, \mathbf{X}\}$, and $\mathcal{A}_3 := \mathscr{P}(\mathbf{X}), \emptyset \neq A \neq \mathbf{X}$, are all σ-algebras on the set $\mathbf{X}$. Here, $\mathcal{A}_1$ is the *coarsest* and $\mathcal{A}_3$ the *finest* σ-algebra.

Definition 4.66 (Measurable Space) Suppose that $\mathbf{X}$ is a nonempty set and $\mathcal{A}$ a σ-algebra over $\mathbf{X}$. The pair $(\mathbf{X}, \mathcal{A})$ is called a *measurable space*. The elements of $\mathcal{A}$ are termed *measurable sets*.

Given any collection of subsets $\mathcal{M}$ of a nonempty set $\mathbf{X}$, there exists always a σ-algebra containing $\mathcal{M}$, namely the power set $\mathscr{P}(\mathbf{X})$. The intersection of all σ-algebras that contain $\mathcal{M}$ is the smallest σ-algebra that contains $\mathcal{M}$. (The reader is encouraged to verify this statement.) It is denoted by $\sigma(\mathcal{M})$ and is called the *σ-algebra generated by* $\mathcal{M}$. In symbols,

$$\sigma(\mathcal{M}) = \bigcap_{\mathcal{M} \subseteq \mathcal{A}} \mathcal{A}.$$

Examples 4.67 Assume that $\emptyset \neq \mathbf{X}$ is a set. Then $\sigma(\{\emptyset\}) = \{\emptyset, \mathbf{X}\}$ and $\sigma(\{A\}) = \{\emptyset, A, \complement A, \mathbf{X}\}, \emptyset \neq A \neq \mathbf{X}$.

Definition 4.68 (Borel σ-Algebra) Suppose that $(\mathbf{X}, d)$ is a nonempty metric space and $\mathcal{O} := \{O \subseteq \mathbf{X} \mid O \text{ open in } \mathbf{X}\}$ the collection of all open subsets of $\mathbf{X}$. Then the σ-algebra $\sigma(\mathcal{O})$ generated by $\mathcal{O}$ is called the *Borel σ-algebra over* $\mathbf{X}$. It is denoted by $\mathcal{B}(\mathbf{X})$.

Note that the Borel σ-algebra over the real numbers $\mathbb{R}$ contains, among others, all open and closed intervals.

Definition 4.69 (Measure) Assume that $\emptyset \neq X$ is a set and $\mathcal{A}$ a σ-algebra over $\mathbf{X}$. A nonnegative function $\mu : \mathcal{A} \to [0, +\infty]$ is called a *measure on* $(\mathbf{X}, \mathcal{A})$ provided the following conditions hold.

1. $\mu(\emptyset) = 0$;

2. Let $\{A_k \mid k \in \mathbb{N}\} \subseteq \mathcal{A}$ be an arbitrary collection of pairwise disjoint sets. Then

$$\mu\left(\bigcup_{k \in \mathbb{N}} A_k\right) = \sum_{k \in \mathbb{N}} \mu(A_k).$$

holds. The triple (X, A, μ) is called a *measure space*. If $A = B(X)$, then the measure μ is usually termed a *Borel measure*. A measure is *finite* if it only attains finite values, i.e., if it maps A into $[0, \infty)$. A measure is referred to as a *probability measure* if $\mu(X) = 1$. In this case, the triple (X, A, μ) is called a *probability space*.

Examples 4.70   Let X denote a nonempty set and A a σ-algebra over X. The following are examples of measures.

(1) **Zero measure:** $\mu(A) := 0$, for all $A \in A$.

(2) **Counting measure:** Assume that X is countable. Then $\mu(A) := \mathrm{card}(A)$.

(3) **Dirac measure:** Let $x \in X$. $\delta_x(A) := \begin{cases} 1, & x \in A; \\ 0, & x \notin A. \end{cases}$

(4) **Borel measure on** $(\mathbb{R}, B(\mathbb{R}))$**:** $\mu([a, b]) := b - a$.

(5) **Hausdorff measure on** $(\mathbb{R}^n, B(\mathbb{R}^n))$**:** The quantity $\mathscr{H}^s$ defined in (4.2) is called *s-dimensional Hausdorff measure*.

(6) **Probability measure on** (X, A, μ)**:** Here, the space X is called the *sample space*, its elements *outcomes*, and the elements of the σ-algebra A *events*. The measure μ assigns to each event $A \in A$ a number between 0 ("impossibility") and 1 ("certainty").

Definition 4.71 (Support of a Measure)   Let $(X, A(X), \mu)$ be a measure space. The *support* of μ is defined as the set

$$\mathrm{supp}\,\mu := \{x \in X \mid \text{for all open sets } O_x \in A(X) \text{ containing } x, \mu(O_x) > 0\}.$$

Example 4.72   Let μ be a Borel measure on $(\mathbb{R}, B(\mathbb{R}))$ given by $\mu([a, b]) := b - a$. Then $\mathrm{supp}\,\mu = \mathbb{R}$.

Example 4.73   Let $X := \mathbb{R}$ and let $x \in \mathbb{R}$. Then the support of the Dirac measure δ_x is the singleton $\{x\}$.

Definition 4.74 (μ–null set)   Let $(X, A(X), \mu)$ be a measure space. A set $N \in A(X)$ is called a μ–*null set* or a *set of μ-measure zero* if $\mu(N) = 0$.

Example 4.75   Let μ be the Borel measure on $(\mathbb{R}, B(\mathbb{R}))$ given by $\mu([a, b]) := b - a$. Then $\mu(\mathbb{Q}) = 0$, i.e., $\mathbb{Q}$ is a μ-null set.

In the exercise section at the end of this chapter, the reader is encouraged to prove several properties of measures. These will be used in the sequel. Of particular importance is the next theorem.

Theorem 4.76 Assume that $(X, \mathcal{B}(X), \mu)$ is a Borel measure space. Further assume that $f : X \to X$ is continuous. Then there exists a unique Borel measure ν such that

$$\nu(A) = \mu(f^{-1}(A)), \quad \forall A \in \mathcal{B}(X).$$

The measure ν is usually denoted by $\nu := f^{\sharp}\mu := \mu \circ f^{-1}$, and is called the *image of the measure μ under f*.

Proof Exercise! $\square$

The final new concept that we need to introduce is that of an *invariant Borel measure.*

Definition 4.77 (Invariant Borel Measure) Suppose (X, d) is a metric space and $(X, \mathcal{B}(X), \mu)$ the measure space of Borel measurable sets. Further suppose that $f : X \to X$ is a continuous function. A Borel measure μ is called *invariant under f* provided that

$$\mu(A) = (f^{\sharp}\mu)(A) = \mu \circ f^{-1}(A), \quad \forall A \in \mathcal{B}(X).$$

Example 4.78 Assume $X := \mathbb{R}$ and that $f : \mathbb{R} \to \mathbb{R}$, $x \mapsto x + v$, $v \in \mathbb{R}$, is a translation. Then any Borel measure μ is invariant under translations.

Now suppose that $f : X \to \mathbb{R}$ is a continuous function defined on a metric space (X, d) and that we are given a Borel measure μ on the measurable space $(X, \mathcal{B}(X))$. We can extend the concept of *step function* introduced in Subsection 1.4.8 to the current setting. To this end, we call a function $s : X \to \mathbb{R}$ a *simple function on X* if its range contains only finitely many values in $[0, \infty)$. If we denote these finitely many values by a_k, $k = 1, \ldots, n$, and set $A_k := \{x \in X \mid s(x) = a_k\}$, then s has the form

$$s(x) = \sum_{k=1}^{n} a_k \, \chi_{A_k}(x).$$

Note that every step function is a simple function but not vice versa.

If for all $k = 1, \ldots, n$, the set $A_k \in \mathcal{B}(X)$, then the *integral of s with respect to the Borel measure μ* is defined by

$$\int_X s(x)d\mu(x) := \sum_{k=1}^{n} a_k \, \mu(A_k).$$

The next result provides us with a means to define the integral of a continuous function $f : X \to [0, \infty]$ with respect to the Borel measure μ.

Theorem 4.79 Suppose that $f : X \to [0, \infty]$ is a continuous function. Then there exists a sequence $\{s_n \mid n \in \mathbb{N}\}$ of simple functions on X such that

1. $0 \leq s_1 \leq s_2 \leq \cdots \leq f$.

2. $\forall x \in \mathsf{X}: s_n(x) \to f(x)$ as $n \to \infty$.

Proof Exercise! $\quad\square$

Definition 4.80 (Integral of a Continuous Function with Respect to a Borel Measure) Assume that $f : \mathsf{X} \to [0, \infty]$ is a continuous function. Then

$$\int_{\mathsf{X}} f(x)\, d\mu(x) := \sup\left\{ \int_{\mathsf{X}} s(x) d\mu(x) \,\Big|\, s \text{ is a simple function with } 0 \le s \le f \right\}.$$

For a continuous function $g : \mathsf{X} \to \mathbb{R}$, we denote by $g^+ := \max\{g, 0\}$ the *positive part of* g and by $g^- := -\min\{g, 0\}$ the *negative part of* g. Then $g = g^+ - g^-$ and the integral of g over μ is defined by

$$\int_{\mathsf{X}} g(x)\, d\mu(x) := \int_{\mathsf{X}} g^+(x)\, d\mu(x) - \int_{\mathsf{X}} g^-(x)\, d\mu(x).$$

For more details about integrals with respect to measures, we refer the interested reader to the vast literature on real analysis and measure theory of which [20, 131, 132], and [146] are but a few examples.

4.7.2 IFSs with Probabilities and Fractal Measures

In the definition of IFS, all the contractive mappings f_i, $i = 1, \ldots, N$, $N \in \mathbb{N}$, are used to obtain the fractal set. In this section, we extend the concept of IFS to include probabilities; the contractions f_i are chosen according to some probability distribution. We will shortly see that this allows us to construct fractal measures whose support are the fractal sets generated by IFS with probabilities.

Definition 4.81 (IFS with Probabilities) Let $1 < N \in \mathbb{N}$. An *IFS with probabilities* is an IFS $(\mathsf{X}; \mathcal{F})$ together with a set of probabilities $\mathcal{P} := \{p_i \,|\, i = 1, \ldots, N\}$, i.e., a collection of nonnegative real numbers p_i such that $\sum_{i=1}^{N} p_i = 1$. The probability p_i is associated with the contraction f_i, $i = 1, \ldots, N$. We denote an IFS with probabilities by $(\mathsf{X}; \mathcal{F}, \mathcal{P})$.

Example 4.82 Suppose that $\mathsf{X} := \mathbb{R}^2$ and f_1 and f_2 are two contractions on X. Suppose that a die is rolled before one of the two maps is applied to some initial set $E_0 \subset \mathbb{R}^2$. If a "1", "2", "3", or "4" is rolled, map f_1 is applied; otherwise f_2 is chosen. Hence, we associate the probability $p_1 = \frac{2}{3}$ with contraction f_1 and the probability $p_2 = \frac{1}{3}$ with contraction f_2. Therefore, $\left(\mathbb{R}^2; \{f_1, f_2\}, \{\frac{2}{3}, \frac{1}{3}\}\right)$ is an IFS with probabilities. It is a natural question to ask what is the limit set in this probabilistic setting, when the contractions are chosen according to the given probability distribution. The answer to this question will be provided below.

Next, we will address the question that was posed in the above example. To this end, we denote the set of all normalized Borel measures μ on a measurable set $(X, \mathcal{B}(X))$, i.e., all Borel measures that satisfy $\mu(X) = 1$, by $P(X)$. It is possible to endow the set $P(X)$ with a metric d_P, called the *Monge*[7] *–Kantorovich*[8] *metric.*

Theorem 4.83 Suppose that (X, d) is a compact metric space and $P(X)$ the set of all normalized Borel measures. The function

$$d_P : P(X) \times P(X) \to \mathbb{R},$$

$$d_P(\mu, \nu) := \sup \left\{ \int_X f \, d\mu - \int_X f \, d\nu \,\middle|\, f \in \mathrm{Lip}^1(X) \right\}$$

is a metric on $P(X)$. Moreover, the metric space $(P(X), d_P)$ is compact.

Proof As the proof employs techniques that will not be developed in this textbook, we refer the reader to the literature for a proof. (See, for instance, [9].) $\square$

Theorem 4.76 shows how mappings from X to itself affect measures defined on the measurable space $(X, \mathcal{B}(X))$. The next result is the analogue of Lemma 4.22 for normalized Borel measures.

Lemma 4.84 Let (X, d) be a compact metric space and let $g : X \to X$ be a contraction with contractivity constant $-1 < \lambda < 1$. Then the induced mapping $g^\sharp : P(X) \to P(X)$, $\mu \mapsto \mu \circ g^{-1}$, is contractive in the Monge–Kantorovich metric d_P with the same contractivity constant.

Proof First, notice that if $f \in \mathrm{Lip}^1(X)$, then

$$\left|(f \circ g)(x) - (f \circ g)(x')\right| \leq |\lambda| \, |f(x) - f(x')| \leq |\lambda| \, d(x, x'), \quad \forall x, x' \in X.$$

Thus, for any $\mu, \nu \in P(X)$, and $\lambda \neq 0$,

$$d_P(g^\sharp \mu, g^\sharp \nu) = \sup \left\{ \int_X f \, d(g^\sharp \mu) - \int_X f \, d(g^\sharp \nu) \,\middle|\, f \in \mathrm{Lip}^1(X) \right\}$$

$$= \sup \left\{ \int_X f \circ g \, d\mu - \int_X f \circ g \, d\nu \,\middle|\, f \in \mathrm{Lip}^1(X) \right\}$$

---

7 GASPARD MONGE, 10 May 1746–28 July 1818. French mathematician and physicist who is primarily known in mathematics for his work on descriptive geometry.
8 LEONID VITALIYEVICH KANTOROVICH, 19 January 1912–7 April 1986. Russian mathematician and economist. He worked in functional analysis and developed the theory of linear programming. In 1975, Kantorovich received the Nobel prize in economics.

$$= |\lambda| \sup\left\{ \int_X |\lambda|^{-1} f \circ g \, d\mu - \int_X |\lambda|^{-1} f \circ g \, d\nu \,\Big|\, f \in \mathrm{Lip}^1(X) \right\}$$

$$\leq |\lambda| \sup\left\{ \int_X h \, d\mu - \int_X h \, d\nu \,\Big|\, h \in \mathrm{Lip}^1(X) \right\} = |\lambda| \, d_P(\mu, \nu).$$

The case when $\lambda = 0$ is clear. $\qquad\square$

Example 4.85 Assume that $X := [0, 1]$ and that $g : [0, 1] \to [0, 1]$ is given by $g(x) = \frac{1}{2}x + \frac{1}{3}$. Then g is contractive on $[0, 1]$ and the induced mapping $g^\sharp$ is contractive on $P([0, 1])$. Since $g^\sharp$ is a contraction on a compact metric space, it has a unique fixed point $\mu_0 \in P([0, 1])$. The unique fixed point μ_0 equals the Dirac measure $\delta_{\frac{2}{3}}$. (The reader is encouraged to verify this statement.)

The following is the analogue of Theorem 4.25 for measures.

Theorem 4.86 Suppose that $1 < N \in \mathbb{N}$, that (X, d) is a compact metric space, and $(X; \mathcal{F}, \mathcal{P})$ an IFS with probabilities. The mapping $F^\sharp : P(X) \to P(X)$ defined by

$$F^\sharp(\mu) := \sum_{i=1}^{N} p_i f_i^\sharp(\mu) \tag{4.11}$$

is a contraction on the compact metric space $(P(X), d_P)$ with contractivity constant $\overline{\lambda} := \sum_{i=1}^{N} p_i |\lambda_i|$, where $\lambda_i := \mathrm{Lip}(f_i)$, $i = 1, \ldots, N$.

Proof We show the result for $N := 2$ and leave the general case to the reader. To this end, let $\mu, \nu \in P(X)$. Then

$$d_P(F^\sharp(\mu), F^\sharp(\nu)) = \sup\left\{ \int_X g \, d(F^\sharp \mu) - \int_X g \, d(F^\sharp \nu) \,\Big|\, g \in \mathrm{Lip}^1(X) \right\}$$

$$= \sup\left\{ p_1 \left(\int_X g \, d(f_1^\sharp \mu) - \int_X g \, d(f_1^\sharp \nu) \right) \right.$$

$$\left. - p_2 \left(\int_X g \, d(f_2^\sharp \mu) - \int_X g \, d(f_2^\sharp \nu) \right) \,\Big|\, g \in \mathrm{Lip}^1(X) \right\}$$

$$\leq \sup\left\{ p_1 \left(\int_X g_1 \, d(f_1^\sharp \mu) - \int_X g_1 \, d(f_1^\sharp \nu) \right) \right.$$

$$\left. - p_2 \left(\int_X g_2 \, d(f_2^\sharp \mu) - \int_X g_2 \, d(f_2^\sharp \nu) \right) \,\Big|\, g_1, g_2 \in \mathrm{Lip}^1(X) \right\}$$

$$= \sum_{i=1}^{2} p_i \, d_P(f_i^\sharp \mu, f_i^\sharp \nu) \leq \overline{\lambda} \, d_P(\mu, \nu). \qquad\square$$

Definition 4.87 (Fractal Measure)   The unique fixed point of the mapping $F^\sharp$ defined in Equation (4.11) is called a *fractal measure* or $F^\sharp$-*invariant measure*, and is denoted by m.

Note that the fractal measure $\mathfrak{m} \in P(\mathbf{X})$ satisfies the fixed point order self-referential equation

$$\mathfrak{m} = F^\sharp(\mathfrak{m}) = \sum_{i=1}^{N} p_i f_i^\sharp(\mathfrak{m}), \tag{4.12}$$

or, equivalently,

$$\mathfrak{m}(A) = F^\sharp(\mathfrak{m})(A) = \sum_{i=1}^{N} p_i f_i^\sharp(\mathfrak{m})(A) = \sum_{i=1}^{N} p_i \mathfrak{m}(f_i^{-1}(A)), \quad \forall A \in \mathcal{B}(\mathbf{X}).$$
$$\tag{4.13}$$

The Banach Fixed Point Theorem 1.35 implies again that the fractal measure $\mathfrak{m}$ can be computed by choosing an initial Borel measure μ_0 and constructing iterates μ_k of μ_0 according to the self-referential equation (4.12) or (4.13): $\mu_k = F^\sharp(\mu_{k-1})$ or $\mu_k(A) = F^\sharp(\mu_{k-1})(A) = \mu_{k-1}(F^{-1}(A))$, $A \in \mathcal{B}(\mathbf{X})$, $k \in \mathbb{N}$.

Remark 4.88   Equation (4.12) may be interpreted as the *expected value* or *average* of the measure-valued contractions $f_i^\sharp$ with respect to the probability distribution given by $\mathcal{P}$.

Example 4.89   We consider the compact metric space $(\square, d)$, where d denotes the Euclidean metric on $\mathbb{R}^2$ restricted to $\square$. Let contractions $f_i :$ $\square \to \square$, $i = 1, 2$, be given by

$$f_1 \begin{pmatrix} x \\ y \end{pmatrix} := \begin{pmatrix} \frac{1}{2} & 0 \\ 0 & \frac{1}{2} \end{pmatrix} \begin{pmatrix} x \\ y \end{pmatrix} + \begin{pmatrix} \frac{1}{2} \\ \frac{1}{4} \end{pmatrix},$$

$$f_2 \begin{pmatrix} x \\ y \end{pmatrix} := \frac{3}{4} \begin{pmatrix} \cos\frac{\pi}{4} & \sin\frac{\pi}{4} \\ -\sin\frac{\pi}{4} & \cos\frac{\pi}{4} \end{pmatrix} \begin{pmatrix} x \\ y \end{pmatrix} + \begin{pmatrix} \frac{1}{2} \\ \frac{3}{4} \end{pmatrix}.$$

Let μ_0 be the Borel measure whose support is the rectangle $[0, \frac{1}{2}] \times [0, 1]$. In Figure 4.7, the sequence μ_1, μ_2, μ_3, and μ_5 of Borel measures as well as the (pre)fractal measure m (of rank 20) are displayed.

Remark 4.90   Given an IFS with probabilities $(\mathbf{X}; \mathcal{F}, \mathcal{P})$, it can be shown (cf. [11, 87, 113]) that the support of the fractal measure m is the attractor $\mathfrak{A}$ of the IFS $(\mathbf{X}; \mathcal{F})$, $\operatorname{supp}\mathfrak{m} = \mathfrak{A}$, provided all probabilities are greater than zero.

The discussions in the previous sections have shown that, given an IFS $(\mathbf{X}; \mathcal{F}, \mathcal{P})$ with probabilities, one can associate with it a set-valued mapping

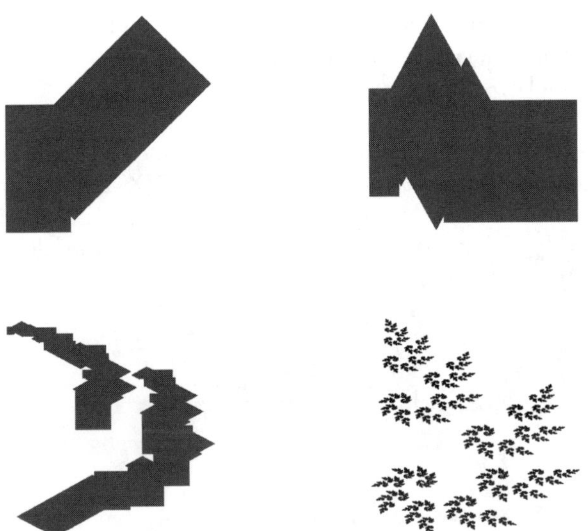

Figure 4.7   A sequence of Borel measures on $(\square, d)$.

$$F : \mathcal{H}(\mathbf{X}) \to \mathcal{H}(\mathbf{X}),$$

$$F(A) = \bigcup_{i=1}^{N} \overline{f}_i(A), \quad A \in \mathcal{H}(\mathbf{X}), \tag{4.14}$$

as well as a measure-valued mapping $F^\sharp : \mathbf{P}(\mathbf{X}) \to \mathbf{P}(\mathbf{X})$

$$F^\sharp(\mu) := \sum_{i=1}^{N} p_i f_i^\sharp(\mu), \quad \mu \in \mathbf{P}(\mathbf{X}). \tag{4.15}$$

Setting $\lambda := \max\{\|\mathrm{Lip}(f_i)\| \mid i = 1, \ldots, N\}$, Theorems 4.25 and 4.86 immediately give the following result.

Theorem 4.91   Let $(\mathbf{X}; \mathcal{F}_N, \mathcal{P})$ be an IFS with nonzero probabilities. The mappings $F : \mathcal{H}(\mathbf{X}) \to \mathcal{H}(\mathbf{X})$ and $F^\sharp : \mathbf{P}(\mathbf{X}) \to \mathbf{P}(\mathbf{X})$ given by (4.14) and (4.15), respectively, are both contractions on their respective metric spaces with the same constant of contractivity λ. Hence, the IFS $(\mathbf{X}; \mathcal{F}_N, \mathcal{P})$ has a unique set attractor $\mathfrak{A} \in \mathcal{H}(\mathbf{X})$ satisfying the self-referential equation $F(\mathfrak{A}) = \mathfrak{A}$ as well as a unique measure attractor $\mathfrak{m} \in \mathbf{P}(\mathbf{X})$ satisfying the self-referential equation $F^\sharp(\mathfrak{m}) = \mathfrak{m}$. Moreover, $\mathrm{supp}\,\mathfrak{m} = \mathfrak{A}$.

Proof   Only the last statement needs to be shown. A proof of it can be found in [11, 87, 113]. $\square$

The next theorem, whose proof can be found in, for instance, [12, 65], is fundamental for a fast algorithmic generation of fractal sets. The result has also been popularized under the name "chaos game."

Theorem 4.92 Suppose that $(\mathbf{X}, d)$ is a compact metric space and $(\mathbf{X}; \mathcal{F}, \mathcal{P})$ an IFS with probabilities. Further assume that $m \in \mathbf{P}(\mathbf{X})$ is the invariant fractal measure. Let $x_0 \in \mathbf{X}$ be arbitrary, and let $x_k := f_i(x_{k-1})$, $k \in \mathbb{N}$, where $f_i \in \mathcal{F}$ is chosen with probability $p_i \in \mathcal{P}$. Then, for almost all random sequences $\{x_k \mid k \in \mathbb{N}\}$ the following equality holds.

$$m(A) = \lim_{k \to \infty} \frac{\mathcal{N}(A \cap \{x_\ell \mid \ell = 0, 1, \ldots, k\})}{k+1}, \tag{4.16}$$

for all $A \in \mathcal{B}(\mathcal{H}(\mathbf{X}))$ with $m(\partial A) = 0$. Here, $\mathcal{N}(A)$ denotes the number of points in set A.

Remark 4.93 The expression "for almost all random sequences" in Theorem 4.92 refers a natural measure ν on the code space (Σ_N, d_F) (N: number of maps in $\mathcal{F}$) associated with the IFS $(\mathbf{X}; \mathcal{F})$. The application of the random algorithm described in the theorem above corresponds to the generation of codes $\sigma = \sigma_1 \cdots \sigma_i \cdots \in \Sigma_N$. Equation (4.16) then holds for all codes in Σ_N except for those lying in a ν-null set. In this context, the reader is encouraged to work out Exercise 45.

Remark 4.94 Let $x_0 \in \mathbf{X}$ be arbitrary and let $\{x_k \mid k \in \mathbb{N}\}$ be the sequence of iterates of x_0 as defined in Theorem 4.92. If one considers the sequence $\left\{ \Delta_k := \frac{\delta_{x_0} + \delta_{x_1} + \cdots + \delta_{x_x}}{k+1} \,\middle|\, k \in \mathbb{N} \right\}$ of Dirac measures associated with $\{x_k \mid k \in \mathbb{N}_0\}$, then it can be shown that the sequence $\{\Delta_k \mid k \in \mathbb{N}_0\}$ converges to the fractal measure m, independent of the choice of x_0. (A reference for this result is, for instance, [53].)

The interpretation of Theorem 4.92 is now immediately at hand: Choosing an arbitrary starting point $x_0 \in \mathbf{X}$, and selecting iterates x_k of x_0 according to $x_k := f_i(x_{k-1})$ where f_i is randomly chosen with probability p_i, the sequence of iterates converges to the fractal attractor $\mathfrak{A} = \text{supp}\, m$. Note that instead of applying *all* the mappings in the IFS at each iteration, it suffices to choose *one* mapping at random (according to the probability distribution given by $\mathcal{P}$) at each iteration step. For the algorithmic generation of fractal sets this results in low memory allocations and requirements, faster computation times, and higher accuracy. We refer the interested reader to [8, 9] for numerous applications of the chaos game to computer graphics, image and video generation.

Exercises

1. Show that the classical (ternary) Cantor set $\mathfrak{C}$ has length zero and the Sierpiński triangle has area zero.

2. Use the fact that the sets $\mathbb{N}$, $\mathbb{Z}$, and $\mathbb{Q}$ are countable, to show that they have topological dimension 0.

3. Verify the statements made in Remark 4.11.

4. Show that the Hausdorff–Besicovitch dimension of the classical Cantor set $\mathfrak{C}$ equals $\log 2/\log 3$.

5. Prove Proposition 4.21.

6. This exercise provides an alternate definition of Hausdorff metric. For this purpose, we need the concept of ε-parallel body of a nonempty compact set.

   Definition 4.95 (ε-parallel Body)   Let $(\mathbf{X}, d)$ be a complete metric space and let $K \in \mathcal{H}(\mathbf{X})$. For an $\varepsilon > 0$, denote by

   $$K_\varepsilon := \{x \in \mathbf{X} \mid d(x, k) \leq \varepsilon \text{ for some } k \in K\}$$

   the ε-parallel body of K.

   Show that the Hausdorff metric h can also be defined by

   $$d_{\mathcal{H}}(A, B) := \inf\{\varepsilon > 0 \mid A \subseteq B_\varepsilon \text{ and } B \subseteq A_\varepsilon\},$$

   where $A, B \in \mathcal{H}(\mathbf{X})$.

7. Complete the proof of Lemma 4.23.

8. Use Lemma 4.23 to prove that

   $$d_{\mathcal{H}}(A \cup B, C \cup D) \leq \max\{d_{\mathcal{H}}(A, C), d_{\mathcal{H}}(B, D)\},$$

   for all $A, B, C, D \in \mathcal{H}(\mathbf{X})$.

9. Complete the proof of Lemma 4.24.

10. Collect natural objects such as leaves, plants, etc., and determine the IFSs that provide fractal models of these natural objects.

11. Give the proof of Proposition 4.36.

12. Consider the geometric Figure 4.8 (left) where $\{A, B, C\} \cap \{A', B', C'\} = \emptyset$. Define the ratios $r := \frac{\overline{AB'}}{\overline{AC}}$, $s := \frac{\overline{BC'}}{\overline{BA}}$, and $t := \frac{\overline{CA'}}{\overline{CB}}$. Define an IFS $(\triangle(A, B, C); \{f_1, f_2, f_3\})$ by requiring that $(A, B, C) \overset{f_1}{\mapsto} (C', A, B')$, $(A, B, C) \overset{f_2}{\mapsto} (C', A', B)$, and $(A, B, C) \overset{f_3}{\mapsto} (C, A', B')$. Choosing $A := (0, 0)$, $B := (1, 0)$, and $C := (0.5, 1)$, derive formulae for the maps f_1, f_2, and f_3. The attractor $\mathfrak{S}(r, s, t)$ is a three-parameter family of

Sierpiński triangles. A representative of this family is depicted on the right.

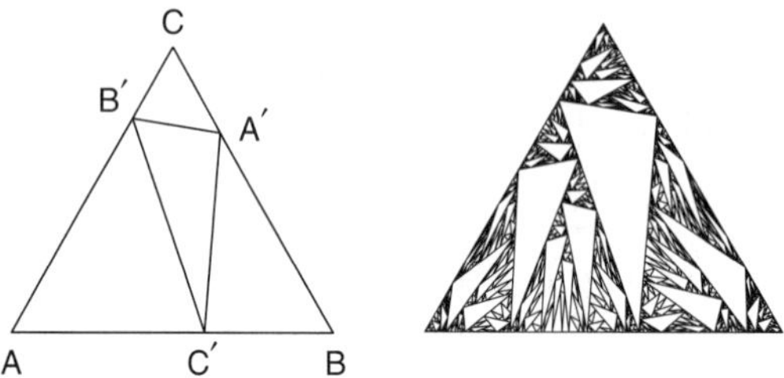

Figure 4.8 The construction of a generalized Sierpiński triangle.

13. Verify the statement made in Remark 4.39.

14. Show that the classical Cantor set $\mathfrak{C}$ and the Sierpiński triangle $\mathfrak{S}$ satisfy the OSC.

15. Prove that (Σ_N, d_F) is a compact metric space and thus complete.

16. Verify that the classical Cantor set $\mathfrak{C}$ is totally disconnected.

17. Show that the relation $\prec$ on Σ_N is irreflexive and transitive.

18. Prove that the tops function $\tau : \mathfrak{A} \to \Omega$ is bijective and thus a right-inverse to the address function $\gamma : \Sigma_N \to \mathfrak{A}$. Also, show that the inverse of the tops function $\tau^{-1} : \Omega \to \mathfrak{A}$ is continuous.

19. Find an example showing that the tops function τ may not be continuous.

20. Let $(\mathbf{X}; \mathcal{F})$ be an IFS with attractor $\mathfrak{A}$ and let $\overline{f}_i \in \mathcal{F}$, $i \in \{1, \ldots, N\}$.

    (a) Show that the following diagram commutes:

$$
\begin{array}{ccc}
\mathfrak{A} & \xrightarrow{\ \overline{f}_i\ } & \mathfrak{A} \\[4pt]
\gamma_{\mathcal{F}} \Big\uparrow & & \Big\uparrow \gamma_{\mathcal{F}} \\[4pt]
\Sigma_N & \xrightarrow{\ s_i\ } & \Sigma_N
\end{array}
$$

    where $s_i : \Sigma_N \to \Sigma_N, \sigma \mapsto i\sigma$, is the inverse shift map on Σ_N.

    (b) Show that for all $i \in \{1, \ldots, N\}$ and for all $\mathfrak{a} \in \mathfrak{A}$ the identity
    $f_i(\mathfrak{a}) = \gamma_{\mathcal{F}}(s_i(\tau_{\mathcal{F}}(\mathfrak{a})))$ holds.

21. What can be said about the sets $\gamma^{-1}(\mathfrak{a}) \cap \gamma^{-1}(\mathfrak{b})$, where $\mathfrak{a}, \mathfrak{b} \in \mathfrak{A}$ are distinct points on the attractor $\mathfrak{A}$? Is $\{\gamma^{-1}(\mathfrak{a}) \mid \mathfrak{a} \in \mathfrak{A}\}$ a partition of Σ_N?

22. Find an example of an IFS for which $\tau_{\mathcal{F}}$ fails to be continuous.

23. In 1890, Guiseppe Peano gave an example of a so-called *space-filling curve*, i.e., a continuous surjective mapping $f : [0, 1] \to [0, 1] \times [0, 1]$. Use iterated function systems to construct such space-filling curves. (*Hint: For some examples of such constructions, you may want to consult [109] or [113].*)

24. Find an IFS that generates the so-called *Koch* [9] *curve* $\mathfrak{K}$ shown in Figure 4.9.

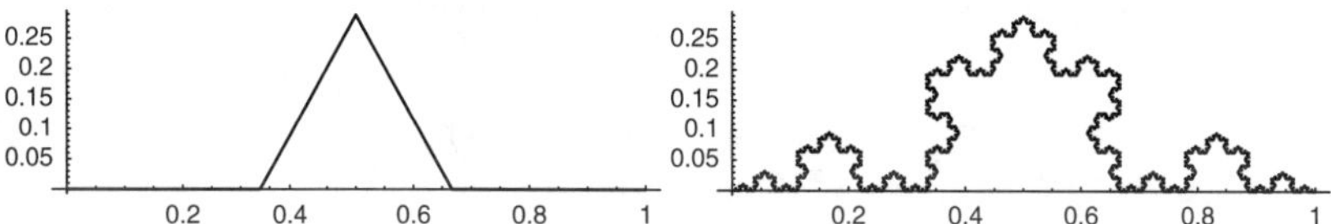

Figure 4.9 The generator of the Koch curve (left) and a prefractal of $\mathfrak{K}$ (right).

(*Hint: Map the unit interval onto the generator triangle!*) The vertices of the generator triangle are $(\frac{1}{3}, 0)$, $(\frac{1}{2}, \frac{\sqrt{3}}{6})$, and $(\frac{2}{3}, 0)$. What is the box dimension of $\mathfrak{K}$?

25. Construct the Koch snowflake by applying the procedure from Exercise 24 to each of the three sides of an equilateral triangle.

26. Create a multitude of Koch islands by taking any closed polygon Π and applying the construction from Exercise 24 to each side of Π. Note that the generator triangle does not have to be equilateral!

27. Establish that the set $\mathfrak{A}_{\mathcal{K}}$ given in Equation (4.9) is indeed the attractor of the IFS $(\mathbf{X} \times \mathbf{Y} \times \Sigma_N; \mathcal{K})$ by showing that $\mathfrak{A}_{\mathcal{K}}$ is nonempty, compact, and satisfies

$$\mathfrak{A}_{\mathcal{K}} = \bigcup_{i=1}^{N} k_i(\mathfrak{A}_{\mathcal{K}}).$$

28. Verify the claim that the set given by Equation (4.10) is equal to graph $\mathfrak{T}_{\mathcal{F}\mathcal{G}}$.

29. Verify the statements made in Example 4.57.

30. Work out the details in Example 4.58. (See also Exercise 12 in this section!)

31. Prove Theorem 4.62.

---

9 Helge von Koch, 25 January 1870–11 March 1924. Swedish mathematician who is primarily known for the fractals that are named after him: the Koch curve, the Koch snow-flake, and the Koch island.

32. Show that every σ-algebra is an algebra, but that not every algebra is a σ-algebra. (*Hint: For the second statement, let* $X := \mathbb{N}$ *and* $\mathcal{A} := \{a \subseteq \mathbb{N} \mid A \text{ or } \complement A \text{ is finite.}\}$)

33. Suppose that $\{\mathcal{A}_i \mid i \in I\}$, I arbitrary index set, is a family of σ-algebras over a set X. Then $\bigcap_{i \in I} \mathcal{A}_i$ is a σ-algebra over X.

34. Show that every finite algebra is a σ-algebra.

35. Suppose that $\{\mu_k \mid k \in \mathbb{N}\}$ is a family of measures on a measurable space $(X, \mathcal{A})$ and $\{c_k \mid k \in \mathbb{N}\}$ a sequence of nonnegative real numbers. Show that

$$\mu(A) := \sum_{k \in \mathbb{N}} c_i \, \mu_k(A)$$

defines a measure on $(X, \mathcal{A})$.

36. Consider the measurable space $(\mathbb{R}, \mathcal{B}(\mathbb{R}))$. Suppose that $w : \mathbb{R} \to \mathbb{R}_0^+$ is a continuous function. Show that

$$\mu(A) := \int_A w(x)\, dx, \quad A \in \mathcal{B}(\mathbb{R}),$$

defines a Borel measure on $(\mathbb{R}, \mathcal{B}(\mathbb{R}))$.

37. Let $\mathcal{B}(X)$ be the Borel σ-algebra on a metric space X. Denote by F_σ the union of countably many closed sets and by G_δ the countable intersection of open sets in X.[10] Interpreting the subscripts σ and δ as operations on closed and open sets, respectively, one defines in analogous fashion the sets $F_{\sigma\delta}$ as the countable intersection of F_σ sets and $G_{\delta\sigma}$ as the countable union of G_δ sets. Show that F_σ, G_δ, $F_{\sigma\delta}$, and $G_{\delta\sigma}$ sets are Borel sets. Note that this justifies the definition of the Borel measure in Example 4.72.

38. Prove the claims made in Example 4.72 and Example 4.73.

39. Assume that $(X, \mathcal{A}, \mu)$ is a measure space and $A, B \in \mathcal{A}$. Show that

    (a) $\mu(A \cup B) \leq \mu(A) + \mu(B)$;
    (b) $A \subseteq B \Longrightarrow \mu(A) \leq \mu(B)$;
    (c) $A \subseteq B$ and μ finite $\Longrightarrow \mu(B \setminus A) = \mu(B) - \mu(A)$.

40. **Continuity of a finite measure.** Let $(X, \mathcal{A}, \mu)$ be a measure space and μ a finite measure. Suppose that $\{A_k \mid k \in \mathbb{N}\} \subseteq \mathcal{A}$ is a sequence of measurable sets. Prove that

---

10 In the expression F_σ, F stands for the French word for closed, *fermé*, and the subscript σ for *sum* or *union*, and the G in G_δ represents the German word *Gebiet*, an open and connected set, and the δ the German word *Durchschnitt*, i.e., intersection.

(a) if $\forall k \in \mathbb{N} : A_k \subseteq A_{k+1}$, then $\mu\left(\bigcup_{k \in \mathbb{N}} A_k\right) = \lim_{k \to \infty} \mu(A_k)$;

(b) if $\forall k \in \mathbb{N} : A_k \supseteq A_{k+1}$, then $\mu\left(\bigcap_{k \in \mathbb{N}} A_k\right) = \lim_{k \to \infty} \mu(A_k)$.

41. Prove Theorem 4.76.

42. Establish the metric properties of the Monge–Kantorovich metric d_{P}.

43. Establish Theorem 4.79. (*Hint: For $n \in \mathbb{N}$ and $k = 1, \ldots, n2^n$, let*

$$A_{nk} := f^{-1}\left(\left[\frac{k-1}{2^n}, \frac{k}{2^n}\right]\right) \quad \text{and} \quad B_n := f^{-1}([n, \infty]).$$

Then, set

$$s_n := \sum_{k=1}^{n2^n} \frac{k-1}{2^n} \chi_{A_{nk}} + n\chi_{B_n}.)$$

44. Verify that $x_0 = \frac{2}{3}$ is the fixed point of the mapping g given in Example 4.85 and that the Dirac measure $\delta_{\frac{2}{3}}$ is the unique fixed point of the induced mapping $g^\sharp$.

45. Let $(\mathbf{X}; \mathcal{F}, \mathcal{P})$ be an IFS with probabilities, and let (Σ_N, d_F) be the code space on N symbols associated with this IFS. Denote by $Z(\sigma_1, \ldots, \sigma_k)$ the *cylinder set* $\{\omega \in \Sigma_N \mid \sigma_j = \omega_j, \ \forall j = 1, \ldots, k\}$. Denote by $\mathcal{B}(\Sigma_N)$ the Borel σ-algebra generated by the cylinder sets.

   (a) Show that v defined by

   $$v(Z(\sigma_1, \ldots, \sigma_k)) := \prod_{j=1}^{k} p_{\sigma_j}, \quad p_{\sigma_j} \in \mathcal{P},$$

   for each $k \in \mathbb{N}$ and all $\sigma_j \in \{1, \ldots, N\}$, defines a unique Borel measure on $(\Sigma_N, \mathcal{B}(\Sigma_N))$.

   (b) Let $s_i : \Sigma_N \to \Sigma_N$ be defined by

   $$s_i(\sigma) := i\sigma, \quad i = 1, \ldots, N.$$

   Show that $(\Sigma_N; \mathcal{S})$ where $\mathcal{S} := \{s_i \mid i = 1, \ldots, N\}$ is an IFS for any set of probabilities $\mathcal{P}$.

   (c) Exhibit the relationship between the $\mathcal{S}^\sharp$-invariant measure ρ of the IFS $(\Sigma_N; \mathcal{S})$ and the $\mathcal{F}^\sharp$-measure m of the IFS $(\mathbf{X}; \mathcal{F}, \mathcal{P})$. (*Hint: Use Theorem 4.44!*) $\square$

5

Fractal Functions

The history of fractal functions goes back to Weierstrass's nowhere differentiable function and beyond. However, it was not until the publication of B. Mandelbrot's book (cf. [105]) where the concept of fractal sets was introduced and common characteristics of such sets, like non-integral dimension and geometric self-similarity, identified that the theory of fractal functions developed into an area of its own. Seemingly different types of nowhere differentiable functions, such as those investigated by Besicovitch, Ursell, Knopp[1], and Kiesswetter, to mention only a few, were unified under the fractal point of view, and this unification led to new mathematical methods and applications in areas that include dimension theory, dynamical systems and chaotic dynamics, image analysis, and wavelet theory.

The type of fractal function considered in this chapter was first systematically considered in [7] and [87]. The construction of fractal functions presented here will follows these lines. In addition to the consideration of the geometric structure of fractal functions, we also introduce the so-called Read–Bajraktarević (RB) operator acting on properly chosen function spaces. The fixed points of these RB operators are then fractal functions naturally associated with the function space onto which the operator acts.

After presenting the more geometric definition of fractal functions originally found in [7], we give their construction via an RB operator. Polynomial-type fractal functions are considered next as are their bases. As in

---

1 KONRAD KNOPP, 22 July 1882–20 April 1957. German mathematician who is noted for his contributions to generalized limits and complex functions.

the case of fractal sets, there exists a code space naturally associated with fractal functions. It is also shown that every continuous function f gives rise to an entire family of fractal functions $f[\lambda]$ parameterized by a certain vector quantity λ where $f[0] = f$. In addition, we also construct fractal functions of class C^k and define fractal-type B-splines.

The approximation of classical functions by classes of fractal functions is presented and some error estimates are derived. Operations on fractal functions, such as indefinite integrals and the Fourier transform, are considered next. The connection with wavelets is also exhibited. Finally, we give some conditions on fractal functions to belong to certain Besov and Triebel–Lizorkin spaces.

As references for this chapter serve [8] and [112], and the original articles [7, 15, 21, 22, 49, 72, 81, 84, 113, 114, 115, 116, 123, 124], and [125].

5.1 Some Examples of Nowhere Differentiable Functions

In 1872, Karl Weierstrass investigated the continuous function $W : \mathbb{R} \to \mathbb{R}$, given by

$$W(x) := \sum_{i=1}^{\infty} \lambda^{s(i-1)} \sin \lambda^i x,$$

where $\lambda > 1$ and $1 < s < 2$, and proved that it is nowhere differentiable. It is conjectured that the Hausdorff–Besicovitch dimension of graph W equals s, but so far it is only known that $\dim_H W \le s$.

Another example of a continuous nowhere differentiable function was introduced in 1903 by T. Takagi [149]. The Takagi function $\tau : [0, 1] \to \mathbb{R}$ is defined by

$$\tau(x) := \sum_{i=0}^{\infty} \frac{\operatorname{dist}(2^n x, \mathbb{Z})}{2^n}.$$

It is known that $\dim_H \operatorname{graph} \tau = \dim_B \operatorname{graph} \tau = 1$.

Besicovitch and Ursell [25] defined in 1937 a class of continuous nowhere differentiable functions $f_\phi : [0, 1] \to \mathbb{R}$ of the form

$$f_\phi(x) := \phi(x) + \sum_{i=1}^{\infty} b_i^{\alpha-2} \, \phi(b_i x),$$

where the function $\phi : \mathbb{R} \to \mathbb{R}$ is given by

$$\phi(x) := \begin{cases} 2x, & 0 \le x \le \tfrac{1}{2}; \\ \phi(-x) = \phi(x+1), & \text{otherwise.} \end{cases}$$

Here, $1 < \alpha < 2$ and $b_i > 0$, for all $i \in \mathbb{N}$. Under certain conditions on the parameters α and b_i, the exact Hausdorff–Besicovitch dimension of graph f_ϕ can be computed.

Proposition 5.1 Let f_ϕ be a Besicovitch–Ursell function. If

$$\lim_{i \to \infty} \frac{b_i}{b_{i+1}} = 0 \quad \text{and} \quad \lim_{i \to \infty} \frac{\log b_{i+1}}{\log b_i} = 1,$$

then $\dim_H \operatorname{graph} f_\phi = \alpha$.

5.2 Fractal Interpolation Functions

In 1986, M. Barnsley [7] introduced a class of fractal functions whose graphs are generated by IFSs consisting of affine mappings. Since these fractal functions were constructed to interpolate a set of data points, he termed them *fractal interpolation functions*. The construction of these functions is presented next.

To this end, suppose that $[a, b] \subset \mathbb{R}$ is a nonempty interval and $1 < N \in \mathbb{N}$. Let $\mathbf{X} := [a, b] \times \mathbb{R}$ and let $Y := \{(x_\nu, y_\nu) \in \mathbf{X} \mid a = x_0 < x_1 < \cdots < x_{N-1} < x_N = b\}$. Moreover, let $\lambda_i \in (-1, 1)$, $i = 1, \ldots, N$, and let $A \in \mathcal{H}(\mathbf{X})$ be given by $A := [a, b] \times [a, b]$. For $i = 1, \ldots, N$, denote by A_i the unique parallelogram with vertices (x_{i-1}, y_{i-1}), (x_i, y_i), $(x_i, y_i + \lambda_i(b - a))$, and $(x_{i-1}, y_{i-1} + \lambda_i(b - a))$. Then there exists a unique affine mapping $\overline{f}_i : \mathcal{H}(\mathbf{X}) \to \mathcal{H}(\mathbf{X})$ so that $\overline{f}_i(A) = A_i$, $i = 1, \ldots, N$. The set-valued map $\overline{f}_i$ is given via the numerical function

$$f_i \begin{pmatrix} x \\ y \end{pmatrix} = \begin{pmatrix} a_i & 0 \\ c_i & \lambda_i \end{pmatrix} \begin{pmatrix} x \\ y \end{pmatrix} + \begin{pmatrix} \alpha_i \\ \beta_i \end{pmatrix}, \quad i = 1, \ldots, N, \tag{5.1}$$

where

$$a_i := \frac{x_i - x_{i-1}}{b - a}, \qquad c_i := \frac{y_i - y_{i-1} - \lambda_i(y_N - y_0)}{b - a}, \tag{5.2a}$$

$$\alpha_i := \frac{bx_{i-1} - ax_i}{b - a}, \qquad \beta_i := \frac{by_{i-1} - ay_i - \lambda_i(by_0 - ay_N)}{b - a}. \tag{5.2b}$$

Since, $\operatorname{Lip}(f_i) = |\lambda_i| < 1$, the IFS $(\mathbf{X}; \mathcal{F}_N)$ with $\mathcal{F}_N := \{f_i \mid i = 1, \ldots, N\}$ possesses a unique fixed point $\mathfrak{G} \in \mathcal{H}(\mathbf{X})$.

Barnsley [7] showed that $\mathfrak{G}$ is the graph of a continuous function $\mathfrak{f} : [a, b] \to \mathbb{R}$ interpolating the set Y: $\mathfrak{f}(x_\nu) = y_\nu$, $\nu = 0, 1 \ldots, N$. The graph of $\mathfrak{f}$ is, in general, a fractal set and Barnsley called these functions *fractal interpolation functions*. Below, we will give a slightly different definition that is based on a certain operator.

Since $\mathfrak{G} = \text{graph}\,\mathfrak{f}$ and satisfies the fixed point equation (4.4), one has

$$\text{graph}\,\mathfrak{f} = \bigcup_{i=1}^{N} \overline{f}_i(\text{graph}\,\mathfrak{f})$$

and, therefore,

$$\left.\begin{pmatrix} \bar{x} \\ \mathfrak{f}(\bar{x}) \end{pmatrix}\right|_{\bar{x}\in[x_{i-1},x_i]} = \left.\begin{pmatrix} a_i & 0 \\ c_i & \lambda_i \end{pmatrix}\begin{pmatrix} x \\ \mathfrak{f}(x) \end{pmatrix}\right|_{x\in[a,b]} + \begin{pmatrix} \alpha_i \\ \beta_i \end{pmatrix}.$$

Now, define functions

$$u_i : [a, b] \to [x_{i-1}, x_i],$$

$$x \mapsto a_i x + \alpha_i,$$

and

$$p_i : [a, b] \to \mathbb{R},$$

$$x \mapsto c_i x + \beta_i,$$

and note that

$$u_i(a) = x_{i-1} \quad \text{and} \quad u_i(b) = x_i, \quad i = 1, \ldots, N.$$

Then, to each $\bar{x} \in [x_{i-1}, x_i]$ there exists exactly one $x \in [a, b]$ so that $\bar{x} = u_i(x)$. Hence,

$$\mathfrak{f}(\bar{x}) = c_i u_i^{-1}(\bar{x}) + \lambda_i \mathfrak{f}(u_i^{-1}(\bar{x})) + \beta_i = (p_i \circ u_i^{-1})(\bar{x}) + \lambda_i(\mathfrak{f} \circ u_i^{-1})(\bar{x}),$$

$\forall \bar{x} \in [x_{i-1}, x_i]$, or, equivalently, since $\mathfrak{f}(x_i{-}) = \mathfrak{f}(x_i{+})$,

$$\mathfrak{f}(x) = \sum_{i=1}^{N} \left[(p_i \circ u_i^{-1})(x) + \lambda_i(\mathfrak{f} \circ u_i^{-1})(x) \right] \chi_{[x_{i-1},x_i]}, \quad \forall x \in [a, b]. \quad (5.3)$$

Let

$$p(x) := \sum_{i=1}^{N} (p_i \circ u_i^{-1})(x)\chi_{[x_{i-1},x_i]}.$$

Then, since $p_i \circ u_i^{-1} : [x_{i-1}, x_i] \to [a, b]$, one has that $p|_{[x_{i-1},x_i]} = p_i \circ u_i^{-1} \in \Pi^2[x_{i-1}, x_i]$ and the question arises under what conditions is p continuous on $[a, b]$, i.e., $p \in S^2(X_N)$, where $X_N := \{x_0, x_1, \ldots, x_N\}$. To this end, observe that for p to be continuous on $[a, b]$ the following conditions have to be satisfied:

$$p_i(b) = p_{i+1}(a), \quad \forall i = 1, \ldots, N - 1. \quad (5.4)$$

Thus, since $p_i(b) = y_i - \lambda_i y_N$ and $p_{i+1}(a) = y_i - \lambda_{i+1} y_0$, one obtains four cases.

1. If $y_0 := y_N := 0$, then p is continuous on $[a, b]$, i.e., $p \in S^2(X)$.

2. If $y_0 y_N \neq 0$, then (5.4) is satisfied for $\lambda_i y_N = \lambda_{i+1} y_0$, i.e., when $\lambda_{i+1} = \lambda_1 \cdot \left(\dfrac{y_N}{y_0}\right)^i$, $i = 1, \ldots, N - 1$. Thus, there is only one free parameter $\lambda_1 \in (-1, 1)$.

3. If $y_0 = 0$ but $y_N \neq 0$, then $\lambda_i = 0$, for all $i = 1, \ldots, N$. In this case, $\mathfrak{f} = p$ on $[a, b]$.

4. If $y_N = 0$ and $y_0 \neq 0$, then $\lambda_1 \neq 0$, but all other $\lambda_i = 0$, $i = 2, \ldots, N - 1$.

In general, none of these cases occurs for general interpolation sets and p is not globally, i.e., on $[a, b]$, continuous.

Considering again the fixed point equation, namely,

$$\mathfrak{f}(x) = p(x) + \sum_{i=1}^{N} \lambda_i (\mathfrak{f} \circ u_i^{-1})(x) \chi_{[x_{i-1}, x_i]}, \quad \forall\, x \in [a, b],$$

one can interpret the fractal interpolation function $\mathfrak{f}$ as constructed by adding to the piecewise polynomial p finitely many vertically compressed affine copies of $\mathfrak{f}$.

Example 5.2 Let $X := [0, 1] \times \mathbb{R}$, $N := 3$, and $Y := \{(0, 0), (0.5, 0.7), (0.7, -0.1), (1.0, 0.3)\}$. Moreover, choose as constants $(\lambda_1, \lambda_2, \lambda_3) := (0.6, -0.5, 0.75)$. The construction of the fractal interpolation function $\mathfrak{f} : [0, 1] \to \mathbb{R}$ is graphically depicted in Figure 5.1.

Remark 5.3 The fixed point $\mathfrak{f}$ depends on the two N-tuples $\boldsymbol{\lambda} := (\lambda_1, \ldots, \lambda_N) \in (-1, 1)^N$ and $\boldsymbol{p} := (p_1, \ldots, p_N) \in \underset{i=1}{\overset{N}{\mathsf{X}}} \Pi[a, b]$ and, when necessary, this dependence is sometimes expressed in the form $\mathfrak{f} = \mathfrak{f}[\boldsymbol{\lambda}, \boldsymbol{p}]$. For a fixed tuple $\boldsymbol{p}$, one may think of $\mathfrak{f}[\boldsymbol{\lambda}]$ as a family of linear splines parametrized by the vector $\boldsymbol{\lambda}$.

The right-hand side of the fixed point Equation (5.3) can be interpreted as an operator acting on continuous functions defined on $[a, b]$. To this end, define an operator $T : C[a, b] \to C[a, b]$ by

$$(Tg)(x) := p(x) + \sum_{i=1}^{N} \lambda_i (g \circ u_i^{-1})(x) \chi_{[x_{i-1}, x_i]}. \tag{5.5}$$

Theorem 5.4 The operator T is a contraction on the Banach space $(C[a, b], \|\cdot\|_{\infty, [a, b]})$ with Lipschitz constant $\lambda := \max\{|\lambda_i| \,|\, i = 1, \ldots, N\}$.

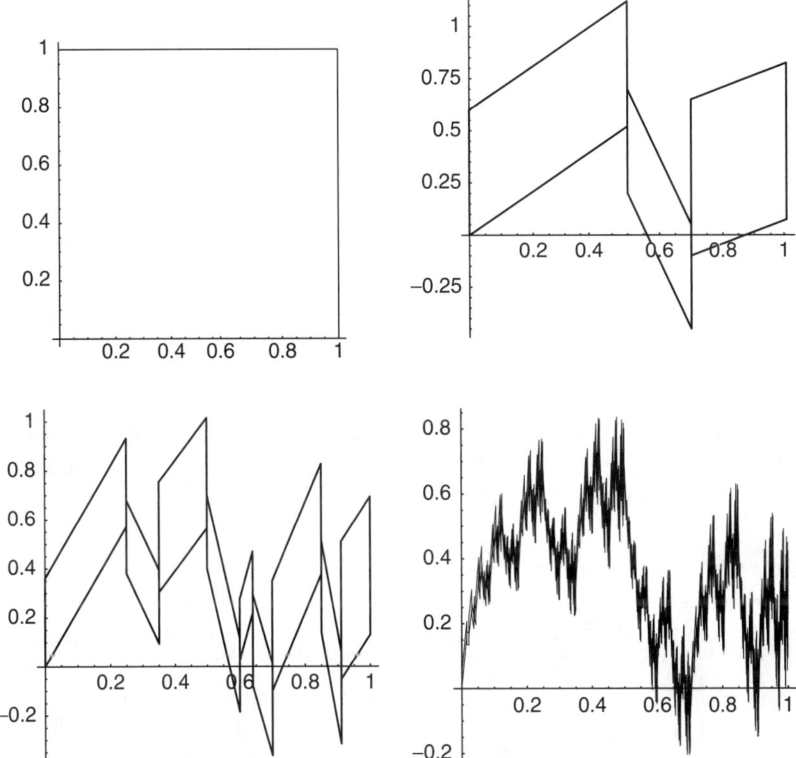

Figure 5.1   The construction of a fractal interpolation function.

Proof   Let $x \in [a, b]$ and let $g, h \in C[a, b]$. Then, there exists a $j \in \{1, \ldots, N\}$ so that $x \in [x_{j-1}, x_j]$. Thus,

$$
\begin{aligned}
|(Tg)(x) - (Th)(x)| &= \left| \sum_{i=1}^{N} \lambda_i (g \circ u_i^{-1})(x) \chi_{[x_{i-1}, x_i]}(x) \right. \\
&\quad \left. - \sum_{i=1}^{N} \lambda_i (h \circ u_i^{-1})(x) \chi_{[x_{i-1}, x_i]}(x) \right| \\
&\leq |\lambda_j| \left| (g \circ u_j^{-1})(x) - (h \circ u_j^{-1})(x) \right| \leq \lambda \, \|g - h\|_{\infty}.
\end{aligned}
$$

Taking the supremum over all $x \in [a, b]$ gives the statement. $\square$

By the completeness of $(C[a, b], \| \ \|_{\infty, [a,b]})$, there exists a unique fixed point $\mathfrak{g}$ of T that, by the Banach Fixed Point Theorem 1.35, can be obtained via the sequence $\{g_n \mid n \in \mathbb{N}\} \subset C[a, b]$ given by

$$
g_n := T g_{n-1},
$$

where $g_0 \in C[a, b]$ satisfies $g_0(x_\nu) = y_\nu$, $\nu = 0, 1, \ldots, N$, but is otherwise arbitrary. Note that since $u_i^{-1}(x_i) = b$, $u_i^{-1}(x_{i-1}) = a$, and $p(x_i) = y_i - \lambda_i y_N$,

$$g_1(x_i) = p(x_i) + \lambda_i (g_0 \circ u_i^{-1})(x_i) = y_i - \lambda_i y_N + \lambda_i g_0(b) = y_i,$$

$i = 1, \ldots, N$. Also, $g_1(a) = y_0$. Thus, g_1, and therefore all g_n, $n \in \mathbb{N}$, interpolate the set Y. Hence, the limit

$$\mathfrak{g} = \lim_{n \to \infty} g_n$$

also interpolates Y. By the uniqueness of the fixed point, we have $\mathfrak{g} \equiv \mathfrak{f}$. This gives rise to the following definition of a fractal function, which uses the operator T as the primary object.

Definition 5.5 (Affine Fractal Interpolation Function)   Let $g_0 \in C[a, b]$ with $g(x_\nu) = y_\nu$, $\nu = 0, 1, \ldots, N$, where $a := x_0$ and $b := x_N$. The unique function $\mathfrak{g} := \lim_{n \to \infty} T^n g_0$ is called an *affine fractal interpolation function* over $[a, b]$.

Remark 5.6   The operator T, as defined in (5.5) is an example of a *Read–Bajraktarević (RB) operator*, and plays a fundamental role in the theory of fractal functions and fractal surfaces.

Example 5.7   Let $\mathsf{X} := [0, 1] \times \mathbb{R}$, let $N := 2$, and let $Y := \left\{(0, 0), (\frac{1}{2}, \frac{1}{2}), (1, 0)\right\}$. Then, $u_1 : [0, 1] \to [0, 1]$, $x \mapsto \frac{1}{2}x$, and $u_2 : [0, 1] \to [0, 1]$, $x \mapsto \frac{1}{2}x + \frac{1}{2}$. Moreover, $p_1 : [0, 1] \to \mathbb{R}$ and $p_2 : [0, 1] \to \mathbb{R}$ are given by $p_1(x) = \frac{1}{2}x$ and $p_2(x) = \frac{1}{2}(x + 1)$, respectively. Choose $\lambda_1 := \lambda_2 := \frac{1}{4}$. Then the RB operator T is given explicitly by

$$(Tf)(x) = \begin{cases} x + \frac{1}{4}f(2x), & x \in [0, \frac{1}{2}]; \\ 1 - x + \frac{1}{4}f(2x - 1), & x \in [\frac{1}{2}, 1]. \end{cases} \tag{5.6}$$

Also note that $p(x) := p_1(x)\chi_{[0, \frac{1}{2}]}(x) + p_2(x)\chi_{[\frac{1}{2}, 1]}(x) \in S^2(\{0, \frac{1}{2}, 1\})$. The function $\mathfrak{f}(x) = 2x(1 - x)$ satisfies the fixed point equation for T and thus, by uniqueness, is the fixed point of T. (The reader is encouraged to verify this statement.)

Example 5.8   Let $\mathsf{X} := [0, 1] \times \mathbb{R}$, let $N := 2$, and let $Y := \left\{(0, 0), (\frac{1}{2}, \frac{1}{2}), (1, 0)\right\}$. Choose $\lambda_1 := \lambda_2 := \frac{1}{2}$. Then T is given by

$$(Tf)(x) = \begin{cases} x + \frac{1}{2}f(2x), & x \in [0, \frac{1}{2}]; \\ 1 - x + \frac{1}{2}f(2x - 1), & x \in [\frac{1}{2}, 1]. \end{cases} \tag{5.7}$$

The Takagi function τ satisfies the fixed point equation for T. (The reader is encouraged to verify this statement.)

Example 5.9 (Kiesswetter's Fractal Function [98])   Let $\mathsf{X} := [0, 1] \times \mathbb{R}$, let $N := 4$, and let $Y := \left\{(0,0), (\frac{1}{4}, -\frac{1}{2}), (\frac{1}{2}, 0), (\frac{3}{4}, \frac{1}{2}), (1, 1)\right\}$. Define numerical functions $f_i : [0, 1] \times [-1, 1] \to \mathbb{R}$, $i = 1, \ldots, 4$, by

$$f_1(x, y) := \begin{pmatrix} \frac{1}{4} & 0 \\ 0 & -\frac{1}{2} \end{pmatrix} \begin{pmatrix} x \\ y \end{pmatrix},$$

$$f_j(x, y) := \begin{pmatrix} \frac{1}{4} & 0 \\ 0 & \frac{1}{2} \end{pmatrix} \begin{pmatrix} x \\ y \end{pmatrix} + \begin{pmatrix} \frac{j-1}{4} \\ \frac{j-3}{4} \end{pmatrix}, \quad j = 2, 3, 4.$$

Then,

$$(Tf)(x) = \begin{cases} -\frac{1}{2} f(4x), & x \in [0, \frac{1}{4}]; \\ -\frac{1}{2} + \frac{1}{2} f(4x - 1), & x \in [\frac{1}{4}, \frac{1}{2}]; \\ \frac{1}{2} f(4x - 2), & x \in [\frac{1}{2}, \frac{3}{4}]; \\ \frac{1}{2} + \frac{1}{2} f(4x - 3), & x \in [\frac{3}{4}, 1]. \end{cases}$$

The unique fixed point $\mathfrak{k}$ is called Kiesswetter's fractal function. (See Figure 5.2.)

Example 5.10 (Casino Functions)   Let $\mathsf{X} := [0, 1] \times \mathbb{R}$, let $N \in \mathbb{N}$, and let $Y := \{(x_\nu, y_\nu) \mid 0 =: x_0 < x_1 < \cdots < x_{N-1} < x_N := 1; \ 0 =: y_0 < y_1 < \cdots < y_{N-1} < y_N := 1\}$. Define an IFS by

$$f_i(x, y) := \begin{pmatrix} x_i - x_{i-1} & 0 \\ 0 & y_i - y_{i-1} \end{pmatrix} \begin{pmatrix} x \\ y \end{pmatrix} + \begin{pmatrix} x_{i-1} \\ y_{i-1} \end{pmatrix}, \quad i = 1, \ldots, N.$$

The associated RB operator T is contractive and its unique fixed point is called a *Casino function* $\mathfrak{c} : [0, 1] \to [0, 1]$. These functions are monotone

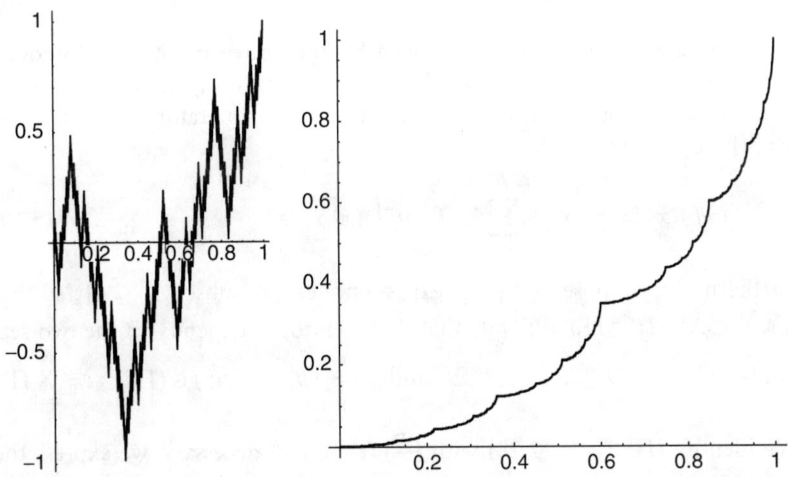

Figure 5.2   Kiesswetter's fractal function (left) and a Casino function (right).

increasing and therefore $\dim_H \operatorname{graph} \mathfrak{c} = \dim_B \operatorname{graph} \mathfrak{c} = 1$. Casino functions play an important role in the theory of games [50]. For a graph of a Casino function, see Figure 5.2.

5.3 Polynomial Fractal Functions

In this section, we generalize the approach that produced affine fractal interpolation functions by using the RB operator as the primary means of defining fractal functions. Initially, we construct fractal functions modeled on polynomials, but later we take any continuous function defined on a compact interval. In particular, we show that every continuous function is a special case of an associated class of fractal functions.

To this end, let $[a, b] \subset \mathbb{R}$ be a nonempty interval and let $\mathbf{X} := [a, b] \times \mathbb{R}$. Suppose that $(\mathbf{F}[a, b], d)$ is a complete metric space of functions defined on $[a, b]$. Moreover, let $1 < N \in \mathbb{N}$ and let $u_i : [a, b] \to [a, b]$, $i = 1, \ldots, N$, be a fixed collection of nonzero contractive homeomorphisms inducing a partition of $[a, b]$ in the following sense:

$$\forall i, j = 1, \ldots, N; \; i \neq j : u_i(a, b) \cap u_j(a, b) = \emptyset, \tag{5.8a}$$

$$\bigcup_{i=1}^{N} u_i[a, b] = [a, b]. \tag{5.8b}$$

Note that (5.8) implies that $u_i(b) = u_{i+1}(a)$, $i = 1, \ldots, N - 1$. Define $x_{i-1} := u_i(a)$ and $x_i := u_i(b)$, $i = 1, \ldots, N$. In addition, let

$$I_i := u_i[a, b), \quad i = 1, \ldots, N - 1, \quad \text{and} \quad I_N := u_N[a, b]. \tag{5.9}$$

Now assume that for some $k \in \mathbb{N}$ and all $i = 1, \ldots, N$, $p_i^k \in \Pi^k := \Pi^k[a, b]$, and set $p^k := \sum_{i=1}^{N} (p_i^k \circ u_i^{-1}) \chi_{I_i}$. Note that $p^k \in \Pi_{\xi}^k$, where $\xi := (x_0 < x_1 < \cdots < x_N)$ is the ordered knot sequence induced by the above partition. Moreover, assume that $\lambda := (\lambda_1, \ldots, \lambda_N) \in \mathbb{R}^N$ is an arbitrary N-tuple. (We sometimes refer to λ also as the *scaling vector*.) Define an operator $T : \mathbf{F}[a, b] \to \operatorname{Map}([a, b], \mathbb{R})$ by

$$(Tf)(x) := p^k(x) + \sum_{i=1}^{N} \lambda_i (f \circ u_i^{-1})(x) \chi_{I_i}(x), \quad x \in [a, b]. \tag{5.10}$$

Furthermore, suppose that there exists a nonempty subset $\Lambda^k \subseteq \Pi_{\xi}^k$ so that for $p^k \in \Lambda^k$, $Tf \in \mathbf{F}[a, b]$. Note that the operator T depends on the two sets of N-tuples $\lambda = (\lambda_1, \ldots, \lambda_N) \in \mathbb{R}^N$ and $\boldsymbol{p}^k := (p_1^k, \ldots, p_N^k) \in (\Pi^k)^N := \underset{i=1}{\overset{N}{\mathsf{X}}} \Pi^k$.

We identify $(\Pi^k)^N$ via (2.28) with $\bigoplus_{i=1}^{N} \Pi^k$. When necessary, we express the dependence of T upon λ and $\boldsymbol{p}^k$ by $T[\lambda, \boldsymbol{p}^k]$.

Definition 5.11 (Polynomial Fractal Function of Class **F**)   If there exists a $\lambda \in \mathbb{R}^N$ and a subset $\emptyset \neq \Lambda^k \subseteq \bigoplus_{i=1}^{N} \Pi^k$ so that the operator T defined in (5.10) is a contraction on the complete metric space of functions $(\mathbf{F}[a, b], d)$, then its unique fixed point $\mathfrak{f} = \mathfrak{f}[\lambda, \boldsymbol{p}^k] \in \mathbf{F}[a, b]$ is called a *polynomial fractal function of class* **F**.

Example 5.12   Let $\mathbf{X} := [0, 1] \times \mathbb{R}$ and let $\mathbf{F} := L^p$, with $p \in [1, \infty)$. Choose $1 < N \in \mathbb{N}$ and define $u_i : [0, 1] \to [0, 1]$ by

$$u_i(x) := \frac{x + i - 1}{N}, \quad i = 1, \dots, N.$$

Let $\Lambda^k := \bigoplus_{i=1}^{N} \Pi^k([0, 1])$, choose $p_i^k \in \Pi^k([0, 1])$, and let $\lambda_i \in \mathbb{R}$, $i = 1, \dots, N$. With these particular choices for λ and $\boldsymbol{p}^k$ it is easy to see that the operator

$$T : \mathbb{R}^N \times \Lambda^k \times L^p[0, 1] \to \mathrm{Map}([0, 1], \mathbb{R})$$

given by

$$(Tf)(x) := (T[\lambda, \boldsymbol{p}^k]f)(x) := \sum_{i=1}^{N} \left[p_i^k(Nx - i + 1) + \lambda_i f(Nx - i + 1) \right] \chi_{I_i}(x)$$

$$= p^k(x) + \sum_{i=1}^{N} \lambda_i f(Nx - i + 1) \chi_{I_i}(x)$$

has range contained in $L^p[0, 1]$.

Now, let $f, g \in L^p[0, 1]$ and consider

$$\| Tf - Tg \|_{L^p}^p = \int_{[0,1]} |(Tf)(x) - (Tg)(x)|^p \, dx$$

$$= \int_{[0,1]} \left| \sum_{i=1}^{N} (\lambda_i f(Nx - i + 1) - \lambda_i g(Nx - i + 1)) \chi_{I_i}(x) \right|^p dx$$

$$\leq \sum_{i=1}^{N} \int_{[\frac{i-1}{N}, \frac{i}{N}]} |\lambda_i|^p |f(Nx - i + 1) - g(Nx - i + 1)|^p \, dx$$

$$= \sum_{i=1}^{N} \frac{|\lambda_i|^p}{N} \int_{[0,1]} |f(x) - g(x)|^p \, dx = \left(\sum_{i=1}^{N} \frac{|\lambda_i|^p}{N} \right) \| f - g \|_{L^p}^p,$$

where in the second to last equation the change of variables $x \mapsto Nx - i + 1$ was used.

If the λ_i, $i = 1, \ldots, N$, are now chosen so that

$$\sum_{i=1}^{N} \frac{|\lambda_i|^p}{N} < 1, \tag{5.11}$$

then T is a contraction on $L^p[0, 1]$. Hence, its unique fixed point $\mathfrak{f}$ is an element of $L^p[0, 1]$ and referred to as a fractal function of class L^p, $p \geq 1$.

Remark 5.13 Note that for a given set $\mathcal{U}_N := \{u_1, \ldots, u_N\}$ of contractive homeomorphisms $u_i : [a, b] \to [a, b]$, $i = 1, \ldots, N$, of the form (5.8), $([a, b]; \mathcal{U}_N)$ is an IFS with attractor $\mathfrak{U} = [a, b]$. By virtue of Theorem 4.44, every point $x \in [a, b] \setminus \{x_i \,|\, i = 1, \ldots, N - 1\}$ has a unique representation as an N-adic expansion. (To get uniqueness at $x = a$ and $x = b$, we take the right N-ary expansion at the former and the left N-ary expansion at the latter.)

For given $\alpha, \beta \in \mathbb{R}$ and a nonempty interval $[a, b] \subset \mathbb{R}$, we introduce the set of functions

$$C_{\alpha,\beta}[a, b] := \{f \in C[a, b] \,|\, f(a) = \alpha \text{ and } f(b) = \beta\}. \tag{5.12}$$

We also define $C_\alpha[a, b] := C_{\alpha,\alpha}[a, b]$. Endowed with the metric

$$d_{\infty,[a,b]} : C[a, b] \times C[a, b] \to \mathbb{R},$$

$$d_{\infty,[a,b]}(f, g) := \|f - g\|_{\infty,[a,b]}, \tag{5.13}$$

where $\| \, \|_{\infty,[a,b]}$ is the Chebyshev norm, $(C_{\alpha,\beta}[a, b], d_{\infty,[a,b]})$ becomes a complete metric space. Let $k \in \mathbb{N}$ and let

$$\Lambda_{\alpha,\beta}^k := \Big\{ p \in \Pi_\xi^k \,\Big|\, p|_{I_i}(x_i) + \lambda_i \beta = p|_{I_{i+1}}(x_i) + \lambda_{i+1} \alpha, \; \forall i = 1, \ldots, N - 1,$$

$$p(a) = (1 - \lambda_1)\alpha, \; p(b) = (1 - \lambda_N)\beta \Big\}, \tag{5.14}$$

where ξ represents the ordered knot sequence $(a = x_0 < x_1 < \cdots < x_{N-1} < x_N = b)$ and $p \in \Pi_\xi^k$ was written in the form $p = \sum_{i=1}^{N} p_i \chi_{I_i}$. Note that $\dim \Lambda_{\alpha,\beta}^k = (k - 1)N - 1$. (The reader is encouraged to verify this statement.)

Now, define an operator $T[\boldsymbol{\lambda}, \boldsymbol{p}^k] : (-1, 1)^N \times \Lambda_{\alpha,\beta}^k \times C_{\alpha,\beta}[a, b] \to \mathrm{Map}([a, b], \mathbb{R})$ by

$$T[\boldsymbol{\lambda}, \boldsymbol{p}^k]f := p^k + \sum_{i=1}^{N} \lambda_i \,(f \circ u_i^{-1})\, \chi_{I_i}. \tag{5.15}$$

Theorem 5.14 Suppose that $\lambda := \max\{|\lambda_i| \,|\, i = 1, \ldots, N\} < 1$ and $\Lambda_{\alpha,\beta}^k$ as in (5.14). Then the operator $T := T[\boldsymbol{\lambda}, \boldsymbol{p}^k]$ defined in (5.15) is a contraction on the complete metric space $(C_{\alpha,\beta}[a, b], \| \, \|_{\infty,[a,b]})$ and its unique fixed point $\mathfrak{f} = \mathfrak{f}[\boldsymbol{\lambda}, \boldsymbol{p}^k]$ is a fractal function of class $C_{\alpha,\beta}$.

Proof We need to show that $Tf \in C_{\alpha,\beta}[a, b]$ for $f \in C_{\alpha,\beta}[a, b]$ and that T is contractive provided $\lambda := \max\{|\lambda_i| \mid i = 1, \ldots, N\} < 1$. To establish the former, note that the join-up conditions for continuity on $[a, b]$ are

$$(Tf)(x_i-) = (Tf)(x_i+), \quad \forall i = 1, \ldots, N-1,$$

which reduce to

$$p_i^k(b) + \lambda_i \beta = p_{i+1}^k(a) + \lambda_{i+1}\alpha, \quad i = 1, \ldots, N-1. \tag{5.16}$$

Furthermore, we need that

$$\alpha = (Tf)(a) = p_1^k(a) + \lambda_1\alpha \quad \text{and} \quad \beta = (Tf)(b) = p_N^k(b) + \lambda_N\beta.$$

Since $f \in C_{\alpha,\beta}[a, b]$ and $p^k \in \Lambda_{\alpha,\beta}^k$, this shows that $Tf \in C_{\alpha,\beta}[a, b]$.

The proof that T is a contraction on $(C_{\alpha,\beta}[a, b], \|\ \|_{\infty,[a,b]})$ for $\lambda < 1$ is straight-forward and left to the reader. $\square$

Note that the fractal functions constructed above do possess free parameters since $\dim \Lambda_{\alpha,\beta}^k > 0$. If we impose *interpolation conditions* at the interior knots $x_1, \ldots, x_{N-1}$, then this number of free parameters is reduced to $\dim \Lambda_{\alpha,\beta}^k - (N-1) = (k-1)N - 1 - (N-1) = (k-2)N$, which in the case that $k := 2$ leaves no free parameters. We thus obtain the setting for affine fractal interpolation functions.

Example 5.15 Let $N := 2$ and let $[a, b] := [0, 1]$. Assume that $u_1 := \frac{1}{2}(\bullet)$ and $u_2 := \frac{1}{2}(\bullet + 1)$. Choose $\lambda_1 := -0.7$ and $\lambda_2 := 0.5$. Moreover, let $\alpha := 1$ and $\beta := 2$. Let $k := 2$ and select $a, b, c, d \in \mathbb{R}$ in $p_1^2 := ax + b$ and $p_2^2 := cx + d$ so that $p_1^2, p_2^2 \in \Lambda_{1,2}^2$. This yields $a = d + (\lambda_1 + \lambda_2 - 1)\alpha - \lambda_1\beta$, $b = (1 - \lambda_1)\alpha$ and $c = -d + (1 - \lambda_2)\beta$. The value of d is arbitrary. (Since $\dim \Lambda_{\alpha,\beta}^2 = 1$ in the current setting, this was to be expected.) Hence, there exists an entire class of fractal functions $\mathfrak{f} = \mathfrak{f}[d] \in C_{1,2}$, for given $p_1^2, p_2^2 \in \Lambda_{1,2}^2$. The graph of the fractal function $\mathfrak{f}[0]$ of class $C_{1,2}$ is displayed in Figure 5.3. It is easy to verify that $\mathfrak{f}(\frac{1}{2}) = \frac{1}{2} + d$. Imposing an interpolation condition at $x = \frac{1}{2} = u_1(1) = u_2(0)$, determines the value of d and $\mathfrak{f}$ becomes a fractal interpolation function.

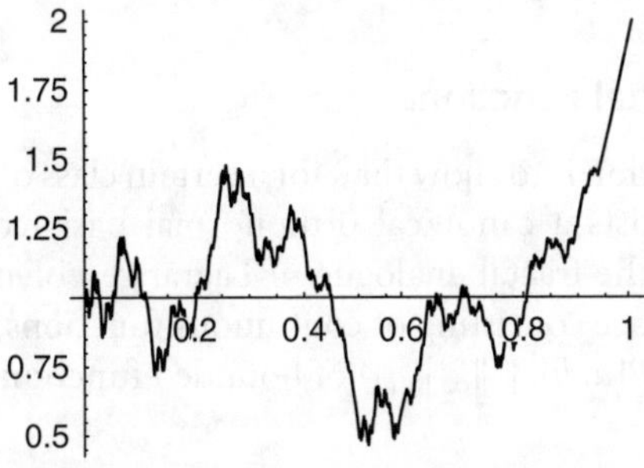

Figure 5.3 A fractal function of class $C_{1,2}$.

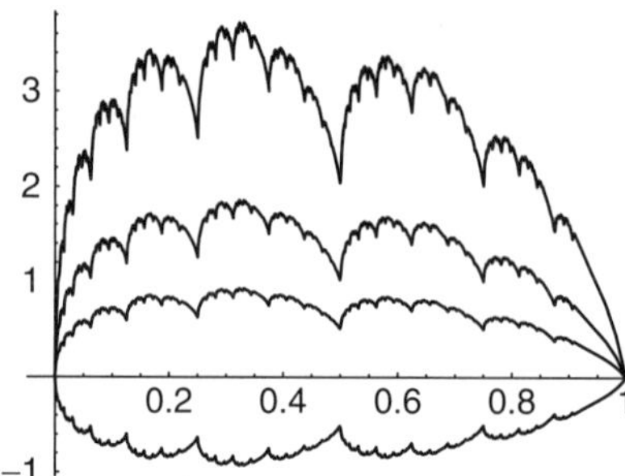

Figure 5.4   Fractal functions of class C_0.

An important special case in the current setting is the following. Suppose that $\alpha = \beta := 0$ and consider the linear space $C_0[a, b]$. The join-up conditions (5.16) now simplify to

$$p_i^k(b) = p_{i+1}^k(a), \quad i = 1, \ldots, N-1.$$

In other words, the function $p^k : [a, b] \to \mathbb{R}$, $p^k := \sum_{i=1}^{N}(p_i^k \circ u_i^{-1})\, \chi_{I_i}$, needs to be an element of $C_0[a, b]$. (It is an easy consequence of the fixed point equation that $p_1(a) = p_N(b) = 0$. The reader is encouraged to verify this statement.) The spaces $\Lambda_{0,0}^k$ can then be taken to be the spline spaces $S_0^k(X_N) := \{s \in S^k(X_N) \mid s(a) = s(b) = 0\}$, where $X_N := \{a = x_0 < x_1 < \ldots < x_{N-1} < x_N = b\}$. We illustrate this particular case by hand of an example and refer the reader to Section 5.9, where this setting will be explored further.

Example 5.16   We again set $N := 2$ and $[a, b] := [0, 1]$. Moreover, let $u_1 := \frac{1}{2}(\bullet)$ and $u_2 := \frac{1}{2}(\bullet + 1)$, and choose $\lambda_1 := 0.75$ and $\lambda_2 := 0.5$. Selecting p^2 from $S_0^2(X_N)$, with $X_N = \{0, \frac{1}{2}, 1\}$, we obtain $p_1^2 = d(\bullet)$ and $p_2^2 = d(1 - \bullet)$, where $d \in \mathbb{R}$ is a free parameter. In Figure 5.4, the graphs of the fractal functions $\mathfrak{f}[d]$ of class C_0 for $d := -0.5, 0.5, 1, 2$ (bottom to top) are shown.

5.4 Bases of Fractal Functions

Our goal in this section is to show that, for a certain class of polynomial fractal functions, there exists a canonical orthonormal basis consisting of fractal functions that are the fractal analogue of Lagrange polynomials. Instead of using a function space consisting of continuous functions, we will work with the Banach space $(B[a, b], \| \ \|_{\infty,[a,b]})$ of bounded functions over a nonempty interval $[a, b]$:

$$B[a, b] := \{f \in \mathrm{Map}([a, b], \mathbb{R}) \mid \exists\, M > 0 \text{ so that } \|f\|_{\infty,[a,b]} \leq M\}.$$

Clearly, $C[a, b] \subsetneq B[a, b]$. Throughout this section we make the following standing assumptions:

Assumptions 5.17

1. $1 < N \in \mathbb{N}$ and $[a, b] \subset \mathbb{R}$, $a < b$.

2. A set of homeomorphisms $\{u_i \mid i = 1, \ldots, N\}$ satisfying (5.8) is given.

3. $x_{i-1} := u_i(a)$, $x_i := u_i(b)$, and I_i is defined as in (5.9) $i = 1, \ldots, N$.

4. $\Lambda^k := \bigoplus_{i=1}^{N} \Pi^k = \underset{i=1}{\overset{N}{\times}} \Pi^k.$

5. A vector $\lambda = (\lambda_1, \ldots, \lambda_N) \in \mathbb{R}^N$ is given.

6. $\mathsf{F}[a, b] := B[a, b]$ endowed with the Chebyshev norm $\| \ \|_{\infty, [a,b]}$.

Based on previously established arguments, the next result is straightforward to prove.

Theorem 5.18 Suppose that $\lambda := \max\{|\lambda_i| \mid i = 1, \ldots, N\} < 1$. Define an operator $T := T[p^k] : \Lambda^k \times B[a, b] \to B[a, b]$ by

$$Tf := \sum_{i=1}^{N} \left[(p_i^k \circ u_i^{-1}) + \lambda_i (f \circ u_i^{-1}) \right] \chi_{I_i} = p^k + \sum_{i=1}^{N} \lambda_i (f \circ u_i^{-1}) \chi_{I_i}.$$

Then T is a contraction on the Banach space $(B[a, b], \| \ \|_{\infty, [a,b]})$ and thus possesses a unique fixed point $\mathfrak{f} \in B[a, b]$. Moreover,

$$\mathfrak{f}(a) = \frac{p_1^k(a)}{1 - \lambda_1} \quad \text{and} \quad \mathfrak{f}(b) = \frac{p_N^k(b)}{1 - \lambda_N}. \tag{5.17}$$

If, in addition, T satisfies the join-up conditions

$$p_{i+1}^k(b) + \lambda_{i+1} \mathfrak{f}(b) = p_i^k(a) + \lambda_i \mathfrak{f}(a), \quad \forall i = 1, \ldots, N - 1, \tag{5.18}$$

then the fixed point $\mathfrak{f}$ is an element of $C[a, b]$.

Proof Exercise! $\square$

Example 5.19 Suppose that $N := 2$ and that $[a, b] := [0, 1]$. Moreover, choose $k := 2$ and set $p_1^2(x) := ax + b$ and $p_2^2(x) := cx + d$, for real constants $a, b, c,$ and d. Imposing the join-up conditions (5.18) on this set of polynomials, we obtain the single requirement that

$$p_2(0) + \frac{\lambda_2}{1 - \lambda_1} p_1(0) = p_1(1) + \frac{\lambda_1}{1 - \lambda_2} p_2(1).$$

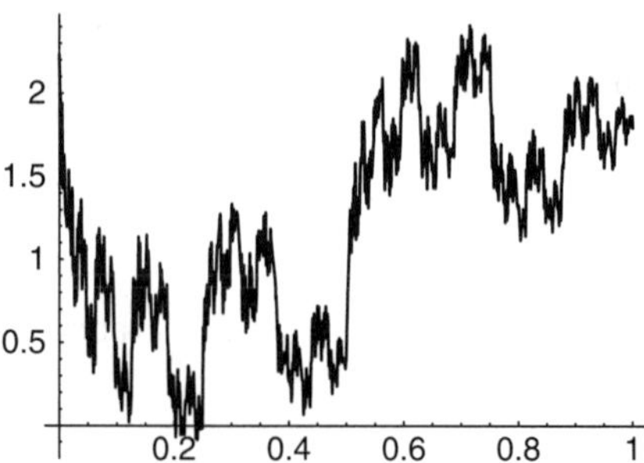

Figure 5.5 A continuous fractal function on $[0, 1]$.

(Here, we used Equations (5.17).) Substituting the expressions for p_1^2 and p_2^2 into the above equation, we obtain that $b, c,$, and d are arbitrary but that

$$a = \left(\frac{\lambda_1 + \lambda_2 - 1}{1 - \lambda_1}\right) b - \left(\frac{\lambda_1}{1 - \lambda_2}\right) c + \left(\frac{1 - \lambda_1 - \lambda_2}{1 - \lambda_2}\right) d.$$

Hence, the space of linear continuous fractal functions on $[0, 1]$ is three-dimensional. The graph of a linear continuous fractal function is depicted in Figure 5.5 for $\lambda_1 := 0.8$ and $\lambda_2 := -0.65$.

Note that the sequence of iterates $\{f_n := Tf_{n-1} \mid n \in \mathbb{N}\}$, with $f_0 \in B[0, 1]$, is not a set of continuous functions, but does converge to a continuous limit, namely the continuous linear fractal function $\mathfrak{f}$.

Our next goal is to establish a correspondence between the linear space $\Lambda^k = \bigoplus_{i=1}^{N} \Pi^k = \underset{i=1}{\overset{N}{\times}} \Pi^k$ and the class of fractal functions $\mathfrak{f} \in B[a, b]$ generated by its elements.

Theorem 5.20 The mapping $\boldsymbol{p}^k \overset{\mathfrak{F}}{\mapsto} \mathfrak{f}[\boldsymbol{p}^k]$, is a linear isomorphism from Λ^k onto $\mathfrak{F}(\Lambda^k) \subseteq B[a, b]$.

Proof Let $\alpha, \beta \in \mathbb{R}$ and let $\boldsymbol{p}^k, \boldsymbol{q}^k \in \Lambda^k$. Then $\alpha \boldsymbol{p}^k + \beta \boldsymbol{q}^k \in \Lambda^k$. Suppose that $\mathfrak{f}[\alpha \boldsymbol{p}^k + \beta \boldsymbol{q}^k]$ is the fractal function generated by the polynomial vector $\alpha \boldsymbol{p}^k + \beta \boldsymbol{q}^k$. Then,

$$\mathfrak{f}[\alpha \boldsymbol{p}^k + \beta \boldsymbol{q}^k] = (\alpha p^k + \beta q^k) + \sum_{i=1}^{N} \lambda_i \, \mathfrak{f}[\alpha \boldsymbol{p}^k + \beta \boldsymbol{q}^k] \circ u_i^{-1} \, \chi_{I_i}. \tag{5.19}$$

However,

$$\alpha \mathfrak{f}[\boldsymbol{p}^k] + \beta \mathfrak{f}[\boldsymbol{q}^k] = \alpha \left(\boldsymbol{p}^k + \sum_{i=1}^{N} \lambda_i \mathfrak{f}[\boldsymbol{p}^k] \circ u_i^{-1} \chi_{I_i} \right) + \beta \left(\boldsymbol{q}^k + \sum_{i=1}^{N} \lambda_i \mathfrak{f}[\boldsymbol{q}^k] \circ u_i^{-1} \chi_{I_i} \right)$$

$$= (\alpha \boldsymbol{p}^k + \beta \boldsymbol{q}^k) + \sum_{i=1}^{N} \lambda_i (\alpha \mathfrak{f}[\boldsymbol{p}^k] + \beta \mathfrak{f}[\boldsymbol{q}^k]) \circ u_i^{-1} \chi_{I_i}. \tag{5.20}$$

The uniqueness of the fixed point implies that the right sides of (5.19) and (5.20) agree, and thus $\mathfrak{f}[\alpha \boldsymbol{p}^k + \beta \boldsymbol{q}^k] = \alpha \mathfrak{f}[\boldsymbol{p}^k] + \beta \mathfrak{f}[\boldsymbol{q}^k]$.

To show injectivity, note that, again by the uniqueness of the fixed point, one has $\mathfrak{f}[\boldsymbol{p}^k] \equiv 0$ iff $\boldsymbol{p}^k = \boldsymbol{0}$. Now, let $\mathfrak{f} \in \mathfrak{F}(\Lambda^k)$ and set $\boldsymbol{p}^k := (\mathfrak{f} \circ u_1 - \lambda_1 \mathfrak{f}, \ldots, \mathfrak{f} \circ u_N - \lambda_N \mathfrak{f})$. Then $\mathfrak{f}[\boldsymbol{p}^k] = \mathfrak{f}$. $\quad\square$

Of particular interest is the space $\mathfrak{F}(\Lambda^k) \cap C[a, b]$ of all continuous polynomial fractal functions over $[a, b]$, i.e., all those fractal functions in $B[a, b]$ that satisfy the join-up conditions (5.18). We denote this space by $\mathfrak{S}^k[a, b]$, or simply by $\mathfrak{S}^k$, when the interval is understood.

Corollary 5.21 The linear space $\mathfrak{S}^k[a, b]$ is $N(k - 1) + 1$-dimensional.

Proof The statement follows immediately from the theorem and the fact that $\dim \overset{N}{\underset{i=1}{\mathsf{X}}} \Pi^k = kN$ and that there are $N - 1$ join-up conditions. $\quad\square$

Let us consider the case of affine fractal interpolation functions, i.e., $k := 2$. Suppose that an interpolation set Y given as $Y := \{(\nu, y_\nu) \in [0, N] \times \mathbb{R} \mid \nu = 0, 1, \ldots, N\}$. For simplicity, set $\boldsymbol{p} := \boldsymbol{p}^2 \in \overset{N}{\underset{i=1}{\mathsf{X}}} \Pi^2$. Then $\boldsymbol{p}$ is uniquely determined by Y, i.e., there exists a linear isomorphism

$$\mathbb{R}^{N+1} \ni \boldsymbol{y} := (y_0, y_1, \ldots, y_N)^\top \mapsto \boldsymbol{p} = (p_1, \ldots, p_N) \in \overset{N}{\underset{i=1}{\mathsf{X}}} \Pi^2.$$

Hence, by Theorem 5.20, this extends to a linear isomorphism

$$\mathbb{R}^{N+1} \ni \boldsymbol{y} := (y_0, y_1, \ldots, y_N)^\top \mapsto \mathfrak{f}[\boldsymbol{p}] \in \mathfrak{S}^2[0, N], \tag{5.21}$$

that is, $\mathfrak{f}[\boldsymbol{p}]$ is uniquely determined by $\boldsymbol{y}$ and one can write $\mathfrak{f}[\boldsymbol{p}] = \mathfrak{f}[\boldsymbol{y}]$. (This is nothing but the construction of affine fractal interpolation functions given a set of interpolation points!)

Every vector $\boldsymbol{y} := (y_0, y_1, \ldots, y_N)^\top \in \mathbb{R}^{N+1}$ can be uniquely written in the form $\boldsymbol{y} = \sum_{\nu=0}^{N} y_\nu \, \boldsymbol{e}_\nu$, where $\boldsymbol{e}_\nu = (0, \ldots, 1, \ldots, 0)^\top \in \mathbb{R}^{N+1}$, with "1" being in the ν-th position. Denote by $\mathfrak{e}_\nu$ the unique affine fractal function generated by e_ν, i.e., the affine fractal interpolation function corresponding to the interpolation set $Y_\nu := \{(\mu, \delta_{\mu\nu}) \,|\, \mu = 0, 1, \ldots, N\}$, $\nu = 0, 1, \ldots, N$. Then, the linear isomorphism (5.21) and the uniqueness of the fixed point imply that

$$\mathfrak{f}[y] = \sum_{\nu=0}^{N} y_\nu\, \mathfrak{e}_\nu.$$

Example 5.22   Let $N := 3$ and let $\lambda_1 := \lambda_2 := \lambda_3 =: 0.5$, let $p_1 := (\tfrac{1}{3} - \lambda_1)x$, $p_2 := (-\tfrac{1}{6} - \lambda_2)x + 1$, and let $p_3 := (\tfrac{1}{3} - \lambda_3)x + \tfrac{1}{2}$. The graph of the affine fractal function and the graphs of the four basis fractal functions are shown in Figure 5.6.

The case $k := 2$ is summarized in the next proposition.

Proposition 5.23   Assume that $\boldsymbol{p} \in \overset{N}{\underset{i=1}{\times}} \Pi^2$. Then the linear space $\mathfrak{S}^2[0, N]$ has dimension $N + 1$. Moreover, every element $\mathfrak{f} \in \mathfrak{S}^2[0, N]$ can be written as a unique linear combination of the form

$$\mathfrak{f} = \sum_{\nu=0}^{N} y_\nu\, \mathfrak{e}_\nu,$$

where $\mathfrak{e}_\nu$ is the unique affine fractal function interpolating the set $Y_\nu := \{(\mu, \delta_{\mu\nu}) \,|\, \mu = 0, 1, \ldots, N\}$, $\nu = 0, 1, \ldots, N$.

We call the unique affine fractal function $\mathfrak{e}_\nu$ interpolating the set $Y_\nu := \{(\mu, \delta_{\mu\nu}) \,|\, \mu = 0, 1, \ldots, N\}$, $\nu = 0, 1, \ldots, N$, an *affine fractal function of Lagrange type*.

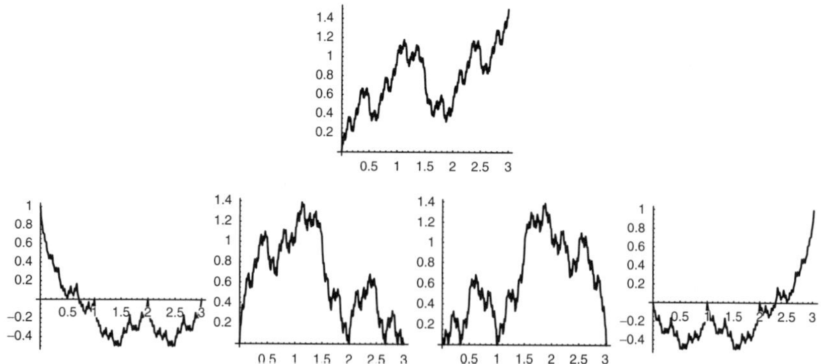

Figure 5.6   A fractal function (top) and its four basis functions (bottom).

The basis $\mathcal{E}_N := \{\mathfrak{e}_\nu \mid \nu = 0, 1, \ldots, N\} \subset \mathfrak{S}^2[0, N]$ can be orthonormalized using the Gram[2]–Schmidt[3] Orthonormalization Procedure. For this purpose, we need to derive formulae for the integral, the moments, and the inner product of affine fractal functions.

Proposition 5.24 Let $\mathfrak{f} \in \mathfrak{S}^2[a, b]$. Then

$$\int_a^b \mathfrak{f}(x)\,dx = \frac{\displaystyle\int_a^b p(x)\,dx}{1 - \displaystyle\sum_{i=1}^N a_i \lambda_i}. \qquad (5.22)$$

Proof By means of the fixed point equation for $\mathfrak{f}$

$$\mathfrak{f} = p + \sum_{i=1}^N \lambda_i \mathfrak{f} \circ u_i^{-1}\, \chi_{I_i},$$

we obtain, after a change of variables,

$$\int_a^b \mathfrak{f}(x)\,dx = \int_a^b p(x)\,dx + \sum_{i=1}^N \lambda_i \int_a^b (\mathfrak{f} \circ u_i^{-1})(x)\,(\chi_{I_i})(x)\,dx$$

$$= \int_a^b p(x)\,dx + \sum_{i=1}^N \lambda_i \int_{x_{i-1}}^{x_i} (\mathfrak{f} \circ u_i^{-1})(x)\,dx$$

$$= \int_a^b p(x)\,dx + \sum_{i=1}^N a_i \lambda_i \int_a^b \mathfrak{f}(x)\,dx.$$

Since $\displaystyle\sum_{i=1}^N a_i \lambda_i < 1$ (the reader is encouraged to verify this), the statement follows by transposition and division by $\displaystyle\sum_{i=1}^N a_i \lambda_i$. $\qquad\square$

To state the next proposition, we need to define the moments of a (Riemann) integrable function $f : K \subseteq \mathbb{R} \to \mathbb{R}$. To this end, let $n \in \mathbb{N}_0$.

2 Jørgen Petersen Gram, 27 June 1850–29 April 1916. Danish mathematician who is known for the Gram–Schmidt Orthogonalization Method, the Gram matrix, and his series representation of the Riemann Zeta function.
3 Erhard Schmidt, 13 January 1876–6 December 1959. German mathematician who is considered to be one of the founders of functional analysis. He also worked in analytic number theory.

Then the *n-th moment* $M_n(f)$ of f is defined by

$$M_n(f) := \int_K x^n f(x)\,dx.$$

Proposition 5.25 Suppose that $f \in \mathfrak{S}^2[a, b]$ and that $n \in \mathbb{N}$. Then the n-th moment $M_n(f)$ can be computed explicitly and recursively in terms of the lower order moments $M_m(f)$, $m = 0, 1, \ldots, n-1$.

Proof Exercise! $\square$

For $[a, b] := [0, N]$ and $u_i(x) := \frac{1}{N}(x - i + 1)$, $i = 1, \ldots, N$, the formula for the n-th moment of an affine fractal function f simplifies to

$$M_n(f) = \frac{M_n(p) + \sum_{m=0}^{n-1} \binom{n}{m} \dfrac{(i-1)^{n-m}}{N^m} M_m(f)}{1 - \sum_{i=1}^{N} \lambda_i N^{-(n+1)}}. \tag{5.23}$$

Proposition 5.26 Let $f, g \in \mathfrak{S}^2[a, b]$. Then

$$\langle f, g \rangle := \int_a^b f(x) g(x)\,dx = \frac{\langle p_f, p_g \rangle + \sum_{i=1}^{N} \langle p_f \circ u_i, g \rangle + \langle p_g \circ u_i, f \rangle}{1 - \sum_{i=1}^{N} a_i \lambda_i^2},$$

where p_f and p_g are the piecewise linear functions generating the affine fractal functions f and g, respectively.

Proof Exercise! $\square$

Remark 5.27 The inner products $\langle p_g \circ u_i, f \rangle$ and $\langle p_f \circ u_i, g \rangle$ can be explicitly computed using Proposition 5.25.

Combining the above propositions and the Gram–Schmidt Orthonormalization Method, we immediately arrive at the next result.

Proposition 5.28 The linear space $\mathfrak{S}^2[a, b]$ has an orthonormal basis consisting of $N + 1$ affine fractal functions of Lagrange type.

5.5 The Box Dimension of Affine Fractal Functions

In this section, we state and prove a dimension result for a class of affine fractal functions [82]. The arguments used in the proof are specific to the form of the functions u_i and p_i, $i = 1, \ldots, N$, and do not necessarily transfer to more complicated settings.

To this end, let $1 < N \in \mathbb{N}$, let $I := [0, 1]$ be the unit interval, and suppose that $Y := \{(x_\nu := \nu/N, y_\nu) \mid \nu = 0, 1, \ldots, N\}$ is a given set of interpolation points. Then the functions $u_i : I \to I$ are given by $u_i := \frac{1}{N}(\bullet + i - 1)$ (see (5.2a)), $i = 1, \ldots, N$. Define affine functions $p_i : I \to \mathbb{R}$ by $p_i := c_i(\bullet) + \beta_i$, where the coefficients c_i and β_i, $i = 1, \ldots, N$, are given by (5.2b). Denote by $\mathfrak{f}$ the associated affine fractal interpolation function with scaling vector $(\lambda_1, \ldots, \lambda_N) \in (-1, +1)^N$. In order to derive a formula for the box dimension of graph $\mathfrak{f}$, we first require some notation and a lemma.

Note that by Remark 4.15, it suffices to consider covers of the graph of $\mathfrak{f}$ whose elements are squares of side N^{-r}, $r \in \mathbb{N}_0$. Denote by $\mathcal{C}_0(r)$ a cover of graph $\mathfrak{f}$ consisting of a finite number of squares of side N^{-r}, $r \in \mathbb{N}_0$. Now consider a specific cover $\mathcal{C}(r)$ of graph $\mathfrak{f}$ of the form

$$\mathcal{C}(r) := \left\{ \left[\frac{k-1}{N^r}, \frac{k}{N^r} \right] \times \left[a, a + \frac{1}{N^r} \right] \;\middle|\; r \in \mathbb{N}_0; \; k \in \{1, \ldots, N^r\}; \; a \in \mathbb{R} \right\}.$$

$$(5.24)$$

By the compactness of graph $\mathfrak{f}$ and by Zorn's[4] Lemma, there exists a minimal cover $\mathcal{C}_0^*(r)$ of graph $\mathfrak{f}$ and also a minimal cover $\mathcal{C}^*(r)$ of graph $\mathfrak{f}$ of the form (5.24). Denote by $\mathcal{N}_0(r)$, respectively, $\mathcal{N}(r)$ the cardinality of these minimal covers. Since covers of the form (5.24) are more restrictive, we have $\mathcal{N}_0(r) \leq \mathcal{N}(r)$. On the other hand, every square of side N^{-r} in a cover of the form $\mathcal{C}_0^*(r)$ can be covered by at most two squares of side N^{-r} from a cover of the form (5.24). Thus, $\mathcal{N}(r) \leq 2\mathcal{N}_0(r)$. Hence, when computing the box dimension of graph $\mathfrak{f}$, it suffices to consider covers of the form (5.24).

For the proof of the main theorem in this section, we require the following lemma.

Lemma 5.29 Suppose that $\mathfrak{f}$ is an affine fractal interpolation function. Denote by $\mathcal{N}(r)$ the cardinality of a minimal cover of graph $\mathfrak{f}$ of the from (5.24). If $\displaystyle\sum_{i=1}^{N} |\lambda_i| > 1$ and if the set Y is not collinear, then

$$\lim_{r \to \infty} \frac{N^r}{\mathcal{N}(r)} = 0.$$

Proof The non-collinearity of the set Y implies the existence of an index $i_0 \in \{1, \ldots, N\}$ so that

$$\delta := |y_{i_0} - (y_N - y_0)x_{i_0} - y_0| > 0.$$

(In other words, there exists at least one interpolation point that lies above the line connecting $(0, y_0)$ and $(1, y_N)$.) Moreover, $\delta \leq \max\{|y_{i_0} - y_0|, |y_{i_0} - y_N|\}$.

---

4 MAX AUGUST ZORN, 6 June 1906–9 March 1993. German-American mathematician who worked primarily in algebra, set theory, and the theory of groups.

Since $\mathfrak{f}$ is continuous over $[0, 1]$, we have that $\mathcal{N}(r) \geq \delta N^r$. Since the RB operator T maps the line segment (= affine function) $\overline{(0, y_0), (1, y_N)}$ to the line segments $\overline{(x_{i-1}, y_{i-1}), (x_i, y_i)}$, we also have that $\mathcal{N}(r) \geq \sum_{i=1}^{N} |\lambda_i| \delta N^r$. By induction, this yields

$$\mathcal{N}(r) \geq \sum_{i_1=1}^{N} \cdots \sum_{i_k=1}^{N} |\lambda_{i_1} \cdots \lambda_{i_k}| \delta N^r,$$

provided $r \geq k \in \mathbb{N}$. Thus,

$$\mathcal{N}(r) \geq \left(\sum_{i=1}^{N} |\lambda_i| \right)^r \delta N^r.$$

As, by assumption, $\sum_{i=1}^{N} |\lambda_i| > 1$, the statement follows. $\square$

Now, we are in a position to state the main result in this section.

Theorem 5.30 Suppose that $\mathfrak{f} : I \to \mathbb{R}$ is an affine fractal interpolation function defined as above and $\mathfrak{G}$ its graph. If $\sum_{i=1}^{N} |\lambda_i| > 1$ and if the interpolation set Y is not collinear, then the box dimension of $\mathfrak{G}$ is given by the formula

$$\dim_B \mathfrak{G} = 1 + \frac{\log \sum_{i=1}^{N} |\lambda_i|}{\log N}; \tag{5.25}$$

otherwise, $\dim_B \mathfrak{G} = \dim_H \mathfrak{G} = 1$.

Proof Let $C(r)$ be a minimal cover of $\mathfrak{G}$ of cardinality $\mathcal{N}(r)$, $r \in \mathbb{N}_0$, consisting of squares of side N^{-r} whose interiors are disjoint. We denote by $C(r, k)$ the collection of all squares in $C(r)$ that lie between $x = \frac{k-1}{N^r}$ and $x = \frac{k}{N^r}$, $k = 1, \ldots, N^r$. Set $\mathcal{N}(r, k) := \mathrm{card}\, C(r, k)$ and denote by $\mathcal{R}(r, k)$ the set

$$\mathcal{R}(r, k) := \bigcup_{C_i \in C(r, k)} C_i.$$

As $C(r)$ is a cover of $\mathfrak{G}$ of minimal cardinality, every square in $C(r)$ must meet $\mathfrak{G}$, and since $\mathfrak{f}$ is continuous over I, the set $\mathcal{R}(r, k)$ must be a rectangle of width N^{-r} and height $N^{-r} \mathcal{N}(r, k)$. Note that $\mathcal{N}(r) = \sum_{k=1}^{N^r} \mathcal{N}(r, k)$. The main idea of the proof is to estimate $\mathcal{N}(r + 1)$ in terms of $\mathcal{N}(r)$ and to use Remark 4.14.

To this end, apply the set-valued mappings $\overline{f}_i$, $i = 1, \ldots, N$, defined via the numerical functions (5.1), to the rectangle $\mathcal{R}(r, k)$. The image of $\mathcal{R}(r, k)$ under the mapping $\overline{f}_i$ is a parallelogram contained in the strip $\left[\dfrac{l(k, i) - 1}{N^{r+1}}, \dfrac{l(k, i)}{N^{r+1}} \right] \times \mathbb{R}$, where we set $l(k, i) := k + (i - 1)N^r$. Observe that

$$\mathcal{N}(r + 1) = \sum_{i=1}^{N} \sum_{k=1}^{N^r} \mathcal{N}(r + 1, l(k, i)).$$

Using the fixed point equation for the graph $\mathfrak{G}$, namely, $\mathfrak{G} = \sum_{i=1}^{N} \overline{f}_i(\mathfrak{G})$, we deduce that

$$\mathfrak{G} \subseteq \bigcup_{i=1}^{N} \overline{f}_i \left(\bigcup_{k=1}^{N^r} \mathcal{R}(r, k) \right).$$

Now, each parallelogram $\overline{f}_i(\mathcal{R}(r, k))$ is contained in a rectangle of width $N^{-(r+1)}$ and height

$$\frac{|\lambda_i| \mathcal{N}(r, k) + |\beta_i|}{N^r}.$$

Therefore,

$$\mathcal{N}(r + 1, l(k, i)) \leq \left(\frac{|\lambda_i| \mathcal{N}(r, k) + |\beta_i|}{N^r} \right) / \left(\frac{1}{N^{r+1}} \right) + 1$$

$$= N\left(|\lambda_i| \mathcal{N}(r, k) + |\beta_i|\right) + 1.$$

Summation over $k = 1, \ldots, N^r$ and $i = 1 \ldots, N$ produces

$$\mathcal{N}(r + 1) \leq \gamma \, N \mathcal{N}(r) + c_1 \, N^{r+1},$$

where we set $\gamma := \sum_{i=1}^{N} |\lambda_i|$ and $c_1 := \sum_{i=1}^{N} |\beta_i| + 1$. Induction over r now yields

$$\mathcal{N}(r) \leq \gamma^r N^r \mathcal{N}(0) + c_1 N^r (1 + \gamma + \cdots + \gamma^{r-1}).$$

Depending on the size of γ, two cases need to be considered next.

Case 1: $\gamma \leq 1$. This implies that $\mathcal{N}(r) \leq c_2 \, r \, N^r$, where $c_2 := \mathcal{N}(0) + c_1$. Thus, this gives

$$\dim_B \mathfrak{G} \leq \lim_{r \to \infty} \frac{\log(c_2 \, r \, N^r)}{\log N^r} = 1.$$

Case 2: $\gamma > 1$. Here, we obtain $\mathcal{N}(r) \leq c_3 (\gamma N)^r$, where $c_3 := \mathcal{N}(0) + c_1(1 - \gamma)^{-1}$. Hence,

$$\dim_B \mathfrak{G} \leq \lim_{r \to \infty} \frac{\log c_3 \, (\gamma N)^r}{\log N^r} = 1 + \frac{\log \gamma}{\log N}.$$

Next, we need to find a non-trivial lower bound for $\dim_B \mathfrak{G}$. As $\mathfrak{f}$ is a continuous function, $\dim_B \mathfrak{G} \geq 1$. Thus, if $\gamma \leq 1$, then $\dim_B \mathfrak{G} = 1 = \dim_H \mathfrak{G}$. Also, if the interpolation set Y is collinear, then $\mathfrak{G} = L$, where L is the line joining the interpolation points $(0, y_0)$ and $(1, y_N)$, and again $\dim_B \mathfrak{G} = 1 = \dim_H \mathfrak{G}$.

Now, suppose that $\gamma > 1$ and that the interpolation set Y is not collinear. Since each $C_i \in \mathcal{C}(r, k)$ meets $\mathfrak{G}$, the image of C_i under the maps $\overline{f}_i$, $i = 1, \ldots, N$, must also meet $\mathfrak{G}$. Hence, $\mathcal{C}(r + 1, l(i, k))$ must at least cover a rectangle of width $N^{-(r+1)}$ and height $(|\lambda_i|[\mathcal{N}(r, k) - 2] - |\beta_i|) N^{-r}$, implying that

$$\mathcal{N}(r + 1, l(k, i)) \geq N \left(|\lambda_i| [\mathcal{N}(r, k) - 2] - |\beta_i| \right) - 1.$$

Summation over k and i gives

$$\mathcal{N}(r + 1) \geq (\gamma N)\mathcal{N}(r) - c_4 N^{r+1},$$

with $c_4 := 1 + \sum_{i=1}^{N} (2|a_i| + |b_i|)$. Induction over r yields

$$\mathcal{N}(r + 1) \geq (\gamma N)^m \mathcal{N}(r - m + 1) - c_3 N^{r+1}(1 + \gamma + \cdots + \gamma^{m-1})$$

$$\geq (\gamma N)^{r-m+1} \mathcal{N}(m) - \frac{c_3}{1 - \gamma^{-1}} N^{r+1}$$

$$= (\gamma N)^{r-m+1} \left(\mathcal{N}(m) - \frac{c_3 N^m}{1 - \gamma^{-1}} \right),$$

for all $m \in \mathbb{N}$ with $1 \leq m \leq n$. Now, by Lemma 5.29, one can choose r and m large enough so that

$$\mathcal{N}(m) - \frac{c_4 N^m}{1 - \gamma^{-1}} > 0,$$

and, therefore, $\mathcal{N}(r) \geq c_5 (\gamma N)^r$, for a constant $c_5 > 0$. Hence, $\dim_B \mathfrak{G} \geq 1 + \dfrac{\log \gamma}{\log N}$, which finishes the proof. $\square$

Remark 5.31 We proved Theorem 5.30 in the case where $\mathfrak{f}$ was defined over the unit interval $I = [0, 1]$. It is, however, straight-forward to show that the proof carries over to any nonempty compact interval $[a, b]$. We leave the details to the reader.

For the sake of completeness, we state the more general dimension result regarding affine fractal interpolation functions.

Theorem 5.32 Suppose that $\mathfrak{f} \in \mathfrak{G}^2[a, b]$ and that the interpolation set $Y :=$ $\{(x_\nu, \mathfrak{f}(x_\nu)) \mid \nu = 0, 1, \ldots, N\}$ is not collinear and $\gamma = \sum_{i=1}^{N} |\lambda_i| > 1$. Then the

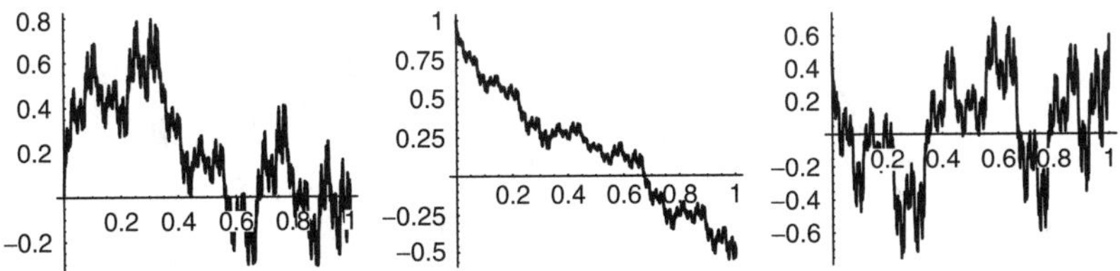

Figure 5.7   Three iso-dimensional affine fractal functions.

box dimension of graph $\mathfrak{f}$ is the unique positive solution of

$$\sum_{i=1}^{N} |\lambda_i|\, a_i^{d-1} = 1;$$

otherwise $\dim_B \operatorname{graph} \mathfrak{f} = \dim_H \operatorname{graph} \mathfrak{f} = 1$.

Proof   See [14] or [107]. $\square$

Remark 5.33   Note that since $a_i = (x_i - x_{i-1})/(b-a)$, $i = 1, \ldots, N$, the box dimension of an affine fractal function depends only on the knot spacings and the scaling factors λ_i. The y-values in the (non-collinear) interpolation set are immaterial. The three fractal functions $\mathfrak{f}, \mathfrak{g}, \mathfrak{h} : [0,1] \to \mathbb{R}$ depicted in Figure 5.7 have knot sequence $x_0 := 0$, $x_1 := 1/3$, $x_2 := 2/3$, and $x_3 := 1$. The scaling vector for each fractal function is $\lambda := (0.6, -0.5, 0.75)$. Thus, they are iso-dimensional with box dimension $\dim_B \operatorname{graph} \mathfrak{f} = \dim_B \operatorname{graph} \mathfrak{g} = \dim_B \operatorname{graph} \mathfrak{h} = 1 + \log 1.85 / \log 3 \approx 1.56$.

If follows from Falconer's result, Theorem 4.42, that for affine fractal functions the box dimension and the Hausdorff–Besicovitch dimension agree except on a set of measure zero in $\mathbb{R}^{2N}$. However, for any two specific affine fractal functions, nothing can be said about the equality of the two dimensions.

As an example of the non-equality of the two dimensions, we consider the following two affine fractal functions. The first is Kiesswetter's fractal function $\mathfrak{f}_K$ (see Example 5.9) whose box dimension is easily shown to be $\dim_B \operatorname{graph} \mathfrak{f}_K = 3/2$, and which agrees with the Hausdorff–Besicovitch dimension. (We refer to [55] for a proof of this fact.)

The second affine fractal function was constructed by Bedford [22] and its specifics are given below. Let $\mathbf{X} := [0,1] \times \mathbb{R}$, $N := 2$, and $Y := \left\{(0,0), (\frac{1}{4}, -\frac{1}{2}), (\frac{1}{2}, 1), (\frac{3}{4}, \frac{1}{2}), (1,1)\right\}$. Define numerical functions $f_i : [0,1] \times [-1,1] \to \mathbb{R}$ by

$$f_1(x,y) := \begin{pmatrix} \frac{1}{4} & 0 \\ 0 & \frac{1}{2} \end{pmatrix} \begin{pmatrix} x \\ y \end{pmatrix},$$

$$f_2(x,y) := \begin{pmatrix} \frac{1}{4} & 0 \\ 0 & \frac{1}{2} \end{pmatrix} \begin{pmatrix} x \\ y \end{pmatrix} + \begin{pmatrix} \frac{1}{4} \\ \frac{1}{2} \end{pmatrix},$$

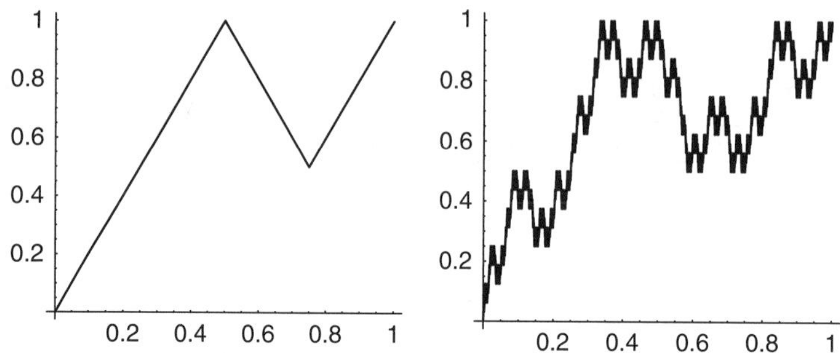

Figure 5.8   Bedford's fractal function.

$$f_3(x, y) := \begin{pmatrix} \frac{1}{4} & 0 \\ 0 & -\frac{1}{2} \end{pmatrix} \begin{pmatrix} x \\ y \end{pmatrix} + \begin{pmatrix} \frac{1}{2} \\ 1 \end{pmatrix},$$

$$f_4(x, y) := \begin{pmatrix} \frac{1}{4} & 0 \\ 0 & \frac{1}{2} \end{pmatrix} \begin{pmatrix} x \\ y \end{pmatrix} + \begin{pmatrix} \frac{3}{4} \\ \frac{1}{2} \end{pmatrix}.$$

The process of generating the graph of Bedford's fractal function f_B and the graph itself are depicted in Figure 5.8.

The box dimension $\dim_B \operatorname{graph} f_B = 3/2$, but $\dim_H \operatorname{graph} f_B \approx 1.45$. The proof of the latter fact uses techniques that go beyond the scope of this text and the interested reader is referred to the original paper by Bedford for details.

Finally, we state a result that relates the box dimension of an affine fractal function to its Lipschitz exponent. We consider again the special case treated in Theorem 5.30, but mention that the result holds true in the more general setting of Theorem 5.32.

Theorem 5.34   Let $f \in \mathfrak{S}^2[a, b]$, suppose that $\gamma > 1$ and that the set $Y := \{(x_v, f(x_v)) \mid v = 0, 1, \ldots, N\}$ is not collinear. Then $f \in \operatorname{Lip}^\alpha[a, b]$, where $\alpha = 2 - \dim_B \operatorname{graph} f$.

Proof   Let $x \in [a, b]$ and let $0 < h < 1$. Then there exists an $r \in \mathbb{N}$ so that both x and $x + h$ lie in $[(k - 1)N^{-r}, kN^{-r}]$, for some $k \in \{1, \ldots, N^r\}$. Then $h \leq N^{-r}$.

Now, the proof of Theorem 5.30 shows that $|f(x + h) - f(x)| \leq \dfrac{\mathcal{N}(r, k)}{N^r}$. Clearly, $\mathcal{N}(r, k) \leq \mathcal{N}(r) \leq c N^{r(d-1)}$, where d denotes the box dimension of $\operatorname{graph} f$ and $c > 0$ is a constant. Thus,

$$|f(x + h) - f(x)| \leq c N^{r(d-2)}.$$

Thus,

$$\log |f(x + h) - f(x)| \leq \log c + r(d - 2) \log N,$$

and, upon division by $\log h < 0$ and employing the inequality $h \le N^{-r}$, we obtain

$$\frac{\log |f(x+h) - f(x)|}{\log h} \ge \frac{\log c}{\log h} + \frac{r(d-2)\log N}{\log h} \ge \frac{\log c}{\log h} - \frac{r(d-2)\log N}{r\log N}$$

$$= \frac{\log c}{\log h} + 2 - d,$$

giving the result. $\square$

5.6 Code Space and Fractal Functions

Let $1 < N \in \mathbb{N}$ and let (Σ_N, d_F) be the code space with N symbols. (See also Section 4.5.) Moreover, let $\emptyset \ne [a, b] \subset \mathbb{R}$, let u_i, $i = 1, \ldots, N$, satisfy (5.8), and let $p_i^k \in \Pi^k[a, b]$ be fixed. Suppose that $T : C[a, b] \to C[a, b]$ be given by

$$T f = \sum_{i=1}^{N} p_i^k \circ u_i^{-1} \chi_{I_i} + \sum_{i=1}^{N} \lambda_i f \circ u_i^{-1} \chi_{I_i}$$

with $\max\{|\lambda_i| \,|\, i = 1, \ldots, N\} < 1$ and

$$(T f)(x_i-) = (T f)(x_i+), \quad i = 1, \ldots, N - 1.$$

Let $i_1 \in \{1, \ldots, N\}$ and let f be the unique fixed point of T. Then

$$f(x) = p_{i_1}^k \circ u_{i_1}^{-1}(x) + \lambda_{i_1} f \circ u_{i_1}^{-1}(x), \quad \forall\, x \in u_{i_1}[a, b].$$

Thus, if $x \in u_{i_1} \circ u_{i_2}[a, b]$, one obtains

$$f \circ u_{i_1}^{-1}(x) = p_{i_2} \circ u_{i_2}^{-1}(x) + \lambda_{i_2} f \circ (u_{i_1} \circ u_{i_2})^{-1}(x),$$

and, therefore,

$$f(x) = (p_{i_1}^k \circ u_{i_1}^{-1})(x) + \lambda_{i_1}(p_{i_2}^k \circ u_{i_2}^{-1})(x) + \lambda_{i_1}\lambda_{i_2} f \circ (u_{i_1} \circ u_{i_2})^{-1}(x),$$

$$\forall\, x \in u_{i_1} \circ u_{i_2}[a, b].$$

Hence, proceeding inductively, yields for $\sigma := (\sigma_1, \sigma_2, \ldots, \sigma_\nu, \ldots) \in \Sigma_N$,

$$f(x) = \sum_{\nu=1}^{n} \lambda_{\sigma(\nu-1)}(p_{\sigma_\nu} \circ u_{\sigma_\nu}^{-1})(x) + \lambda_{\sigma(n)}(f \circ u_{\sigma(n)}^{-1})(x), \quad \forall\, x \in u_{\sigma(n)}[a, b],$$

where $\lambda_{\sigma(0)} := 1$.

Now choose $x \in [a, b]$ arbitrary and let σ be its address, i.e., $x = \gamma(\sigma)$, where γ is the surjection defined in Theorem 4.44. Then, since $\lambda_{\sigma(n)} \to 0$ as $n \to \infty$, we obtain a series representation of f in the form

$$f(x) = \sum_{\nu=1}^{\infty} \lambda_{\sigma(\nu-1)}(p_{\sigma_\nu}^k \circ u_{\sigma_\nu}^{-1})(x).$$

5.7 Continuous Functions as Special Cases of Fractal Functions

Next, we show that any continuous function defined on a compact interval can be regarded as a special case of a class of fractal functions. This was first noticed by Barnsley [7].

To this end, let $\emptyset \neq [a, b] \subset \mathbb{R}$ and let $\alpha, \beta \in \mathbb{R}$ be arbitrary. Recall that

$$C_{\alpha,\beta}[a, b] := \{f \in C[a, b] \mid f(a) = \alpha, \ f(b) = \beta\}$$

endowed with the metric $d_{\infty,[a,b]}$ is a complete metric space.

For a fixed $1 < N \in \mathbb{N}$, let $u_i : [a, b] \to [a, b]$, $i = 1, \ldots, N$, be contractive homeomorphisms satisfying the conditions given in (5.8). Moreover, set $x_{i-1} := u_i(a)$ and $x_i := u_i(b)$, $i = 1, \ldots, N$. Furthermore, suppose that $q_i : [a, b] \to \mathbb{R}$, $i = 1, \ldots, N$, are continuous functions. Define operators

$$T[\lambda, q] : \mathbb{R}^N \times \underset{i=1}{\overset{N}{\times}} C[a, b] \times C_{\alpha,\beta}[a, b] \to C_{\alpha,\beta}[a, b] \text{ via}$$

$$(T[\lambda, q]f)(x) := \sum_{i=1}^{N}(q_i \circ u_i^{-1})(x)\,\chi_{I_i}(x) + \sum_{i=1}^{N}\lambda_i(f \circ u_i^{-1})(x)\,\chi_{I_i}(x), \quad (5.26)$$

where $\lambda := (\lambda_1, \ldots, \lambda_N) \in \mathbb{R}^N$ are arbitrary, $q := (q_1, \ldots, q_N) \in (C[a, b])^N$, and I_i, $i = 1, \ldots, N$, is defined as in (5.9). Additionally, we require that $T[\lambda, q]$ satisfies

$$(T[\lambda, q])(x_i-) = (T[\lambda, q])(x_i+), \quad i = 1, \ldots, N-1.$$

These last conditions guarantee that $T[\lambda, q]f$ is continuous on $[a, b]$. The operator $T[\lambda, q]$ possesses a unique fixed point $f[\lambda, q] \in C_{\alpha,\beta}[a, b]$ provided that $\lambda := \max\{|\lambda_i| \mid i = 1, \ldots, N\} < 1$.

Now suppose that $g \in C[a, b]$ is chosen arbitrarily and held fixed. Set $\alpha := g(a)$ and $\beta := g(b)$. For $i = 1, \ldots, N$, let

$$q_i := g \circ u_i - \lambda_i h,$$

where $h \in C_{\alpha,\beta}[a, b]$ is arbitrary. For this particular choice of q, we denote the above-defined RB operator simply by $T[\lambda]$:

$$(T[\lambda]f)(x) := g(x) + \sum_{i=1}^{N}\lambda_i(f - h) \circ u_i^{-1}(x)\,\chi_{I_i}(x), \quad \forall x \in [a, b]. \quad (5.27)$$

Since $T[\lambda] : \mathbb{R}^N \times C_{\alpha,\beta}[a, b] \to C_{\alpha,\beta}[a, b]$, we compute

$$(T[\lambda]f)(x_i-) = g(x_i-) + \lambda_i(f - h) \circ u_i^{-1}(x_i) = g(x_i) + \lambda_i(f(b) - h(b))$$

$$= g(x_i), \quad (5.28a)$$

$$(T[\lambda]f)(x_i+) = g(x_i+) + \lambda_{i+1}(f - h) \circ u_{i+1}^{-1}(x_i) = g(x_i) + \lambda_{i+1}(f(a) - h(a))$$

$$= g(x_i). \quad (5.28b)$$

Hence, provided that $\lambda < 1$, $\exists\, \mathfrak{g}[\lambda] \in C_{\alpha,\beta}[a, b]$ so that

$$\mathfrak{g}[\lambda] = g + \lambda_i(\mathfrak{g}[\lambda] - h) \circ u_i^{-1}, \quad \text{on } u_i[a, b],\ i = 1, \ldots, N,$$

or, equivalently,

$$\mathfrak{g}[\lambda] = g + \sum_{i=1}^{N} \lambda_i(\mathfrak{g}[\lambda] - h) \circ u_i^{-1}\, \chi_{I_i}, \quad \text{on } [a, b]. \tag{5.29}$$

Notice that if $\lambda := 0$, then $\mathfrak{g}[0] = g$, and one may interpret (5.29) as defining an entire class of fractal functions parametrized by $\lambda \in \mathbb{R}^N$ with g as its germ.

Remark 5.35 The fixed point $\mathfrak{g}[\lambda]$ also depends on the choice of $h \in C_{\alpha,\beta}[a, b]$, although we have not indicated this dependence in our notation. There are, of course, several ways of how to choose h, but a natural and also simple choice is

$$h(x) := \frac{(\beta - \alpha)x + (\alpha b - a\beta)}{b - a}. \tag{5.30}$$

Henceforth, we will use this form of h when we compute the class of fractal functions $\mathfrak{g}[\lambda]$ associated with a given $g \in C[a, b]$.

Other possibilities for h exist as well and may be employed depending on the nature of the fractal modeling problem at hand.

Remark 5.36 Notice that given $g \in C_{\alpha,\beta}[a, b]$ and choosing $h \in C_{\alpha,\beta}[a, b]$, the continuity conditions (5.28a) and (5.28b) imply that

$$\mathfrak{g}[\lambda](x_j) = g(x_j), \quad \forall j = 0, 1, \ldots, N.$$

We can reverse this setting as follows.

Suppose we are given an interpolation set $Y := \{x_\nu, y_\nu \,|\, \nu = 0, 1, \ldots, N\}$, for some $1 < N \in \mathbb{N}$. Set $a := x_0$ and $b := x_N$. We can take any interpolant $g : [x_0, x_N] \to \mathbb{R}$ through Y and any function $h \in C_{\alpha,\beta}[a, b]$, where $\alpha := g(a)$ and $\beta := g(b)$ and use them to construct a fractal function $\mathfrak{g}[\lambda] : [a, b] \to \mathbb{R}$ as outlined above. Then $\mathfrak{g}[\lambda]$ interpolates the set Y for any choice of $\lambda = (\lambda_1, \ldots, \lambda_N)$ subject to the usual constraint $\max\{|\lambda_i| \,|\, i = 1, \ldots, N\} < 1$. This fractal interpolation function is then the "fractal generalization" of the "classical" interpolation function g.

Remark 5.37 From the form of the operator $T[\lambda]$ one obtains immediately the following statements.

1. $T[0] = \mathrm{id}_{C_{\alpha,\beta}[a,b]}$.
2. If h is of the form (5.30) and $g := h$, then $\mathfrak{g}[\lambda] = g$.

Proof   In this case, the fixed point equation for $\mathfrak{g}[\lambda]$ reads

$$\mathfrak{g}[\lambda] = h + \lambda_i(\mathfrak{g}[\lambda] - h) \circ u_i^{-1}, \quad \text{on } u_i[a, b], \, i = 1, \ldots, N,$$

which is satisfied by $\mathfrak{g}[\lambda] = h$. The uniqueness of the fixed point now yields the result. $\square$

3. If $g \equiv \alpha$, then clearly $h \equiv \alpha$ and thus by the above observation, $\mathfrak{g}[\lambda] \equiv \alpha$, $\forall \, \lambda \in (-1, 1)^N$.

Example 5.38   Choose $N := 2$, $[a, b] := [0, 2\pi]$, and let $g := \sin$. Then $\alpha = \beta = 0$ and one may choose $h := 0$. Let $u_1 := \frac{1}{2}(\bullet)$ and $u_2 := \frac{1}{2}(\bullet + \pi)$. For the continuous functions q_i we thus obtain

$$q_1 = \sin \frac{(\bullet)}{2} \quad \text{and} \quad q_2 = \sin \frac{(\bullet) + \pi}{2},$$

for an arbitrary N-tuple $\lambda \in (-1, 1)^2$. We denote the class of fractal functions generated in this way by $\sin[\lambda_1, \lambda_2]$. In Figure 5.9, the graphs of $\sin[0.6, -0.8]$ and $\sin[0.4, 0.75]$ are shown.

Example 5.39   We again choose $N := 2$, but let $[a, b] := [0, 1]$ and $g := \exp$. Thus, $\alpha = 1$ and $\beta = e$, and therefore $h = (e - 1)\,\mathrm{id}_{[0,1]} + 1$. For u_1, u_2 we again choose $u_1 = \frac{1}{2}(\bullet)$ and $u_2 = \frac{1}{2}(\bullet + 1)$. The functions q_1 and q_2 are then given by

$$q_1 = \exp \frac{1}{2}(\bullet) - \lambda_1 e\,\mathrm{id}_{[0,1]} \quad \text{and} \quad q_2 = \exp \frac{1}{2}(\bullet + 1) - \lambda_2 e\,\mathrm{id}_{[0,1]},$$

for arbitrary $\lambda_1, \lambda_2 \in (-1, 1)$.

The fractal functions generated by these mappings will be denoted by $\mathfrak{exp}[\lambda_1, \lambda_2]$. In Figure 5.10 ,we display the graph of $\mathfrak{exp}[0.25, -0.15]$ and the graph of the difference $\mathfrak{exp}[0.25, -0.15] - \exp$ over the interval $[0, 1]$.

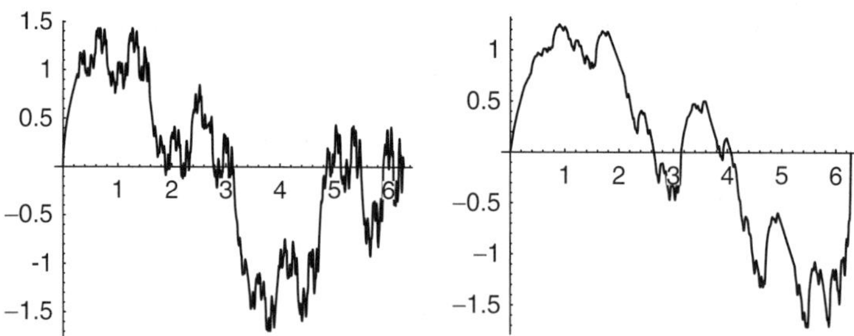

Figure 5.9   The graphs of the fractal functions $\sin[0.6, -0.8]$ and $\sin[0.4, 0.75]$.

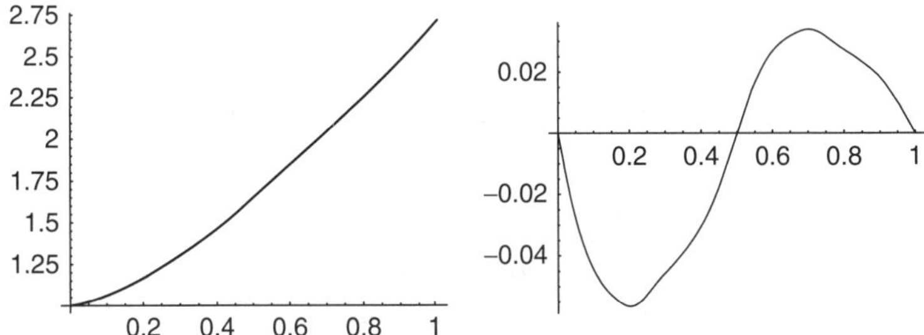

Figure 5.10 The graph of the fractal function $\mathrm{e\mathfrak{r}p}[0.25, -0.15]$ (left) and the graph of $\mathrm{e\mathfrak{r}p}[0.25, -0.15] - \exp$ (right).

The reason why the fractal function $\mathrm{e\mathfrak{r}p}[0.25, -0.15]$ looks so smooth (it turns out that the function $\mathrm{e\mathfrak{r}p}[0.25, -0.15]$ is actually differentiable on $[0, 1]$) will become clear in Section 5.8.

Example 5.40 Let $N := 2$, let $[a, b] := [-2, 3]$, and let $g := (x + 1)(x - 1)$ $(x - 2)$. This gives $\alpha = -12$ and $\beta = 8$. The function h is chosen to be $h = 4(\bullet) - 4$, and for the maps u_1 and u_2 we select $u_1 = \frac{1}{2}(\bullet) - 1$ and $u_2 = \frac{1}{2}(\bullet) + \frac{3}{2}$. Thus,

$$q_1(x) = 2x(2x + 1)(2x + 3) - 4\lambda_1(x - 1)$$

and

$$q_2(x) = 4(x - 1)(x - 2)(2x - 5) - 4\lambda_2(x - 1),$$

for $\lambda_1, \lambda_2 \in (-1, 1)$. We denote the fractal functions generated by these mappings by $\mathfrak{p}[\lambda_1, \lambda_2]$. The graphs of $\mathfrak{p}[-0.85, -0.75]$ and $\mathfrak{p}[-0.25, -0.25]$ are shown in Figure 5.11.

Using Theorems 5.32 and 5.34, it is readily verified that the fractal function $\mathfrak{p}[-0.25, -0.25]$ is in $\mathrm{Lip}^1[-2, 3]$, but not in $C^1[-2, 3]$, as we will see in Section 5.8.

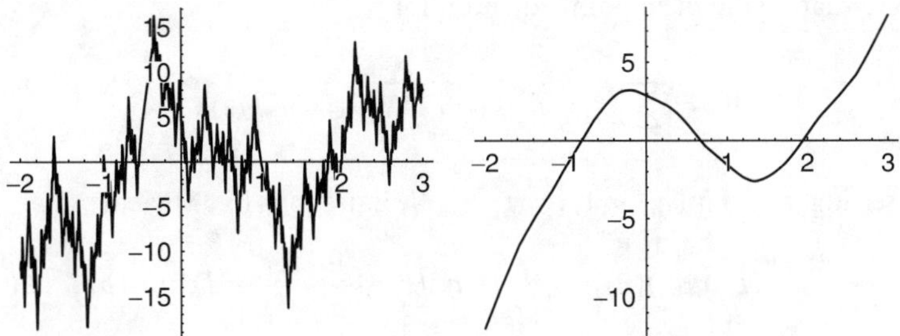

Figure 5.11 The graphs of the fractal functions $\mathfrak{p}[-0.85, -0.75]$ and $\mathfrak{p}[-0.25, -0.25]$.

5.8 Fractal Functions of Class C^k

Examples 5.39 and 5.40 in the previous section gave an indication that fractal functions may actually possess some smoothness, depending on the size of the modulus of the components of the parameter vector $\boldsymbol{\lambda}$. Next, we will construct fractal functions of class $C^k[a, b]$, where $[a, b] \subset \mathbb{R}$ is a nonempty interval.

To this end, we consider the collection of all functions $f : [a, b] \to \mathbb{R}$ that are k-times, $k \in \mathbb{N}_0$, uniformly differentiable on $[a, b]$ and endow it with the norm $\| \; \|_{C^k,[a,b]} : C^k[a, b] \to \mathbb{R}$ (see also Example 1.7) given by

$$\|f\|_{C^k,[a,b]} := \max_{\varkappa=0,1,\dots,k} \left\{ |(D^\varkappa f)(x)| \,|\, x \in [a, b] \right\}. \tag{5.31}$$

It is an easy exercise to show that $(C^k[a, b], \| \; \|_{C^k,[a,b]})$ is a Banach space. (The reader is encouraged to verify this statement.))

Let $1 < N \in \mathbb{N}$ and define contractions $u_i : [a, b] \to [a, b]$, $i = 1, \dots, N$, as in (5.8). Recall the associated partition of $[a, b]$ into subintervals I_i, $i = 1, \dots, N$, given in (5.9). In addition, we require that the mappings u_i, $i = 1, \dots, N$, are diffeomorphisms instead of just homeomorphisms, i.e., u_i and u_i^{-1} are uniformly continuously differentiable on $[a, b]$. As before, we set $x_{i-1} := u_i(a)$ and $x_i := u_i(b)$, $i = 1, \dots, N - 1$. Furthermore, assume that we are given functions $q_i \in C^k[a, b]$ and scalars $\lambda_i \in \mathbb{R}$, $i = 1, \dots, N$.

We define an RB operator $T[\boldsymbol{\lambda}, \boldsymbol{q}] : \mathbb{R}^N \times (\underset{i=1}{\overset{N}{\mathsf{X}}} C^k[a, b]) \times B[a, b] \to B[a, b]$ by

$$T[\boldsymbol{\lambda}, \boldsymbol{q}]f := \sum_{i=1}^{N} (q_i \circ u_i^{-1}) \chi_{I_i} + \sum_{i=1}^{N} \lambda_i (f \circ u_i^{-1}) \chi_{I_i}. \tag{5.32}$$

If $\max\{|\lambda_i| \,|\, i = 1, \dots, N\} < 1$, then the unique fixed point $\mathfrak{f}$ of T is an element of $B[a, b]$. We now require that this fixed point is an element of $C^k[a, b]$, i.e., we impose the following conditions:

$$(D^\varkappa \mathfrak{f})(x_i-) = (D^\varkappa \mathfrak{f})(x_i+), \quad \forall \varkappa = 0, 1, \dots, k; \; \forall i = 1, \dots, N - 1. \tag{5.33}$$

Differentiating the fixed point equation for $\mathfrak{f}$,

$$\mathfrak{f} = \sum_{i=1}^{N} (q_i \circ u_i^{-1}) \chi_{I_i} + \sum_{i=1}^{N} \lambda_i (\mathfrak{f} \circ u_i^{-1}) \chi_{I_i},$$

and setting $a_i := \mathrm{Lip}(u_i) < 1$, $i = 1, \dots, N$, Equations (5.33) yield

$$a_{i+1}^{-\varkappa} (D^\varkappa q_{i+1})(a) - a_i^{-\varkappa} (D^\varkappa q_i)(b) + \frac{\lambda_{i+1}}{a_N^\varkappa - \lambda_N} (D^\varkappa q_N)(b)$$

$$- \frac{\lambda_i}{a_1^\varkappa - \lambda_1} (D^\varkappa q_1)(a) = 0, \tag{5.34}$$

for all $\varkappa = 0, 1, \ldots, k$ and $i = 1, \ldots, N - 1$. Here, we also used that

$$(D^\varkappa \mathfrak{f})(a) = \frac{(D^\varkappa q_1)(a)}{a_1^\varkappa - \lambda_1} \quad \text{and} \quad (D^\varkappa \mathfrak{f})(b) = \frac{(D^\varkappa q_N)(b)}{a_N^\varkappa - \lambda_N}. \tag{5.35}$$

(The reader is encouraged to verify this statement.) If we now define functionals $L_i^\varkappa : \underset{i=1}{\overset{N}{\times}} C^k[a, b] \to \mathbb{R}$, $\varkappa = 0, 1, \ldots, k$ and $i = 1, \ldots, N - 1$, by

$$L_i^\varkappa \begin{pmatrix} q_1 \\ \vdots \\ q_N \end{pmatrix} := a_{i+1}^{-\varkappa} (D^\varkappa q_{i+1})(a) - a_i^{-\varkappa} (D^\varkappa q_i)(b) + \frac{\lambda_{i+1}}{a_N^\varkappa - \lambda_N} (D^\varkappa q_N)(b)$$

$$- \frac{\lambda_i}{a_1^\varkappa - \lambda_1} (D^\varkappa q_1)(a),$$

then conditions (5.33) are equivalent to the requirement that

$$q := \begin{pmatrix} q_1 \\ \vdots \\ q_N \end{pmatrix} \in \bigcap_{\varkappa=0}^{k} \bigcap_{i=1}^{N-1} \ker L_i^\varkappa =: \mathcal{K}^k.$$

In other words, if $q \in \mathcal{K}^k$ then the fixed point $\mathfrak{f}$ of T is an element of $C^k[a, b]$. In order to obtain $\mathfrak{f}$ as the limit of a sequence of C^k-functions, one needs to require that

$$\max\{|\lambda_i| \, a_i^{-k} \,|\, i = 1, \ldots, N\} < 1. \tag{5.36}$$

(The reader is encouraged to verify this statement.) We summarize these results in a theorem.

Theorem 5.41 Suppose we are given diffeomorphisms $u_i : [a, b] \to [a, b]$, $i = 1, \ldots, N$, of the form (5.8), inducing a partition of $[a, b]$ according to (5.9). Furthermore, let $k \in \mathbb{N}$ and suppose that $q \in \mathcal{K}^k$. Moreover, let $\lambda \in \mathbb{R}^N$ be such that $\max\{|\lambda_i| \, a_i^{-k} \,|\, i = 1, \ldots, N\} < 1$, where $a_i := \mathrm{Lip}(u_i) < 1$, $i = 1, \ldots, N$. Then the unique fixed point of the operator $T := T[\lambda, q] : \mathbb{R}^N \times (\underset{i=1}{\overset{N}{\times}} C^k[a, b]) \times C^k[a, b] \to B[a, b]$ is a fractal function of class C^k. Moreover,

$$D^\varkappa \mathfrak{f} = \sum_{i=1}^{N} a_i^{-\varkappa} (D^\varkappa q_i) \circ u_i^{-1} \chi_{I_i} + \sum_{i=1}^{N} a_i^{-\varkappa} \lambda_i (D^\varkappa \mathfrak{f}) \circ u_i^{-1} \chi_{I_i}, \quad \forall \, \varkappa = 0, 1 \ldots, k.$$

In other words, if $\mathfrak{f}$ is generated via the operator $T[\lambda, q]$, then $D^\varkappa \mathfrak{f}$ is generated via the operator $T[a^{-\varkappa}\lambda, a^{-\varkappa} D^\varkappa q]$, or equivalently, $\mathfrak{f}[a^{-\varkappa}\lambda, a^{-\varkappa} D^\varkappa q] = D^\varkappa(\mathfrak{f}[\lambda, q])$. (Here, we defined the component-wise product uv of two vectors $u, v \in \mathbb{R}^N$ by $uv := (u_1 v_1, \ldots, u_N v_N)^\top$.)

Example 5.42 As an example, we consider the case where $N := 2$, $[a, b] := [0, 1]$, and $k := 1$. With $u_1 := \frac{1}{2}(\bullet)$ and $u_2 := \frac{1}{2}(\bullet + 1)$, we take as functions

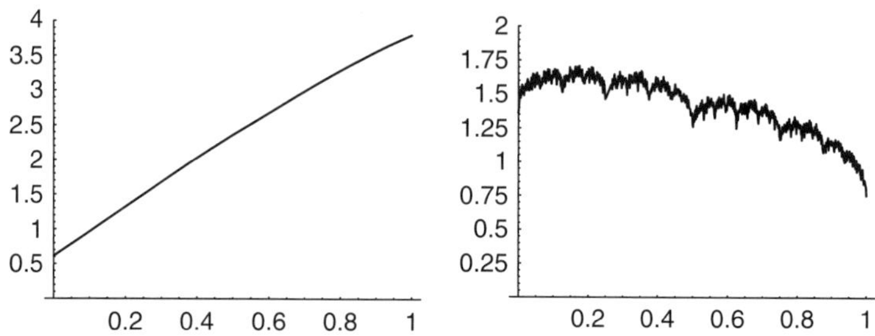

Figure 5.12 A fractal function $\mathfrak{f}$ of class C^1 (left) and its derivative $D\mathfrak{f}$ (right).

$q_1 := \exp(a + b(\bullet))$ and $q_2 := \exp(c + d(\bullet))$, where a, b, c and d are parameters to be determined so that $\mathbf{q} := (q_1, q_2)^\top \in \mathcal{K}^1$. Letting $\lambda_1 = \lambda_2 := \lambda \in \left(-\frac{1}{2}, \frac{1}{2}\right)$, we obtain for the operators $L_1^\varkappa$, $\varkappa = 0, 1$, the following expressions:

$$L_1^0 \begin{pmatrix} q_1 \\ q_2 \end{pmatrix} = (1 - \lambda)(e^c - e^{a+b}) + \lambda(e^a - e^{c+d}),$$

$$L_1^1 \begin{pmatrix} q_1 \\ q_2 \end{pmatrix} = 2\lambda(be^a - de^{c+d}) + (1 - 2\lambda)(de^c - be^{a+b}).$$

Selecting $\lambda := 0.4$, $a := -1$, and $c := 0.75$, and requiring that $\mathbf{q} \in \mathcal{K}^1$, we obtain for b and d the numerical solutions

$$b \approx 0.827899 \quad \text{and} \quad d \approx 0.0744116.$$

The graphs of the fractal function $\mathfrak{f}$ of class C^1 obtained in this way and its derivative $D\mathfrak{f}$ are depicted in Figure 5.12. Note that $\mathfrak{f}(0) = e^a/(1 - \lambda) \approx 0.613132$ and $\mathfrak{f}(1) = e^{c+d}/(1 - \lambda) \approx 3.79933$, and $(D\mathfrak{f})(0) = (be^a)/(0.5 - \lambda) \approx 1.52302$ and $(D\mathfrak{f})(1) = (de^{c+d})/(0.5 - \lambda) \approx 0.843452$.

5.9 Construction of Fractal Functions of Class C^k from Splines

In this section and the next, we explore the connection between splines and a class of fractal functions generated by spaces $\Lambda^k \subseteq \Pi_\xi^k$ whose elements are certain splines.

For this purpose, assume that $u_i := \frac{1}{N}(\bullet + (N - i)a + (i - 1)b)$, $i = 1, \ldots, N$, i.e., the diffeomorphisms u_i divide the interval into N subintervals of equal length $h := (b - a)/N$. Writing out the conditions in (5.33) explicitly and using the specific definition of the u_i, we see that the following conditions, namely,

$$(D^\varkappa q_{i+1})(a) + \lambda_{i+1}(D^\varkappa f)(a) = (D^\varkappa q_i)(b) + \lambda_i(D^\varkappa f)(b), \quad \varkappa = 1, \ldots, k-1,$$

need to hold for all $i = 1, \ldots, N - 1$. Here, we decide to choose the functions q_i to be elements of a – yet to be determined – subset Λ^k of Π_ξ^k, where $\xi := (a = x_0 < x_1 < \cdots < x_{N-1} < x_N = b)$ is the ordered knot sequence associated with the partition induced by the mappings u_i, $i = 1, \ldots, N$. (As a matter of fact, the knots x_i are given by $x_i = a + (i - 1)h$, $i = 1, \ldots, N$.)

The above conditions suggest the investigation of two distinct cases.

Case 1: $D^\varkappa f|_{\partial[a,b]} = 0$, $\varkappa = 0, 1, \ldots, k$.

The vanishing of f and all its derivatives up to order k on the boundary of $[a, b]$, i.e., its endpoints a and b, reduces the above requirements to

$$L_i^\varkappa \begin{pmatrix} q_1 \\ \vdots \\ q_N \end{pmatrix} := (D^\varkappa q_{i+1})(a) - (D^\varkappa q_i)(b) = 0, \ \forall \varkappa = 0, 1, \ldots, k; \ \forall i = 1, \ldots, N-1,$$

$$(5.37)$$

where $L_i^\varkappa : \overset{N}{\underset{i=1}{\times}} C^\varkappa[a, b] \to \mathbb{R}$, $\varkappa = 0, 1, \ldots, k$, $i = 1, \ldots, N - 1$, are linear functionals. In other words, the function $q : [a, b] \to \mathbb{R}$ given by $q = \sum_{i=1}^{N} q_i(N \bullet - (N - i)a - (i - 1)b)\chi_{I_i}$ must be in $C^k[a, b]$. The choice of $\lambda_i \in (-1, 1)$, $i = 1, \ldots, N$, is arbitrary. The natural function space to be considered in this case is

$$C_0^k[a, b] := \{f \in C^k[a, b] \,|\, D^\varkappa f|_{\partial[a,b]} = 0, \ \forall \varkappa = 0, 1, \ldots, k\}.$$

(We also define $C_0^0[a, b] := C_0[a, b]$, see (5.12).) A simple computation shows that the function space $(C_0^k[a, b], \| \ \|_{C^k})$ is a Banach space. Imposing the additional conditions $D^\varkappa q|_{\partial[a,b]} = 0$, $\varkappa = 0, 1, \ldots, k$, on the function q, i.e., $q \in C_0^k[a, b]$, and requiring that $\max_{1 \leq i \leq N} |\lambda_i| < N^{-k}$, implies that the unique fixed point $\mathfrak{f} = \mathfrak{f}[\boldsymbol{\lambda}, \boldsymbol{q}]$ of the RB operator given by

$$T[\boldsymbol{\lambda}, \boldsymbol{q}] : (-N^{-k}, +N^{-k})^N \times \left[\bigcap_{i=1}^{N-1} \bigcap_{\varkappa=0}^{k} \ker L_i^\varkappa \cap \bigcap_{\varkappa=0}^{k} (\ker R_a^\varkappa \cap \ker R_b^\varkappa) \right]$$

$$\times C_0^k[a, b] \to C_0^k[a, b] \qquad (5.38)$$

$$f \mapsto q + \sum_{i=1}^{N} \lambda_i (f \circ u_i^{-1}) \chi_{I_i}$$

where the boundary operators $R_a^\varkappa, R_b^\varkappa : C^\varkappa[a, b] \to \mathbb{R}$, $\varkappa = 0, 1, \ldots, k$, are defined by

$$R_a^\varkappa f := (D^\varkappa f)(a) \quad \text{and} \quad R_b^\varkappa f := (D^\varkappa f)(b),$$

is an element of $C_0^k[a, b]$. (The reader is encouraged to verify this statement.)

Note that the function q has to satisfy a totality of $(k+1)(N-1) + 2(k+1) = (N+1)(k+1)$ conditions. Now, recall that a B-spline of order $k+2$ is a C_0^k-function on its support satisfying join-up conditions of the type stated in (5.37). Hence, one may choose for q a scaled version of the cardinal B-spline $B_{k+2} := B_{0,k+2,\mathbb{Z}}$. More precisely, suppose that $r : [a, b] \to [0, k+2]$,

$$r := \frac{k+2}{b-a}(\bullet - a),$$

is a re-scaling map. Then we select

$$q_i := B_{k+2} \circ r \circ u_i, \quad i = 1, \ldots, N.$$

Since with this selection, $\boldsymbol{q} = (q_1, \ldots, q_N)^\top \in \bigcap_{i=1}^{N-1} \bigcap_{\varkappa=0}^{k} \ker L_i^\varkappa \cap \bigcap_{\varkappa=0}^{k} (\ker R_a^\varkappa \cap \ker R_b^\varkappa)$, it is suggestive to choose

$$\Lambda^{k+2} := \bigcap_{i=1}^{N-1} \bigcap_{\varkappa=0}^{k} \ker L_i^\varkappa \cap \bigcap_{\varkappa=0}^{k} (\ker R_a^\varkappa \cap \ker R_b^\varkappa).$$

Note that since we use B-splines of order $k+2$, we expressed this explicitly in the definition of the spaces Λ^{k+2}. The dependence of the re-scaling map r was, however, suppressed.

The following example will illustrate the current Case 1.

Example 5.43 Let $N := 2$ and let $k := 1$. Moreover, let $[a, b] := [0, 1]$. Then $u_1 = \frac{1}{2}(\bullet)$ and $u_2 = \frac{1}{2}(\bullet + 1)$. The join-up conditions (5.37) now read

$$q_2(0) = q_1(1), \quad q_2'(0) = q_1'(1),$$

and

$$q_1(0) = q_1(1) = q_2(0) = q_2(1) = 0, \quad q_1'(0) = q_1'(1) = q_2'(0) = q_2'(1) = 0.$$

Selecting for the functions q_1 and q_2 the quadratic cardinal B-spline, B_3,

$$q_1 := B_3(3\,u_1) \quad \text{and} \quad q_2 := B_3(3\,u_2)$$

(since the re-scaling function $r = 3(\bullet)$) and positive constants $\lambda_1, \lambda_2 \in \mathbb{R}$ so that $|\lambda_1| + |\lambda_2| < \frac{1}{2}$ produces a contractive RB operator on $C_0^1[0, 1]$ whose fixed point $\mathfrak{b} = \mathfrak{b}[\lambda_1, \lambda_2]$ is a fractal function of class C^1 over $[0, 1]$. In Figure 5.13, we show the graphs of $\mathfrak{b}[0.25, -0.2]$ and $\mathfrak{b}[0.1, 0.1]$.

Case 2: $(D^\varkappa f)(a) = (D^\varkappa f)(b) \neq 0, \varkappa = 0, 1 \ldots, k,$ **and** $\lambda_i = \lambda, \forall i = 1, \ldots, N.$

In this case, (5.33) implies that

$$(D^\varkappa q_{i+1})(a) = (D^\varkappa q_i)(b), \quad \forall \varkappa = 0, 1, \ldots, k; \ \forall i = 1, \ldots, N-1.$$

This suggests using the function space

$$C^k[a, b] := \{f \in C^k[a, b] \mid (D^\varkappa f)(a) = (D^\varkappa f)(b), \ \varkappa = 0, 1 \ldots, k\},$$

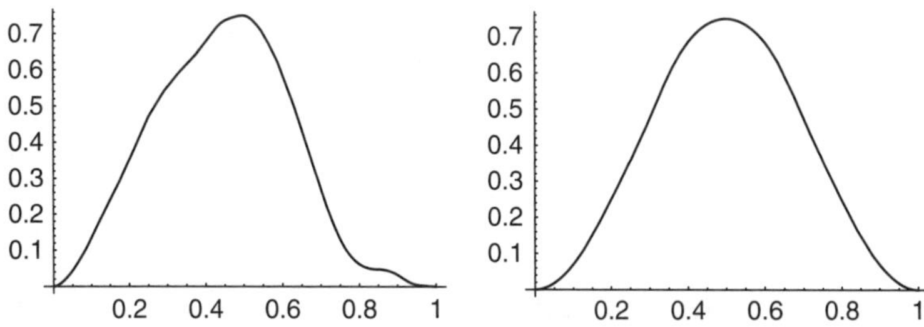

Figure 5.13 The graph of the fractal functions $\mathfrak{b}[0.25, -0.2]$ and $\mathfrak{b}[0.1, 0.1]$.

which, when endowed with the C^k-norm $\| \; \|_{C^k,[a,b]}$, becomes a complete metric space.

As above, set $\boldsymbol{q} := (q_1, \ldots, q_N)^\top$ and, for $\varkappa = 0, 1, \ldots, k$, define linear functionals $\widetilde{L}_i^\varkappa : \underset{i=1}{\overset{N}{\mathsf{X}}} C^\varkappa[a, b] \to \mathbb{R}$ by

$$\widetilde{L}_i^\varkappa \boldsymbol{q} := (D^\varkappa q_{i+1})(a) - (D^\varkappa q_i)(b), \ \forall i = 1, \ldots, N-1,$$

and linear boundary functionals $\widetilde{R}^\varkappa : \underset{i=1}{\overset{N}{\mathsf{X}}} C^\varkappa[a, b] \to \mathbb{R}, \varkappa = 0, 1, \ldots, k$, by

$$\widetilde{R}^\varkappa \boldsymbol{q} := (D^\varkappa q_1)(a) - (D^\varkappa q_N)(b).$$

We leave it to the reader to establish that the RB operator defined by

$$\widetilde{T}[\lambda, \boldsymbol{q}] : (-N^{-k-1}, +N^{-k-1}) \times \left[\bigcap_{i=1}^{N-1} \bigcap_{\varkappa=0}^{k} \ker \widetilde{L}_i^\varkappa \cap \bigcap_{\varkappa=0}^{k} \ker \widetilde{R}^\varkappa \right]$$

$$\times C^k[a, b] \to C^k[a, b] \tag{5.39}$$

$$f \mapsto q + \sum_{i=1}^{N} \lambda \, (f \circ u_i^{-1}) \, \chi_{I_i}$$

has range $C^k[a, b]$ and thus possesses a unique fixed point $\mathfrak{f} = \mathfrak{f}[\lambda, \boldsymbol{q}]$ in $C^k[a, b]$. (Since this fixed point depends only on *one* scaling constant, namely λ, we can write $\boldsymbol{\lambda} = \lambda \mathbf{1}$, where $\mathbf{1} := (1, \ldots, 1)^\top \in \mathbb{R}^N$, and thus express this dependence in the indicated way.)

As we specify $k + 1$ conditions at each of the knots x_ν, $\nu = 0, 1, \ldots, N$, where $a = x_0$ and $b = x_N$, it is suggestive to use for the functions q_i, $i = 1, \ldots, N$, Hermite splines, i.e., elements of the function space $\Pi_{\xi, k+1}^{2(k+1)}$, where ξ is the ordered knot set $\{a = x_0 < x_1 < \cdots < x_{N-1} < x_N = b\}$. (See Section 2.8.) Note that $\dim \Pi_{\xi, k+1}^{2(k+1)} = (N+1)(k+1)$, which is the total number of conditions the functions $q_i \in C^k[a, b]$ have to satisfy. Hence, let $h \in \Pi_{\xi, k+1}^{2(k+1)}$ and set

$$q_i \circ u_i^{-1} := h|_{I_i}, \quad i = 1, \ldots, N.$$

We remark that for this choice of q_i's one has $q \in \bigcap\limits_{i=1}^{N-1} \bigcap\limits_{\varkappa=0}^{k} \ker \widetilde{L}_i^{\varkappa} \cap \bigcap\limits_{\varkappa=0}^{k} \ker \widetilde{R}^{\varkappa}$
and, therefore, we can choose

$$\Lambda^{2(k+1)} := \bigcap_{i=1}^{N-1} \bigcap_{\varkappa=0}^{k} \ker \widetilde{L}_i^{\varkappa} \cap \bigcap_{\varkappa=0}^{k} \ker \widetilde{R}^{\varkappa},$$

where we again expressed the explicit dependence upon the Hermite splines of order $2(k+1)$.

We consider a simple example of this special case for this particular choice of q_i.

Example 5.44 Choose $N := 2$ and select $k := 1$. Moreover, let $[a, b] := [0, 1]$. Then, we obtain $u_1 = \frac{1}{2}(\bullet)$ and $u_2 = \frac{1}{2}(\bullet + 1)$. For the functions q_1 and q_2, we select a cubic Hermite spline $h \in \Pi_{\xi,2}^4$ with ordered knot set $\xi := \{0, \frac{1}{2}, 1\}$:

$$q_1 \circ u_1^{-1} := -10x^3 + 7x^2 + x + 1 \quad \text{and} \quad q_2 \circ u_2^{-1} := 22x^3 - 49x^2 + 33x - 5.$$

Choose a positive constant $\lambda \in \mathbb{R}$ so that $|\lambda| < \frac{1}{4}$. The fixed point $\mathfrak{h} = \mathfrak{h}[\lambda]$ of the underlying contractive RB operator is a fractal function of class C^2 over $[0, 1]$. The graph of $\mathfrak{h}[0.2]$ and the cubic Hermite spline h are shown in Figure 5.14.

Let us now return to the definition of the RB operator $T = T[\lambda, q]$ given in (5.32) and (5.33). For a given $k \in \mathbb{N}_0$ and $\varkappa = 1, \ldots, k$, we define the derivative $D^{\varkappa} T$ of T by

$$
\begin{aligned}
(D^{\varkappa} T)f &:= \sum_{i=1}^{N} D^{\varkappa}(q_i \circ u_i^{-1})\, \chi_{I_i} + \sum_{i=1}^{N} \lambda_i\, D^{\varkappa}(f \circ u_i^{-1})\, \chi_{I_i} \\
&= \sum_{i=1}^{N} a_i^{-\varkappa}(D^{\varkappa} q_i) \circ u_i^{-1}\, \chi_{I_i} + \sum_{i=1}^{N} a_i^{-\varkappa} \lambda_i\, (D^{\varkappa}f) \circ u_i^{-1}\, \chi_{I_i},
\end{aligned}
\tag{5.40}
$$

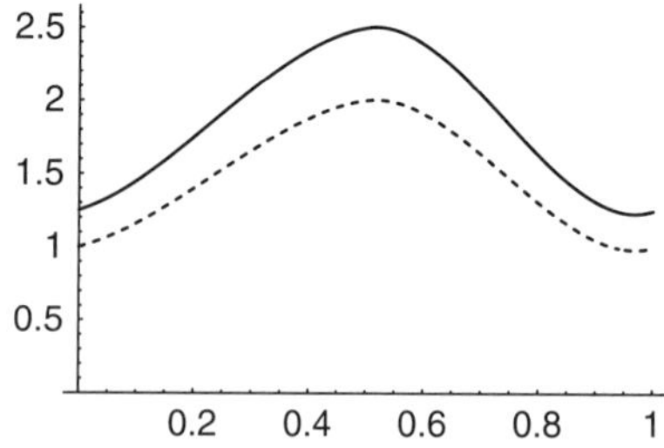

Figure 5.14 The graphs of the fractal function $\mathfrak{h}[0.2]$ (solid line) and the cubic Hermite spline h (dashed line).

where we set $a_i := \mathrm{Lip}(u_i) \in (0,1)$, $i = 1, \ldots, N$. The range of the operator

$$D^{\varkappa} T : \mathbb{R}^N \times (\underset{i=1}{\overset{N}{\textsf{X}}} C^{k-\varkappa}[a,b]) \times C^{k-\varkappa}[a,b] \to \mathrm{Map}([a,b],\mathbb{R})$$

is contained in $C^{k-\varkappa}[a,b]$ provided

$$D^m(D^{\varkappa} T)f(x_i-) = D^m(D^{\varkappa} T)f(x_i+), \quad \forall m \in \{0, \ldots, k-\varkappa\} \text{ and}$$

$$\forall i \in \{1, \ldots, N-1\}.$$

Note that, if we require

$$\max\{|a_i^{-\varkappa} \lambda_i| \mid i = 1, \ldots, N\} < 1, \tag{5.41}$$

then the unique fixed point, $\mathrm{Fix}(D^{\varkappa} T) =: D^{\varkappa}\mathfrak{f}$, of $D^{\varkappa} T$ may be interpreted as the $\varkappa$-th derivative of the fractal function $\mathfrak{f}$ of class C^k. The fractal function $D^{\varkappa}\mathfrak{f}$ is then of class $C^{k-\varkappa}$. If we indicate the dependence of this fixed point on the N-tuples λ and $\boldsymbol{q}$, we see that

$$\mathrm{Fix}(D^{\varkappa} T) = D^{\varkappa}\mathfrak{f}[\boldsymbol{a}^{-\varkappa}\lambda, \boldsymbol{a}^{-\varkappa}D^{\varkappa}\boldsymbol{q}], \quad \varkappa = 0, 1, \ldots, k,$$

where we defined

$$\boldsymbol{a}^{-\varkappa}\lambda := (a_1^{-\varkappa}\lambda_1, \ldots, a_N^{-\varkappa}\lambda_N) \in \mathbb{R}^N$$

and

$$\boldsymbol{a}^{-\varkappa}D^{\varkappa}\boldsymbol{q} := (a_1^{-\varkappa}D^{\varkappa}q_1, \ldots, a_N^{-\varkappa}D^{\varkappa}q_N) \subset \underset{i=1}{\overset{N}{\textsf{X}}} C^{k-\varkappa}[a,b], \quad \varkappa = 0, 1, \ldots, k.$$

In other words, every fractal function $D^{\varkappa}\mathfrak{f}$ is in a lower differentiability class and has a larger scaling vector.

Example 5.45 We will use the setting given in Example 5.44, but with $k := 2$ to illustrate the above findings. Let $\mathfrak{s}[\lambda] \in \Pi^6_{\xi,3}$ be the unique fixed point satisfying

$$\mathfrak{s}[\lambda](x) = \begin{cases} 244x^5 - 309x^4 + 106x^3 - \frac{1}{4}x^2 - x + 1 + \lambda\,\mathfrak{s}[\lambda](2x), & x \in \left[0, \frac{1}{2}\right) \\ -148x^5 + 559x^4 - 814x^3 + \frac{2271}{4}x^2 - \frac{381}{2}x + \frac{107}{4} \\ \quad + \lambda\,\mathfrak{s}[\lambda](2x-1), & x \in \left[\frac{1}{2}, 1\right]. \end{cases}$$

Then $D\mathfrak{s}$ and $D^2\mathfrak{s}$ satisfy the following fixed point equations:

$$D\mathfrak{s}[2\lambda](x) = \begin{cases} 2440x^4 - 2472x^3 + 636x^2 - x - 2 + 2\lambda\,D\mathfrak{s}[2\lambda](2x), & x \in \left[0, \frac{1}{2}\right) \\ -1480x^4 + 4472x^3 - 4884x^2 + 2271x - 381 \\ \quad + 2\lambda\,D\mathfrak{s}[2\lambda](2x-1), & x \in \left[\frac{1}{2}, 1\right], \end{cases}$$

and

$$D^2\mathfrak{s}[4\lambda](x) = \begin{cases} 19520x^3 - 14832x^2 + 2544x - 2\lambda\,D^2\mathfrak{s}[4\lambda](2x), & x \in \left[0, \frac{1}{2}\right) \\ -11840x^3 + 26832x^2 - 19536x + 4542 \\ \quad + 4\lambda\,D^2\mathfrak{s}[4\lambda](2x-1), & x \in \left[\frac{1}{2}, 1\right]. \end{cases}$$

The graphs of the fractal functions $\mathfrak{s}[0.2]$, $D\mathfrak{s}[0.4]$, and $D^2\mathfrak{s}[0.8]$ are depicted in Figure 5.15.

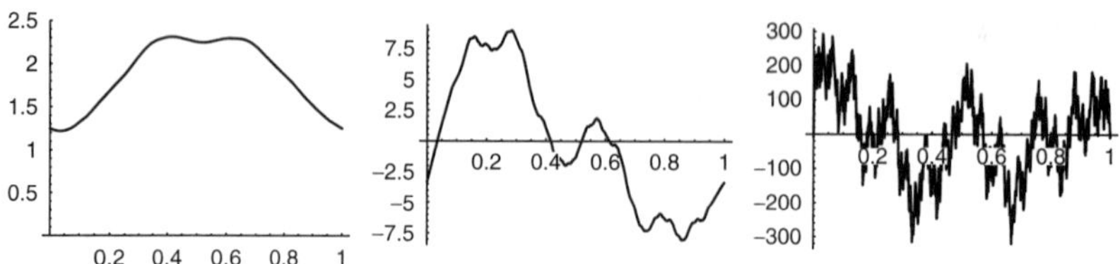

Figure 5.15 The graphs of the fractal functions $\mathfrak{s}[0.2]$, $D\mathfrak{s}[0.4]$, and $D^2\mathfrak{s}[0.8]$.

5.10 Fractal B-Splines

In this short section, we define the fractal analogue of cardinal B-splines. To this end, denote by $B_N := B_{0,N,\mathbb{Z}}$ the cardinal B-spline with support $[0, N]$, where $N \in \mathbb{N}$, and define

$$u_i := \frac{(\bullet) - i + 1}{N}, \quad i = 1, \ldots, N.$$

Then the u_i are contractive diffeomorphisms on $[0, N]$. Note that $B_N \in C_0^{N-2}[0, N]$, the natural function space for this setting, and that the knots of B_N are the images of 0 and N under the diffeomorphisms u_i.

Now define an RB operator $T := T[\lambda] : \mathbb{R}^N \times C_0^{N-2}[0, N] \to C_0^{N-2}[0, N]$ by

$$T[\lambda]f := \sum_{i=1}^{N} (B_N \circ u_i) \circ u_i^{-1} \chi_{I_i} + \sum_{i=1}^{N} \lambda_i (f \circ u_i^{-1}) \chi_{I_i} \qquad (5.42)$$

$$= B_N + \sum_{i=1}^{N} \lambda_i (f \circ u_i^{-1}) \chi_{I_i},$$

where $I_i = u_i[0, N)$, $i = 1, \ldots, N - 1$, and $I_N = u_N[0, N]$. (Show that $T[\lambda]$ indeed maps $C_0^{N-2}[a, b]$ into itself!)

It is straight-forward to show that if $\max\{|\lambda_i|N^\nu\} < 1$, then $T[\lambda]$ has a unique fixed point in $C_0^\nu[0, N]$, $\nu = 0, 1 \ldots, N - 2$. (See also Theorem 5.41.) This unique fixed point is completely determined by the cardinal B-spline B_N and is therefore denoted by $\mathfrak{B}_N[\lambda]$, as its regularity depends on the size of the entries in the scaling vector λ. The fractal function $\mathfrak{B}_N[\lambda]$ will be called a *fractal B-spline of order N*. Its fixed point equation reads

$$\mathfrak{B}_N[\lambda] = B_N + \sum_{i=1}^{N} \lambda_i (\mathfrak{B}_N[\lambda] \circ u_i^{-1}) \chi_{I_i}. \qquad (5.43)$$

Note that $\mathfrak{B}_N[0] = B_N$ and thus $\mathfrak{B}_N[\lambda]$ may be interpreted as an entire class of fractal analogues of the classical cardinal B-spline parametrized by the scaling vector λ.

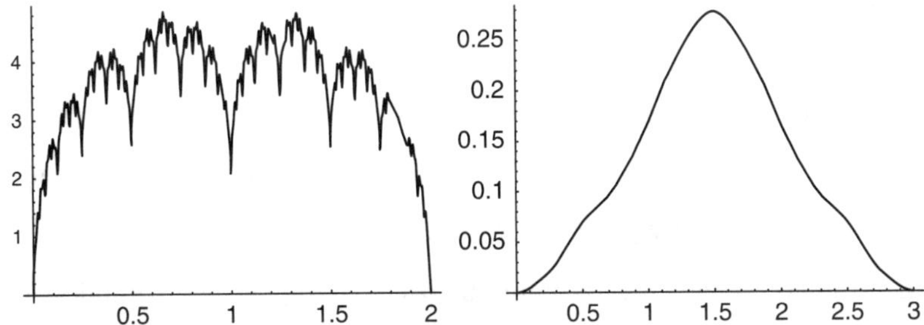

Figure 5.16 The fractal B-splines $\mathfrak{B}_2[0.75, 0.75]$ and $\mathfrak{B}_3[0.25, 0.25, 0.25]$.

Two examples of fractal B-splines are given in Figure 5.16. The first fractal B-splines is of order two and only continuous, whereas the second one is of order three and differentiable.

The graphs of the fractal B-splines in Figure 5.16 suggest the following result, whose proof is an exercise left to the reader.

Proposition 5.46 If $\lambda = \lambda 1$, then the fractal B-splines $\mathfrak{B}_N[\lambda]$ are symmetric about a vertical line through the midpoint of their interval of support.

Proof Exercise! □

We also notice that (5.43) is a special case of the setting considered in Section 5.7 with the endpoint conditions $\alpha := \beta := 0$, thus implying $h \equiv 0$.

Remark 5.47 It should be clear from the above elaborations that any spline, B-spline, cardinal spline, exponential spline, $\mathcal{L}$-spline, etc. can be used to obtain a family of fractal analogues of these classical approximants.

5.11 Approximation with Fractal Functions

In this section, we derive estimates for the approximation of continuous functions by their fractal analogues. For this purpose, we follow the setting given in Section 5.7 and, in particular, recall equation (5.29).

Let $g \in C[a, b]$ and let $\lambda := (\lambda_1, \ldots, \lambda_N) \in \mathbb{R}^N$ be such that $\lambda := \max\{|\lambda_i| \mid i = 1, \ldots, N\} < 1$. Moreover, let $g[\lambda]$ be the family of fractal functions generated from g. Then we have the following estimates.

Proposition 5.48

$$\|g[\lambda] - g\|_{\infty,[a,b]} \leq \frac{\lambda}{1 - \lambda} \|g - h\|_{\infty,[a,b]}$$

Proof Suppose that $g \in C[a, b]$ is given. Choose $x \in [x_{i-1}, x_i]$, $i = 1, \ldots, N$. Then

$$|g[\lambda](x) - g(x)| = |\lambda_i(g[\lambda] \circ u_i^{-1})(x) - \lambda_i(h \circ u_i^{-1})(x)|$$

$$\leq \lambda|(g[\lambda] \circ u_i^{-1}) - (g \circ u_i^{-1})(x)| + \lambda|(g \circ u_i^{-1})(x) - (h \circ u_i^{-1})(x)|$$

$$\leq \lambda\|g[\lambda] - g\|_{\infty,[a,b]} + \lambda\|g - h\|_{\infty,[a,b]},$$

which, after taking the supremum over all $x \in [a, b]$, implies the statement. $\square$

In case we consider functions h of the form (5.30), more can be said about the above estimate.

Corollary 5.49 Suppose that the function h is of the form given in Equation (5.30) and that $g \in C[a, b]$. Then

$$\|g[\lambda] - g\|_{\infty,[a,b]} \leq \frac{\lambda}{1-\lambda}(\max\{|g(a)|, |g(b)|\} + \|g\|_{\infty,[a,b]})$$

$$\leq \frac{2\lambda}{1-\lambda}\|g\|_{\infty,[a,b]}.$$

Proof We only need to estimate the term $\|g - h\|_{\infty,[a,b]}$. But,

$$\|g - h\|_{\infty,[a,b]} \leq \|g\|_{\infty,[a,b]} + \|h\|_{\infty,[a,b]}$$

$$= \|g\|_{\infty,[a,b]} + \left\|\frac{(g(b) - g(a))(\bullet) + (bg(a) - ag(b))}{b - a}\right\|_{\infty,[a,b]}$$

$$\leq \|g\|_{\infty,[a,b]} + \max\{|g(a)|, |g(b)|\} \leq \frac{2\lambda}{1-\lambda}\|g\|_{\infty,[a,b]}. \quad \square$$

The norm of the RB operator associated with the fixed point equation (5.29) can be estimated as well.

Proposition 5.50 Assume that $0 \neq g \in C[a, b]$ and that h is of the form (5.30). Then

$$\|T[\lambda]\|_{\infty,[a,b]} \leq \frac{1+\lambda}{1-\lambda}.$$

Proof Consider the following sequence of inequalities:

$$\|g[\lambda]\|_{\infty,[a,b]} \leq \|g[\lambda] - g\|_{\infty,[a,b]} + \|g\|_{\infty,[a,b]}$$

$$\leq \frac{2\lambda}{1-\lambda}\|g\|_{\infty,[a,b]} + \|g\|_{\infty,[a,b]} = \frac{1+\lambda}{1-\lambda}\|g\|_{\infty,[a,b]}.$$

This yields the result. $\square$

In the case of fractal splines, we can use Theorem 2.72 to obtain an error estimate between a function $f \in C^k[a, b]$ and the fractal splines $\mathfrak{s}[\boldsymbol{\lambda}]$. Since every spline $s \in S^k(X_N)$ is an element of $C^{k-2}[a, b]$, where as usual $a := x_0 < x_1 < \cdots < x_N := b$, we also have that the fractal family $\mathfrak{s}[\boldsymbol{\lambda}]$ generated by s is also an element of $C^{k-2}[a, b]$ provided that $\sum_{i=1}^{N} |\lambda_i| (\Delta x_i)^{k-2} < 1$. (Here, $\Delta x_i := x_i - x_{i-1}$, since we are using linear diffeomorphisms u_i, $i = 1, \ldots, N$.)

Now, suppose that $f \in C^k[a, b]$. Then, by Corollary 2.73 we have

$$E_k(f; \mathcal{S}^k(X_N)) \leq c_{k,j} \|\boldsymbol{x}\|^j \|D^j f\|_{\infty, [a,b]} \leq c_{k,j} \|\boldsymbol{x}\|^j \|f\|_{C^j}, \quad j = 0, 1, \ldots, k-1.$$

For $s \in \mathcal{S}^k(X_N) = S^k(\Xi)$ (on $[a, b]$) and $\mathfrak{s}[\boldsymbol{\lambda}]$, we get

$$\|f - \mathfrak{s}[\boldsymbol{\lambda}]\|_{C^{k-2}} \leq \|f - s\|_{C^{k-2}} + \|s - \mathfrak{s}[\boldsymbol{\lambda}]\|_{C^{k-2}}$$

$$\leq c_{k,k-2} \|\boldsymbol{x}\|^{k-2} \|f\|_{C^{k-2}} + \frac{2(\max\{\Delta x_i \mid i = 1, \ldots, N\})^{k-2} \lambda}{1 - \max\{\Delta x_i \mid i = 1, \ldots, N\})^{k-2} \lambda} \|s\|_{C^{k-2}},$$

given that h is of the form (5.30) and $\lambda := \max\{|\lambda_i| \mid i = 1, \ldots, N\}$.

5.12 Indefinite Integrals of Fractal Functions

In this section, we briefly investigate indefinite integrals of fractal functions and show that the indefinite integral of a fractal interpolation function is again a fractal interpolation function, but with new scaling factors and for different interpolation points. The discussion here is based on the original paper by Barnsley and Harrington [15].

We adopt the setting described in Remark 5.36, and suppose that we are given an interpolation set $Y := \{(x_\nu, y_\nu) \mid \nu = 0, 1, \ldots, N\}$, for some $1 < N \in \mathbb{N}$. Moreover, we assume that $g \in C[x_0, x_N]$ is any interpolant for Y and $h \in C_{y_0, y_N}[x_0, x_N]$ is arbitrary. Furthermore, let $u_i : [x_0, x_N] \to I_i := [x_{i-1}, x_i]$, $i = 1, \ldots, N$, be affine contractions generating a partition of $[x_0, x_N]$.

For a given $\mathbf{0} \neq \boldsymbol{\lambda} \in (-1, 1)^N$, we denote by $\mathfrak{g} := \mathfrak{g}[\boldsymbol{\lambda}] : [a, b] \to \mathbb{R}$ the fractal interpolation function constructed as in Section 5.7. Then, $\mathfrak{g}$ satisfies the fixed point equation (see also (5.29))

$$\mathfrak{g}(x) = g + \sum_{i=1}^{N} \lambda_i (\mathfrak{g} - h) \circ u_i^{-1}(x) \, \chi_{I_i},$$

or, equivalently,

$$\mathfrak{g}(u_i(x)) = q_i(x) + \lambda_i \mathfrak{g}(x), \quad \forall i = 1, \ldots, N; \ \forall x \in [x_0, x_N], \tag{5.44}$$

with $q_i := g(u_i(x)) - \lambda_i h(x)$.

Now consider the indefinite integral $\mathcal{I}(\mathfrak{g})(x) := \int_{x_0}^{x} \mathfrak{g}(t) dt$, for $x \in [x_0, x_N]$. Since $\mathfrak{g}$ is continuous, the integral $\mathcal{I}(\mathfrak{g})$ exists for all $x \in [x_0, x_N]$.

Moreover, $\mathscr{I}(\mathfrak{g}) : [x_0, x_N] \to \mathbb{R}$ and, clearly, $\mathscr{I}(\mathfrak{g})(x_0) = 0$. We like to investigate the question under what condition(s) the function $\mathscr{I}(\mathfrak{g})$ is again a fractal interpolation function. If we can show that $\mathscr{I}(\mathfrak{g})$ satisfies a fixed point equation for some RB operator, then the uniqueness of the fixed point would give us the result.

For this purpose, we apply the integral operator $\mathscr{I}$ to the left-hand side of the fixed point equation (5.44) and obtain, for $i = 1, \ldots, N$,

$$(\mathscr{I}\mathfrak{g})(u_i(t)) = \int_{x_0}^{u_i(x)} \mathfrak{g}(t)\,dt = \int_{x_0}^{x_{i-1}} \mathfrak{g}(t)\,dt + \int_{x_{i-1}}^{u_i(x)} \mathfrak{g}(t)\,dt, =: \widetilde{y}_{i-1}$$

$$+ \int_{x_0}^{x} \mathfrak{g}(u_i(t))\,du_i(t) = \widetilde{y}_{i-1} + a_i \int_{x_0}^{x} \mathfrak{g}(u_i(t))\,dt, \quad \forall x \in [x_0, x_N],$$

where $a_i := \mathrm{Lip}\, u_i$. Defining $\mathfrak{G}(x) := (\mathscr{I}\mathfrak{g})(x) = \int_{x_0}^{x} \mathfrak{g}(t)\,dt$ and employing the fixed point equation (5.44) yields

$$\mathfrak{G}(u_i(x)) = \widetilde{y}_{i-1} + a_i Q_i(x) + a_i \lambda_i \mathfrak{G}(x), \tag{5.45}$$

which is valid for all $x \in [x_0, x_N]$ and for all $i = 1, \ldots, N$. Here we defined

$$Q_i(x) := \int_{x_0}^{x} q_i(t)\,dt, \quad i = 1, \ldots, N.$$

The continuity of the integral operator $\mathscr{I}$ implies that the continuous join-up conditions are satisfied at the interpolation points $(x_\nu, y_\nu) \in Y$. Setting $x = x_N$ in (5.45) produces

$$\mathfrak{G}(x_i) = \widetilde{y}_i = \widetilde{y}_{i-1} + a_i Q_i(x_N) + a_i \lambda_i \widetilde{y}_N,$$

and therefore

$$\widetilde{y}_i = \sum_{j=1}^{i} a_j (\lambda_j \widetilde{y}_N + Q_j(x_N)), \quad \forall i = 1, \ldots, N. \tag{5.46}$$

In addition, we define $\widetilde{y}_0 := 0$. Setting $i = N$ gives an explicit formula for $\widetilde{y}_N$:

$$\widetilde{y}_N = \frac{\displaystyle\sum_{j=1}^{N} a_j Q_j(x_N)}{1 - \displaystyle\sum_{j=1}^{N} a_j \lambda_j}. \tag{5.47}$$

Notice that $0 < \sum_{i=1}^{N} a_i \lambda_i < 1$ (The reader is encouraged to verify this statement.). This formula together with (5.46) allows the recursive computation of the new interpolation values $\widetilde{y}_\nu$, $\nu = 0, 1 \ldots, N$.

The indefinite integral $\mathfrak{G}$ is clearly the fixed point of the $\| \, \|_\infty$-contractive operator

$$f \longmapsto \sum_{i=1}^{N} \left[\tilde{y}_{i-1} + a_i Q_i(x)\right] \circ u_i^{-1} \chi_{I_i} + \sum_{i=1}^{N} a_i \lambda_i (f \circ u_i^{-1}) \chi_{I_i},$$

with scaling vector $a\lambda$. Hence, $\mathfrak{G}$ is a fractal function interpolating the set $\tilde{Y} := \{(x_\nu, \tilde{y}_\nu) \mid \nu = 0, 1, \dots, N\}$. We summarize these results in a theorem.

Theorem 5.51 Let $Y := \{(x_\nu, y_\nu) \mid \nu = 0, 1, \dots, N\}$, $1 < N \in \mathbb{N}$, be a given interpolation set and let $g \in C[x_0, x_N]$ be any interpolant of Y. Furthermore, let $h \in C[x_0, x_N]$ satisfy $h(x_0) = y_0$ and $h(x_N) = y_N$. Let $\mathbf{0} \neq \lambda \in (-1, 1)^N$ and let $\mathfrak{g} := \mathfrak{g}[\lambda] : [x_0, x_N] \to \mathbb{R}$ be the fractal function generated by the operator (5.27). Let $\mathfrak{G}(x) = \mathscr{I}(\mathfrak{g})(x) = \displaystyle\int_{x_0}^{x} \mathfrak{g}(t)\,dt$ be the indefinite integral of $\mathfrak{g}$. Furthermore, let $Q_i(x) = \displaystyle\int_{x_0}^{x} q_i(t)\,dt = \int_{x_0}^{x} (g(u_i(t)) - \lambda_i h(t))\,dt$, $i = 1, \dots, N$. Let $\tilde{Y} = \{(x_\nu, \tilde{y}_\nu) \mid \nu = 0, 1, \dots, N\}$ be the interpolation set with $\tilde{y}_\nu$, $\nu = 0, 1, \dots, N$, given by (5.46) and (5.47).

Then $\mathfrak{G} : [x_0, x_N] \to \mathbb{R}$ is a fractal function interpolating the set $\tilde{Y}$ and is generated by the RB operator $T[a\lambda] : \mathbb{R}^N \times C_{\tilde{y}_0, \tilde{y}_N}[a, b] \to C_{\tilde{y}_0, \tilde{y}_N}[x_0, x_N]$,

$$T[a\lambda]f := \sum_{i=1}^{N} \left[\tilde{y}_{i-1} + a_i Q_i(x)\right] \circ u_i^{-1} \chi_{I_i} + \sum_{i=1}^{N} a_i \lambda_i (f \circ u_i^{-1}) \chi_{I_i},$$

with $a_i = \operatorname{Lip} u_i$, $i = 1, \dots, N$.

As an illustration, we compute the indefinite integral of Kiesswetter's fractal function. (See Example 5.9.)

Example 5.52 Kiesswetter's fractal function $\mathfrak{f}_K$ interpolates on the knot set $\{(0, 0), (\frac{1}{4}, -\frac{1}{2}), (\frac{1}{2}, 0), (\frac{3}{4}, \frac{1}{2}), (1, 1)\}$. Thus, $a_1 = a_2 = a_3 = a_4 = \frac{1}{4}$. Furthermore, the scaling factors are $\lambda_1 = -\frac{1}{2}, \lambda_2 = \lambda_3 = \lambda_4 = \frac{1}{2}$. The functions q_i are given by

$$q_1(x) = q_3(x) = 0, \quad q_2(x) = -\frac{1}{4}, \quad q_4(x) = \frac{1}{4},$$

implying that

$$Q_1(x) = Q_3(x) = 0, \quad Q_2(x) = -\frac{x}{4}, \quad Q_4(x) = \frac{x}{4}.$$

Moreover, $\tilde{y}_0 = \tilde{y}_1 = 0, \tilde{y}_2 = \tilde{y}_3 = -\frac{1}{16}$, and $\tilde{y}_4 = 0$. Setting $\mathfrak{F}_K(x) := \int_0^1 \mathfrak{f}_K(t)\,dt$, the indefinite integral $\mathfrak{F}_K = \mathscr{I}(\mathfrak{f}_K)$ of Kiesswetter's fractal function is thus

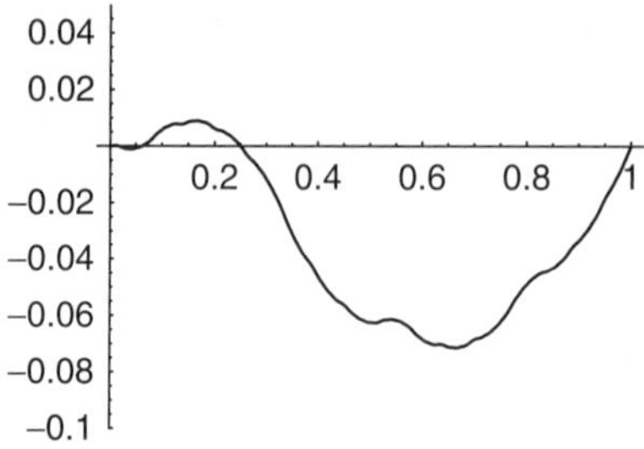

Figure 5.17 The indefinite integral of Kiesswetter's fractal function.

given by the fixed point equation

$$
\mathfrak{F}_K(x) = \begin{cases}
-\frac{1}{8}\mathfrak{F}_K(4x), & x \in [0, 1/4]; \\
-\frac{1}{16}(4x+1) + \frac{1}{8}\mathfrak{F}_K(4x+1), & x \in [1/4, 1/2]; \\
-\frac{1}{16} + \frac{1}{8}\mathfrak{F}_K(4x+2), & x \in [1/2, 3/4]; \\
\frac{1}{16}(4x+3) + \frac{1}{8}\mathfrak{F}_K(4x+3), & x \in [3/4, 1].
\end{cases}
$$

The graph of $\mathfrak{F}_K$ is shown in Figure 5.17.

5.13 Fourier Transform of Fractal Functions

In this section, we show how the Fourier transform of a fractal function can be computed and its value approximated to any arbitrary order. To this end, we assume that $\mathfrak{f} : [a, b] \to \mathbb{R}$ is a continuous fractal function satisfying the fixed point equation

$$
\mathfrak{f}(x) = \sum_{i=1}^{N} \left[(q_i \circ u_i^{-1})(x) + \lambda_i (\mathfrak{f} \circ u_i^{-1})(x) \right] \chi_{I_i}, \quad \forall x \in [a, b], \qquad (5.48)
$$

together with $\mathfrak{f}(a) = \alpha$ and $\mathfrak{f}(b) = \beta$, for fixed $\alpha, \beta \in \mathbb{R}$. Here, we assume that the u_i are affine contractions of the type considered in (5.8) inducing a *uniform* partition of $[a, b]$ of the form (5.9). Then $\mathrm{Lip}\, u_i = \frac{1}{N}$, $i = 1, \ldots, N$. The functions $q_i : [a, b] \to \mathbb{R}$, $i = 1, \ldots, N$, are assumed to be continuous and chosen in such a way that the associated RB operator (5.26) maps $C_{\alpha,\beta}[a, b]$ into itself.

Apply the Fourier transform $\mathscr{F}$ to both sides of (5.48) to obtain

$$
(\mathscr{F}\mathfrak{f})(\omega) =: \widehat{\mathfrak{f}}(\omega) = \sum_{i=1}^{N} \frac{1}{\sqrt{2\pi}} \int_{x_{i-1}}^{x_i} e^{-i\omega x} \left[q_i(u_i^{-1}(x)) + \lambda_i \widehat{\mathfrak{f}}(u_i^{-1}(x)) \right] dx,
$$

$\forall \omega \in \mathbb{R}$.

Since u_i is affine with nonzero Lipschitz constant $\frac{1}{N}$ and translation part b_i, we write $\omega x = \frac{1}{N}\omega\, u_i^{-1}(x) + \omega\, b_i$. Substituting this expression for ωx into the above equation and integrating the left-hand side with respect to $u_i^{-1}(x)$ yields

$$\widehat{f}(\omega) = \frac{1}{N}\sum_{i=1}^{N} e^{-i\omega b_i}\left[\widehat{q_i}(\omega/N) + \lambda_i\widehat{f}(\omega/N)\right], \quad \forall \omega \in \mathbb{R}. \tag{5.49}$$

Note that $\widehat{q_i}$ is the Fourier transform of q_i on the interval $[a, b]$, i.e., we extend q_i to all of $\mathbb{R}$ by setting it identically equal to zero off $[a, b]$.

Setting $\Lambda(\omega) := \frac{1}{N}\sum_{i=1}^{N} e^{-i\omega b_i}\lambda_i$, $\widehat{Q}(\omega) := \frac{1}{N}\sum_{i=1}^{N} e^{-i\omega b_i}\widehat{q_i}(\omega)$, and iterating (5.49) n-times produces the following expression:

$$\widehat{f}(\omega) = \sum_{k=1}^{n}[\Lambda(\omega)]^{k-1}\widehat{Q}\left(\frac{\omega}{N^k}\right) + [\Lambda(\omega)]^n\widehat{f}\left(\frac{\omega}{N^n}\right), \tag{5.50}$$

which is valid for all $\omega \in \mathbb{R}$.

As $n \to \infty$, $\widehat{f}(\omega/N^n) \to \widehat{f}(0)$ by the continuity of the Fourier transform ($f \in L^1(\mathbb{R})$!). Moreover, $|\Lambda(\omega)| < 1$, and thus

$$[\Lambda(\omega)]^n \xrightarrow{\; n\to\infty \;} 0.$$

By the Riemann–Lebesgue Lemma (2.30), the Fourier transform of q_i is uniformly bounded and therefore

$$\left|\widehat{Q}\left(\frac{\omega}{N^k}\right)\right| \le \frac{1}{N}\sum_{i=1}^{N}\left|\widehat{q_i}\left(\frac{\omega}{N^k}\right)\right| \le \max\{\|\widehat{q_i}\|_\infty \mid i = 1, \ldots, N\}, \quad \forall k \in \mathbb{N}.$$

These remarks, together with an application of the root test shows that

$$\sum_{k=1}^{\infty}[\Lambda(\omega)]^{k-1}\widehat{Q}\left(\frac{\omega}{N^k}\right) = \widehat{f}(\omega) \tag{5.51}$$

converges absolutely for all $\omega \in \mathbb{R}$. Also note that the smoothness of the functions q_i determines the rate of decay of the Fourier transform of f. (See Proposition 2.31.)

Formula (5.51) provides a means for computing the Fourier transform of a fractal function to arbitrarily high orders of approximation.

Example 5.53 Let $f : [0, 1] \to \mathbb{R}$ be the affine fractal function generated by the RB operator $T : C_{\alpha,\beta}[0, 1] \to C_{\alpha,\beta}[0, 1]$

$$(T f)(x) := \begin{cases} \frac{7}{10}x + \frac{1}{20} + \frac{3}{5}f(2x), & x \in [0, 1/2]; \\ -\frac{1}{10}x + 1 - \frac{4}{5}f(2x - 1), & x \in [1/2, 1], \end{cases}$$

with $\alpha := \frac{1}{4}$ and $\beta := \frac{1}{2}$. (See Figure 5.18.)

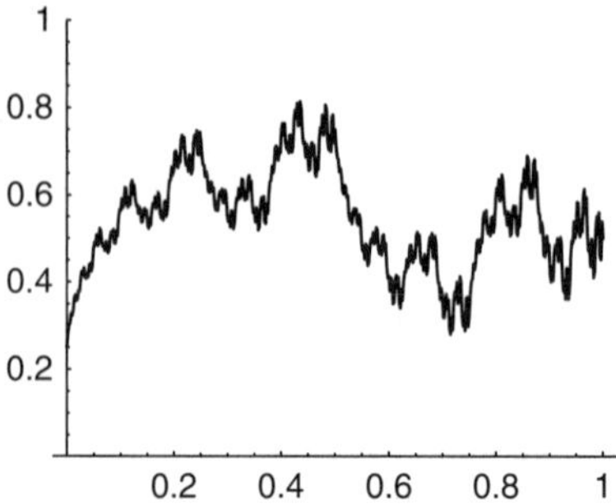

Figure 5.18 The affine fractal function $\mathfrak{f}$ in Example 5.53.

Then,

$$\Lambda(\omega) = \frac{1}{10}\left(3 - 4e^{-i\omega/2}\right)$$

and

$$\widehat{Q}(\omega) = \frac{1}{40\omega^2}\left(-7 - 2\,i\omega + (7 + 9i\omega)\cos\omega + (9\omega - 7\,i)\sin\omega\right.$$

$$\left. + e^{-i\omega/2}\left[1 - 19i\omega + (-1 + 18i\omega)\cos\omega + (18\omega + i)\sin\omega\right]\right).$$

Choosing $k = 20$ produces the following plots of $\mathrm{Re}\,\widehat{\mathfrak{f}}(\omega)$, $\mathrm{Im}\,\widehat{\mathfrak{f}}(\omega)$, and $|\widehat{\mathfrak{f}}(\omega)|$. (See Figures 5.19 and 5.20.) Notice that since q_1 and q_2 are C^1-functions, the rate of decay of the Fourier transform is $\mathcal{O}(\omega^{-1})$.

5.14 Wavelets and Fractal Functions

In Section 2.14 we have seen that cardinal B-splines can be used to define a multiresolution analysis of $L^2(\mathbb{R})$. The spline spaces $S^k(\mathbb{Z})$ formed the basis for this construction. In this section, we show that starting with slightly

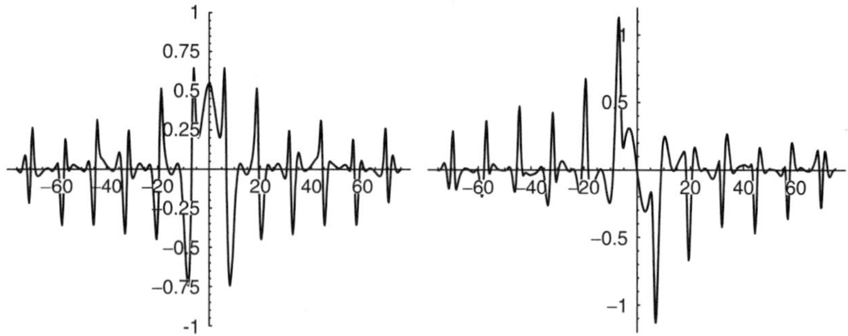

Figure 5.19 The real (left) and imaginary (right) part of the Fourier transform of $\mathfrak{f}$.

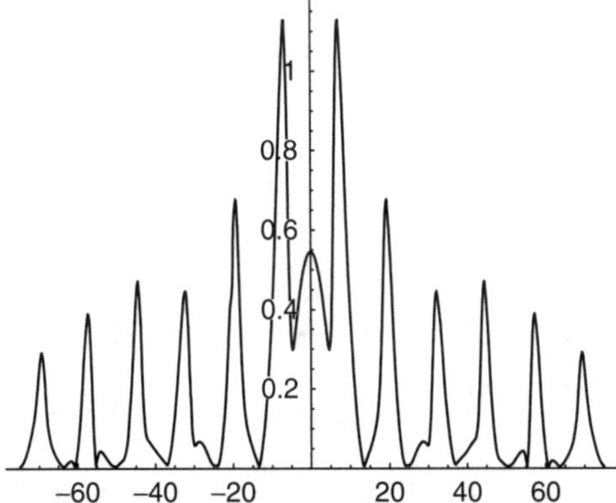

Figure 5.20 The absolute value of the Fourier transform of $\mathfrak{f}$.

different cardinal spline spaces, one can construct a FSI space consisting of two generators, each of which is a (piecewise) affine fractal function, and that this FSI space then generates a multiresolution analysis of $L^2(\mathbb{R})$. This construction commenced in [81] and was carried out fully in [49, 72]. It provided the first example of a compactly supported orthonormal scaling function and wavelet wavelet system outside the Daubechies' system of wavelets. Instead of using the more technical construction of [49] and [72], we follow the ideas presented in [130] to obtain the scaling functions.

For this purpose, define the (symmetric) cardinal spline spaces $\widetilde{S}^k(\mathbb{Z})$ by

$$s \in \widetilde{S}^k(\mathbb{Z}) \quad \Longleftrightarrow \quad s\left(\bullet - \frac{k}{2}\right) \in S^k(\mathbb{Z}).$$

In other words, the basis elements $\widetilde{B}_k := \widetilde{B}_{k,\mathbb{Z}}$ of $\widetilde{S}^k(\mathbb{Z})$ are all centered symmetrically around the origin and thus satisfy $\operatorname{supp}\widetilde{B}_k = [-k/2, k/2]$. Let $k := 2$ and set $h := B_2(\bullet - 1)$. The function h is also called the *hat* or *chapeau function*. The PSI space $S(h)$ is a closed subspace of $L^2(\mathbb{R})$, but the generator h is not orthonormal.

Example 5.54 Consider the functions $f_1 : [0, 1] \to [0, 1]$, $x \mapsto B_2(x - 1)$, and $f_2 : [0, 1] \to [0, 1]$, $x \mapsto 4x(1 - x)\chi_{[-1,1]}$. Then $S(\{f_1, f_2\})$ is an FSI space isomorphic to the linear space of all continuous piece-wise quadratic L^2-functions on integer knots. However, $\mathcal{F} := \{f_1, f_2\}$ is not an orthonormal generator of $S(\mathcal{F})$.

In order to obtain an FSI space with orthonormal generator, we proceed as follows. We take the hat function h together with a yet unknown continuous function u supported on the interval $[0, 1]$ and consider the FSI space $S(\{h, u\})$. The idea is to construct a third function v using a linear combination of h, u, and $u(\bullet + 1)$ so that v becomes orthogonal to u and its integer translates, and also to its own integer translates. Note that since u is defined on $[0, 1]$ it is automatically orthogonal to its own integer translates.

Define $v : [-1, 1] \to \mathbb{R}$ by

$$v := h - \frac{\langle u, h \rangle}{\langle u, u \rangle} u - \frac{\langle u(\bullet + 1), h \rangle}{\langle u, u \rangle} u(\bullet + 1).$$

Here $\langle , \rangle$ denotes the L^2-inner product. Clearly, $\langle v, u \rangle = \langle v, u(\bullet + 1) \rangle = 0$. Requiring that $\langle v, v(\bullet + 1) \rangle = \langle v, v(\bullet - 1) \rangle = 0$ yields

$$\frac{\langle h, u \rangle \langle h(\bullet - 1), u \rangle}{\langle u, u \rangle} = \langle h, h(\bullet - 1) \rangle = \frac{1}{6}. \tag{5.52}$$

Since we also require shift-invariance, the function u must be a linear combination of the functions $h(2 \bullet - 1)$, $u(2\bullet)$, and $u(2 \bullet - 1)$ on the interval $[0, 1]$. Thus, u has to satisfy the following refinement equation

$$u(x) = \lambda_0 \, h(x) + \sum_{i=1}^{2} \lambda_i \, u(2x - i), \tag{5.53}$$

where $\lambda_0, \lambda_1, \lambda_2 \in \mathbb{R}$. Since rescaling u affects neither the definition of v nor (5.52), we may assume that $\lambda_0 = 1$. Equation (5.53) is the fixed point equation for the operator $T : \mathbb{R}^2 \times C[0, 1] \to C[0, 1]$ given by

$$(T[\lambda_1, \lambda_2]f)(x) := h(x) + \sum_{i=1}^{2} \lambda_i f(2x - i). \tag{5.54}$$

To find a solution of (5.53), we note that $T[\lambda_1, \lambda_2]$ is an example of an RB operator and of the type considered earlier in this chapter when we constructed fractal functions. (In particular, see (5.43).) We also know that there exists a unique solution of (5.54) provided $\max\{\lambda_1, \lambda_2\} < 1$. Now, *assuming* that $\max\{\lambda_1, \lambda_2\} < 1$, we are guaranteed a unique solution u and it remains to show whether the unknowns λ_1 and λ_2 can be chosen so that (5.52) also holds. If in addition we set $\lambda_1 = \lambda_2$, then the resulting function u will be symmetric about the line $x = \frac{1}{2}$. Substituting (5.53) into (5.52) and computing the inner products using (5.22), Exercise 8, and (5.23) yields, after a lengthly but straight-forward calculation, as the only acceptable solution

$$\lambda_1 = \lambda_2 = -\frac{1}{5}. \tag{5.55}$$

Hence, there exists a unique symmetric solution u to the fixed point equation (5.53) also satisfying (5.52). This function u is a fractal function with box dimension one. To obtain an orthonormal generator, we define

$$\varphi_1 := \frac{u}{\sqrt{\langle u, u \rangle}} \quad \text{and} \quad \varphi_2 := \frac{v}{\sqrt{\langle v, v \rangle}}.$$

The generator $\Phi := (\varphi_1, \varphi_2)$ is called the *GHM scaling vector* and was constructed via different means in [72]. In Figure 5.21 the graphs of these two functions are displayed.

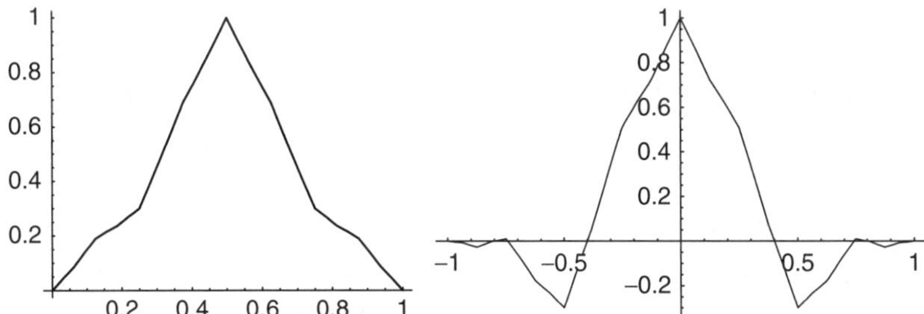

Figure 5.21 The GHM scaling vector.

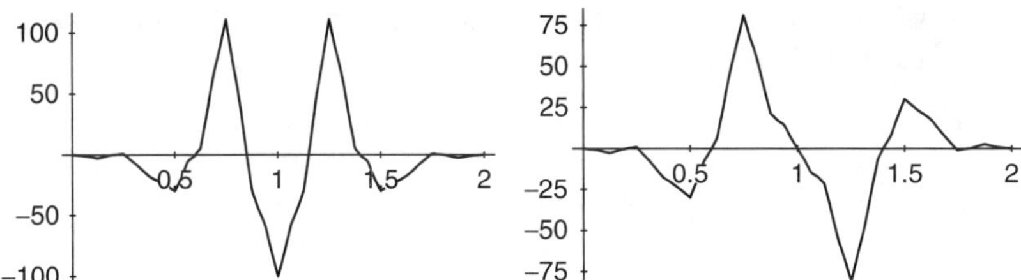

Figure 5.22 The DGHM multiwavelet.

We note that φ_2 is a *piecewise* fractal function consisting of two affine fractal functions supported on $[-1, 0]$ and $[0, 1]$, respectively.

The ladder of spaces $V_k := \mathcal{D}^k S(\Phi)$, $k \in \mathbb{Z}$, forms an MRA of $L^2(\mathbb{R})$, and the detail or wavelet spaces $W_k := V_{k+1} \ominus V_k$ are generated by two functions, $\psi_1, \psi_2 : [-1, 1] \to \mathbb{R}$, each of which is a piecewise affine fractal function. The graphs of these two wavelets are shown in Figure 5.22. The generator $\Psi := (\psi_1, \psi_2)$ is called the *DGHM multiwavelet*. For the construction of Ψ we refer to the original paper [49] or the summary in [113].

At this point, we like to summarize some of the properties of the generators Φ and Ψ. The interested reader is referred to the literature [72, 49, 84, 85, 113, 115, 143, 144] for proofs, detailed discussions, and applications.

- $\operatorname{supp}\phi_1 = [0, 1]$ and $\operatorname{supp}\phi_2 = \operatorname{supp}\psi_1 = \operatorname{supp}\psi_2 = [-1, 1]$.

- The scaling vector Φ and the associated multiwavelet Ψ satisfy the full set of L^2-orthonormality conditions: For all $j, k, \ell, m \in \mathbb{Z}$, let $\Phi_{jk} := 2^{j/2}\,\Phi(2^j \bullet -k)$ and $\Psi_{jk} := 2^{j/2}\,\Psi(2^j \bullet -k)$. Then

$$\langle \Phi_{jk}, \Phi_{j\ell} \rangle = \delta_{k\ell} I_2,$$

$$\langle \Psi_{jk}, \Psi_{m\ell} \rangle = \delta_{jm, k\ell} I_r,$$

$$\langle \Phi_{jk}, \Psi_{j\ell} \rangle = O_2,$$

where I_2 and O_2 denote the 2×2-identity and 2×2-zero matrix, respectively.

Here we defined the inner product of two vector-valued functions

$$f, g : \mathbb{R}^m \to \mathbb{R}^n \text{ by } \langle f, g \rangle := \int_{\mathbb{R}^n} f(x) g^\top(x) \, dx.$$

- The wavelets ψ_1 and ψ_2 are antisymmetric and symmetric, respectively.

- The FSI space $S(\Phi)$ has approximation order two, i.e., $\Pi^2 \subset S[\Phi]$. Since the wavelet space W_0 is orthogonal to the approximation space $V_0 = S(\Phi)$, this condition is equivalent to $W_0 \perp \Pi^2$, which itself is equivalent to the statement that the generator Ψ has vanishing moments of orders 0 and 1: $\langle (\bullet)^n, \Psi \rangle = 0$, $n = 0, 1$.

- $\Phi, \Psi \in C^{0,1}(\mathbb{R}) \times C^{0,1}(\mathbb{R})$. Hence all four component functions possess a weak first derivative.

- The GHM scaling vector is interpolatory: Given a set of interpolation points $X := \{X_i\}$ supported on $\frac{1}{2}\mathbb{Z}$, there exists a set of vector coefficients $\{\alpha_k\}$ such that $\sum_k \alpha_k^\top \Phi(x - k)$ interpolates X. (Note that $\phi_1(1/2) = 1 = \phi_2(0)$.)

- The DGHM multiwavelet system can be easily modified to obtain a multiresolution analysis on $L^2[0, 1]$ *without* the addition of boundary functions. (See, for instance, [117]).

- The length of support of Φ and Ψ, supp $\Phi = 3$, and the approximation order are the same as that of the Daubechies $_2\phi$ scaling function and $_2\psi$ wavelet, but the GHM scaling vector and DGHM multiwavelet have slightly higher regularity. It turns out that the Daubechies wavelet system $(_2\phi, _2\psi)$ and the (GHM, DGHM) wavelet system are the only two wavelet systems that have approximation order two and local dimension three [80].

5.15 Fractal Functions and Function Spaces

In this section, we very briefly address the question under what conditions on the scaling factors a given fractal function belongs to a certain function space. To this end, we recall Example 5.12 where we found conditions on the scaling factors λ_i, $i = 1, \ldots, N$, so that a fractal function $\mathfrak{f}$ is an element of $L^p(\mathbb{R})$, and Section 5.9 where condition (5.41) guaranteed that $\mathfrak{f}$ is an element of the function space $C^k[a, b]$. We will present similar conditions that guarantee that a certain class of fractal functions belongs to so-called *Besov* and *Triebel–Lizorkin spaces*. These two types of function spaces encompass many of the spaces that one encounters in approximation and interpolation theory. For a detailed introduction into the theory of both types of spaces, we refer to [151].

Before presenting the definition of these two function spaces, we first consider Hölder spaces. For this purpose, suppose that $\Omega \subseteq \mathbb{R}^n$, $n \in \mathbb{N}$, and $f : \Omega \to \mathbb{R}$ a function that belongs to the space $C^k(\Omega)$, $k \in \mathbb{N}$.

Definition 5.55 (Hölder space)   Let $0 \leq \alpha < 1$ and $k \in \mathbb{N}$. The linear space

$$C^{k,\alpha}(\Omega) := \left\{ f \in \mathrm{Map}(\Omega, \mathbb{R}) \mid \|f\|_{C^\alpha} := \sum_{|m| \leq k} \|D^m f\|_{\infty, \Omega} + \sum_{|m| = k} |D^m f|_{\mathrm{Lip}^\alpha} < \infty \right\}$$

is called the *Hölder space* of functions over Ω.

Remark 5.56   The Hölder space is the space of all k-times continuously differentiable functions from $\Omega \to \mathbb{R}$ whose k-th partial derivatives are Lipschitz continuous with exponent α. Hence, they extend the notion of Lipschitz space. Thus, one also sets $C^{0,\alpha} := \mathrm{Lip}^\alpha$.

In the literature, one also finds the notation $C^s(\Omega)$, $s > 0$, for Hölder spaces. Defining the *floor function* $\lfloor\,\rfloor : \mathbb{R} \to \mathbb{Z}$ by $\lfloor x \rfloor := \max\{z \in \mathbb{Z} \mid z \leq x\}$, and the *fractional part function* $\{\,\} : \mathbb{R} \to [0, 1)$ by $\{x\} := x - \lfloor x \rfloor$, one has

$$C^s(\Omega) = C^{\lfloor s \rfloor, \{s\}}(\Omega).$$

Remark 5.57   The Hölder spaces $C^{k,\alpha}(\Omega)$ fill in the gaps between the spaces $C^k(\Omega)$ and $C^{k+1}(\Omega)$, $k \in \mathbb{N}_0$.

In Section 5.9, we derived a condition for a fractal function to be of class C^k. Similar arguments can be employed to obtain a condition for a fractal function to belong to a Hölder space $C^{k,\alpha}(\Omega)$. We encourage the reader to derive such a condition.

Now, we are in a position to define Besov and Triebel–Lizorkin spaces. To this end, recall the concept of forward difference operator Δ_h^M of order M and step size $h > 0$ given in Definition 1.75. For the sake of completeness and generality, we also define what is called a *quasi-norm*.

Definition 5.58   Let X be a real or complex vector space. A function $\|\,\| : \mathsf{X} \to \mathbb{R}$ is called a *quasi-norm* if it satisfies all the usual conditions of a norm except for the triangle inequality, which is replaced by

$$\|x + y\| \leq c\,(\|x\| + \|y\|) \tag{5.56}$$

for a constant $c \geq 1$. If $c = 1$, then $\|\,\|$ is a norm. A complete quasi-normed space is called a *quasi-Banach space*.

Example 5.59   The spaces $L^p(\mathbb{R})$ with $0 < p < 1$ are quasi-Banach spaces. (The reader is encouraged to verify this statement.)

Definition 5.60   Let $0 < p, q \leq \infty$ and let $\sigma_p := \dfrac{1}{\min\{p,1\}} - 1$. Suppose that $s \in \mathbb{R}$ is chosen so that $s > \sigma_p$. Furthermore, assume that $M \in \mathbb{N}$ is such that $M > s \geq M - 1$. Then a function $f \in L^p(\mathbb{R})$ belongs to the *Besov space* $B_q^s(L^p)$ iff

$$
\|f\|_{B_q^s(L^p)} := \begin{cases} \|f\|_{L^p} + \left(\displaystyle\int_{\mathbb{R}} |h|^{-sq} \|\Delta_h^M f\|_{L^p}^q \, \frac{dh}{|h|} \right)^{\frac{1}{q}} < \infty, & 0 < q < \infty \\[4mm] \|f\|_{L^p} + \sup_{0 \neq h \in \mathbb{R}} |h|^{-s} \|\Delta_h^M f\|_{L^p} < \infty, & q = \infty. \end{cases}
$$

We remark that $B_q^s(L^p)$ is a Banach space for $1 \leq p, q \leq \infty$ and a quasi-Banach space for $0 < p < 1$.

Definition 5.61   Let $0 < p, q \leq \infty$ and suppose $s > \dfrac{1}{\min\{p,q\}}$. If $M \in \mathbb{N}$ is such that $M > s \geq M - 1$. A function $f \in L^p$ is said to belong to the *Triebel–Lizorkin space* $F_q^s(L^p)$ iff

$$
\|f\|_{F_q^s(L^p)} := \begin{cases} \|f\|_{L^p} + \left\| \left(\displaystyle\int_{\mathbb{R}} |h|^{-sq} |\Delta_h^M f(\bullet)|^q \, \frac{dh}{|h|} \right)^{\frac{1}{q}} \right\|_{L^p} < \infty, & 0 < q < \infty \\[4mm] \left\| \sup_{0 \neq h \in \mathbb{R}} |h|^{-s} |\Delta_h^M f(\bullet)| \right\|_{L^p} < \infty, & q = \infty. \end{cases}
$$

Similar to the situation above, $F_q^s(L^p)$ is a Banach space for $1 \leq p, q \leq \infty$ and a quasi-Banach space for $0 < p < 1$.

The following list shows some commonly used function spaces that are special cases of Besov and Triebel–Lizorkin spaces, thus expressing their versatility.

- **Hölder spaces:** For $s > 0$ and $s \notin \mathbb{N}$, $C^s = B_\infty^s(L^\infty)$.

- **Sobolev spaces:** For $1 < p < \infty$ and $m \in \mathbb{N}$, $W^m(L^p) = F_2^m(L^p)$.

- **Slodeckiĭ spaces:** For $1 \leq p < \infty$ and $s > 0$, $W^s(L^p) = B_p^s(L^p) = F_p^s(L^p)$.

- **Bessel[5] potential spaces:** For $1 < p < \infty$ and $s > 0$: $H^s(L^p) = F_2^s(L^p)$. Here, we defined for $s \in \mathbb{R}$

$$
H^s(L^p) = \{ f \in \mathscr{S}'(\mathbb{R}^n) : \|\mathscr{F}^{-1}[(1+|\xi|^2)^{s/2} \mathscr{F}f(\xi)](\bullet)\|_{L^p} < \infty \}.
$$

For details and proofs, we again refer to [151].

---

5 FRIEDRICH WILHELM BESSEL, 22 July 1784–17 March 1846. German mathematician, astronomer and geodesist. He is known for his work in astronomy using the parallax and Bessel functions.

The main results of this section are the following two theorems, which we state without proof. These two results are special cases, $n := 1$, of Theorem 3 and Theorem 4 in [116], and we refer the interested reader to this reference.

Theorem 5.62 Suppose that $[a, b]$ is a nonempty interval. Further suppose that $1 < N \in \mathbb{N}$, that $\{u_i : [a, b] \to [a, b] \mid i = 1, \ldots, N\}$ is a collection of nonzero contractive homeomorphisms of type (5.8), and that $\mathfrak{f}$ is the associated fractal function of class C^k. Denote the Lipschitz constant of u_i by $a_i, i = 1, \ldots, N$, and the scaling vector by $\lambda = (\lambda_1, \ldots, \lambda_N)$. Then $\mathfrak{f} : [a, b] \to \mathbb{R}$ belongs to the Besov space $B_q^s(L^p)$ provided that

$$\sum_{i=1}^{N} |\lambda_i|^q \, a_i^{q\left(\frac{1}{p}-s\right)} < 1, \quad 0 < p \le \infty, \quad 0 < q < \infty;$$

$$\sum_{i=1}^{N} |\lambda_i| \, a_i^{\frac{1}{p}-s} < 1, \quad 0 < p \le \infty, \quad q = \infty.$$

Example 5.63 In the special case that $a_i = 1/N$, $i = 1, \ldots, N$, $s \in \mathbb{N}$, and $q = p < \infty$, one obtains

$$\sum_{i=1}^{N} |\lambda_i|^p \, N^{ps-1} < 1$$

as a condition for $\mathfrak{f} \in W^s(L^p)$. For $p = q = \infty$, one deduces the known requirement for a fractal function to be in C^s:

$$\sum_{i=1}^{N} |\lambda_i| \, N^s < 1.$$

These conditions were also derived in [114].

Theorem 5.64 Suppose the hypotheses of Theorem 5.62 are satisfied. Then a fractal function $\mathfrak{f} : [a, b] \to \mathbb{R}$ belongs to the Triebel–Lizorkin space $F_q^s(L^p)$ provided

$$\sum_{i=1}^{N} |\lambda_i|^p \, a_i^{1-sp} < 1, \quad 0 < p, q \le \infty;$$

$$\sum_{i=1}^{N} |\lambda_i| \, a_i^{-s} < 1, \quad p = q = \infty.$$

Proof The interested reader is again referred to [116] for a proof of this result. $\square$

Exercises

1. Verify (5.2a) and (5.2b).

2. Verify that the polynomial $\mathfrak{f}(x) := 2x(1-x)$ is the unique fixed point of the operator defined in (5.6).

3. Show that the Takagi function τ satisfies the fixed point equation (5.7).

4. Prove that $(C_{\alpha,\beta}[a, b], d_{\infty,[a,b]})$ is a complete metric space and that the operator T defined in (5.15) is contractive on $(C_{\alpha,\beta}[a, b], d_{\infty,[a,b]})$.

5. Show that $(B[a, b], \|\,\|_{\infty,[a,b]})$ is a Banach space.

6. Give a proof of Theorem 5.18.

7. Prove Proposition 5.25.

8. Suppose that $N := 2$ and $\mathfrak{f} \in \mathfrak{S}[a, b]$. Compute the L^2-inner product $\langle \mathfrak{f}, \mathfrak{f} \rangle$.

9. Prove Proposition 5.26.

10. Let $k \in \mathbb{N}_0$ and let $\emptyset \neq [a, b] \subset \mathbb{R}$. Show that the spaces $(C^k[a, b], \|\,\|_{C^k})$, $k \in \mathbb{N}_0$, are Banach spaces.

11. Verify Equations (5.35) and fill in all details that show the validity of Theorem 5.41.

12. Show that the fixed point $\mathfrak{f} = \mathfrak{f}[\lambda, q]$ of the operator defined by (5.38) is unique and an element of the function space $C_0^k[a, b]$.

13. Verify that the operator $T[\lambda]$ given in (5.42) has range $C_0^{N-2}[a, b]$.

14. Prove Proposition 5.46.

15. Let $\square := [0, 1] \times [0, 1]$ and let Q denote a quadrilateral with vertices A, B, C, and D. Consider the mapping $T : \square \to Q$, $(x, y) \overset{T}{\mapsto} (\alpha x + \beta, ax + by + cxy + d)^\top$, $\alpha, \beta, a, b, c, d \in \mathbb{R}$, defined as follows: $T(0, 0) := A$, $T(1, 0) := B$, $T(1, 1) := C$, and $T(0, 1) := D$.

    (a) Determine the values of α, β, a, b, c, and d.
    (b) Show that the lattice $\Gamma := \{(i/N, y) \mid i = 0, \ldots, N\} \cup \{(x, j/M) \mid j = 0, \ldots, M\} \subseteq \square$ is mapped onto a lattice $\Gamma' \subset Q$ consisting also of straight lines.
    (c) Show that any other line $L \subset \square$ not contained in a lattice Γ of the above form is mapped onto a curve. What kind of curve is it?

16. Consider the unit square $\square$ together with the collection of points $A := (0, \underline{y}_0)$, $B := (0.5, \underline{y}_1)$, $C := (1, \underline{y}_2)$, $D := (1, \overline{y}_2)$, $E := (0.5, \overline{y}_1)$, and $F := (0, \overline{y}_0)$, as depicted in Figure 5.23.

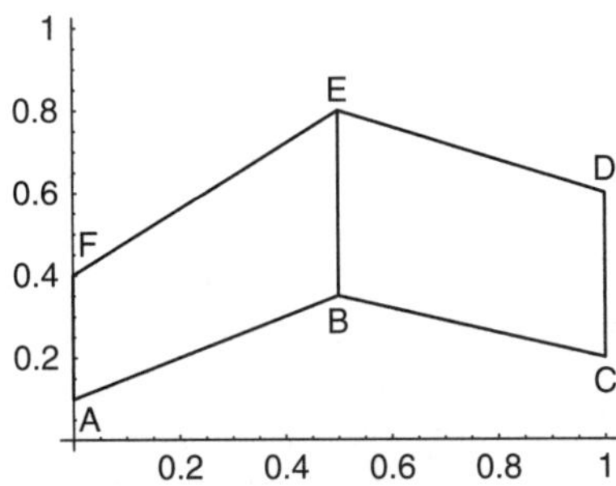

Figure 5.23 Construction of fractal interpolation functions using quadrilaterals.

(a) Using the mapping T defined in Exercise 15, construct a fractal interpolation function through an appropriate set of interpolation points.

(b) Under what conditions does there exist a unique fractal interpolation function?

(c) Generalize the construction to arbitrary increasing knot sequences of length $N+1$ and arbitrary compact intervals.

17. Following the notation of Exercise 16, let $\underline{\lambda}_i := \min\{\overline{y}_{i-1} - \underline{y}_{i-1}\}$ and $\overline{\lambda}_i := \max\{\overline{y}_{i-1} - \underline{y}_{i-1}\}$, $i = 1, \ldots, N$. Show that the box dimension d of a fractal interpolation function generated using quadrilaterals (more precisely, trapezoids) over a uniform knot sequence has the following lower and upper bounds:

$$1 + \log_N \sum_{i=1}^{N} |\underline{\lambda}_i| \leq d \leq 1 + \log_N \sum_{i=1}^{N} |\overline{\lambda}_i|.$$

(*Hint: See the proof of Theorem 5.30.*)

18. Given that $\mathbf{0} \neq \boldsymbol{\lambda} \in (-1, 1)^N$ and $a_i = \operatorname{Lip} u_i$, $i = 1, \ldots, N$, show that

$$0 < \sum_{i=1}^{N} a_i \lambda_i < 1.$$

19. Compute the indefinite integral of the fractal B-spline (5.43).

20. Verify (5.50).

21. Compute the Fourier transform of the fractal B-spline (5.43).

22. Work out the details in Example 5.53.

23. Verify the claims made in Example 5.54.

24. Recall that the n-order Legendre[6] polynomial $P_n : [-1, 1] \to \mathbb{R}$, $n \in \mathbb{N}_0$, can be defined via Rodrigues's[7] formula as

$$P_n(x) := \frac{1}{2^n \, n!} \frac{d^n}{dx^n} (x^2 - 1)^n,$$

and that $\{P_n \mid n \in \mathbb{N}_0\}$ forms an orthogonal system of functions on $[-1, 1]$ with respect to the L^2-inner product $\langle \, , \, \rangle$.

(a) Verify that $\langle P_n, P_m \rangle = \dfrac{2}{2n+1} \delta_{mn}$.

(b) Define $\mathcal{F} := \big\{ P_0(2 \bullet - 1) \chi_{[0,1]}, \, \sqrt{3} \, P_1(2 \bullet - 1) \chi_{[0,1]}, \, \sqrt{5} \, P_2 \, (2 \bullet - 1) \chi_{[0,1]} \big\}$. Show that $\mathcal{F}$ is an orthonormal generator for the FSI space $S(\mathcal{F})$.

(c) Determine to which linear space over $[0, 1]$ is $S(\mathcal{F})$ isomorphic.

(d) Now let $N \in \mathbb{N}_0$, and define

$$\mathcal{F}_N := \left\{ \sqrt{2n+1} \, P_n(2 \bullet - 1) \, \chi_{[0,1]} \; \middle| \; n = 0, 1, \ldots, N-1 \right\}.$$

Based on the previous results, construct an MRA of $L^2(\mathbb{R})$ using the spaces $V_k := \mathcal{D}^k S(\mathcal{F}_N)$, $k \in \mathbb{Z}$.

More details about this generator and the corresponding wavelet spaces can be found in [3].

25. Establish the result given in (5.55).

26. Show that the Hölder spaces $C^{k,\alpha}(\overline{\Omega})$ are complete normed spaces.

27. Find an example of a function f, which belongs to $C^{1,\alpha}(\mathbb{R})$ with $\alpha := \frac{1}{2}$, but not to $C^{1,\beta}$ with $\beta > \frac{1}{2}$.

28. Using the arguments in Section 5.9, derive a condition for a fractal function to belong to the Hölder space $C^{k,\alpha}[a, b]$.

29. Show that for $0 < p < 1$, the spaces $L^p(\mathbb{R})$ are quasi-Banach spaces. Can you determine the value of the constant c in (5.56)?   □

---

6 ADRIEN-MARIE LEGENDRE, 18 September 1752–10 January 1833. French mathematician who made significant contributions to abstract algebra, analysis, number theory, and probability theory.
7 BENJAMIN OLINDE RODRIGUES, 1795–1851. French mathematician and social reformer. He is remembered for two formulae: one regarding the rotation of vectors and the other about series of orthogonal polynomials.

6

Fractal Surfaces

A construction of fractal surfaces with coplanar boundary values using iterated function systems was first presented in [110]. The generalization to fractal surfaces with arbitrary boundary values was carried out shortly thereafter in [71]. A more general class of $\mathbb{R}^m$-valued multivariate fractal functions $\mathfrak{f}: D \subseteq \mathbb{R}^n \to \mathbb{R}^m$ was investigated in [83]. The domain D for these three types of construction consisted of simplices and simplicial complexes, i.e., specific unions of simplices. Several other constructions not necessarily based on the theory of iterated functions systems appeared in the literature, notably those of Dubuc and his co-workers (cf. [51, 52]).

Constructions of fractal surfaces over grids or rectangular domains using iterated function systems can be found in [27] and [28]. These references also consider the case of smoothly defined fractal surfaces.

It is always possible to construct fractal surfaces as tensor products of univariate continuous fractal functions. However, these tensor product fractal surfaces lack most of the exciting features of the aforementioned fractal surfaces and exhibit too much of an isotropic behavior, making them less amenable for practical purposes.

The first section in this chapter introduces tensor product fractal surfaces. The construction of non-tensor product $\mathbb{R}$-valued fractal surfaces over simplicial domains in $\mathbb{R}^2$ generated by affine iterated function systems is presented next. Two cases are considered: coplanar boundary values with arbitrary scaling factors and non-coplanar boundary values with a single scaling factor. The next sections deal with properties of affine fractal surfaces such as the computation of their box dimension and Hölder continuity. In the fifth section, bilinear fractal surfaces over rectangular domains are presented. Finally, fractal surfaces arising from a class of quadratic forms are constructed and the indefinite two-dimensional integral of a class of fractal functions defined over the unit square is computed. This will yield smooth fractal surfaces.

A reference for this chapter is [112] and the original papers [27, 28, 29, 30, 71, 73, 74, 83, 110], and [111], although some go beyond the scope of this textbook.

6.1 Tensor Product Fractal Surfaces

Here, we show how the tensor product can be used to obtain fractal-like surfaces from univariate $\mathbb{R}$-valued continuous fractal functions.

In what follows, the notation and terminology of Section 5.4 is used. To this end, suppose that $1 < N \in \mathbb{N}$ and let $[a, b]$ and $[c, d]$ be fixed nonempty compact intervals of $\mathbb{R}$. Furthermore, for $k, l \in \mathbb{N}$, consider $\mathfrak{S}^k[a, b]$ and $\mathfrak{S}^l[c, d]$, the finite-dimensional subspaces of $C[a, b]$ and $C[c, d]$, respectively, introduced in Section 5.4. Moreover, let $\{\mathfrak{f}^1, \ldots, \mathfrak{f}^n\}$ and $\{\mathfrak{g}^1, \ldots, \mathfrak{g}^m\}$ be bases for these spaces, and let $\mathfrak{f}$ and $\mathfrak{g}$ be fractal functions generated via

$$\mathfrak{f} = \sum_{i=1}^{n} a_i \, \mathfrak{f}^i \qquad \text{and} \qquad \mathfrak{g} = \sum_{j=1}^{m} b_j \, \mathfrak{g}^j,$$

where $\{a_i \mid i = 1, \ldots, n\}$ and $\{b_j \mid = 1, \ldots, m\}$, are sets of real numbers. Then the tensor product of $\mathfrak{f}$ with $\mathfrak{g}$ is the bivariate fractal function $\mathfrak{f} \otimes \mathfrak{g} : [a, b] \times [c, d] \to \mathbb{R}$ given by

$$\mathfrak{f} \otimes \mathfrak{g} = \sum_{i=1}^{n} \sum_{j=1}^{m} a_i \, b_j \, \mathfrak{f}^i \otimes \mathfrak{g}^j. \tag{6.1}$$

The graph of the fractal function $\mathfrak{f} \otimes \mathfrak{g}$ is called a *tensor product fractal surface (of polynomial type)*. We denote the space of all such tensor product fractal surfaces $\mathfrak{f} \otimes \mathfrak{g}$ by $(\mathfrak{S}^k \otimes \mathfrak{S}^l)[a, b] \times [c, d]$.

Example 6.1 Let $N := 2$, let $[a, b] = [c, d] := [0, N]$, and let $k := l := 2$. Then, any fractal function $\mathfrak{f}$ in $\mathfrak{S}^k[0, 2]$ is an affine fractal function. The space $\mathfrak{S}^2[0, 2]$ is thus three-dimensional and has a basis given by, for instance, $\{\mathfrak{f}^1, \mathfrak{f}^2, \mathfrak{f}^3\}$, where $\mathfrak{f}^i$ is the affine fractal function that interpolates the points $(0, \delta_{1i})$, $(1, \delta_{2i})$, and $(2, \delta_{3i})$, i.e., the $\mathfrak{f}^i$ are fractal functions of Lagrange type. (See also Proposition 5.23 and the arguments leading to it.)

If the scaling factors are denoted by λ_1 and λ_2, then the following expressions for the polynomial functions $\boldsymbol{p} \in \Lambda^2$ are obtained:

$$\mathfrak{f}^1 : \quad \boldsymbol{p}^1 = \left((\lambda_1 - 1)\left(\frac{1}{2}\,\mathrm{id}_{\mathbb{R}} - 1\right),\ \lambda_2\left(\frac{1}{2}\,\mathrm{id}_{\mathbb{R}} - 1\right)\right),$$

$$\mathfrak{f}^2 : \quad \boldsymbol{p}^2 = \left(\frac{1}{2}\,\mathrm{id}_{\mathbb{R}},\ -\frac{1}{2}\,\mathrm{id}_{\mathbb{R}} + 1\right),$$

$$\mathfrak{f}^3 : \quad \boldsymbol{p}^3 = \left(-\frac{\lambda_1}{2}\,\mathrm{id}_{\mathbb{R}},\ \left(1 - \frac{\lambda_2}{2}\right)\mathrm{id}_{\mathbb{R}}\right).$$

The tensor products $\mathfrak{f}^i \otimes \mathfrak{f}^j$ satisfy the functional equations

$$\mathfrak{f}^i \otimes \mathfrak{f}^j = \sum_{k,\ell=0}^{1} \left(p_k^i \circ u_k^{-1} + \lambda_k\, \mathfrak{f}^i \circ u_k^{-1}\right)\left(p_\ell^j \circ u_\ell^{-1} + \lambda_\ell\, \mathfrak{f}^j \circ u_\ell^{-1}\right) \chi_{u_k[0,2]}\, \chi_{u_\ell[0,2]},$$

with $u_k = (1/2)\,\mathrm{id}_{\mathbb{R}} + (k - 1)$, $k = 1, 2$.

Now consider the affine fractal functions $\mathfrak{f} := 2\mathfrak{f}^1 - 3\mathfrak{f}^2 + \mathfrak{f}^3$ with $\lambda_{\mathfrak{f}} := (0.3, -0.7)$ and $\mathfrak{g} := \mathfrak{f}^1 + 3\mathfrak{f}^2 - 2\mathfrak{f}^3$ with $\lambda_{\mathfrak{g}} := (-0.5, -0.6)$. In Figure 6.1, the affine tensor product fractal surface $\mathfrak{f} \otimes \mathfrak{g}$ is shown.

Note that if (x, y, z) is a coordinate system in $\mathbb{R}^3$ and if the graph of $\mathfrak{f} \otimes \mathfrak{f}$ is analytically represented by $z = (\mathfrak{f} \otimes \mathfrak{f})(x, y)$, then the intersection of graph $\mathfrak{f} \otimes \mathfrak{f}$ with the hyperplane $y = c$, $c \in [a, b]$, is a rescaled version of graph $\mathfrak{f}$.

We finally remark that tensor-product fractal surfaces have, of course, the same shortcomings as tensor product splines. (See Section 3.3.2.) To overcome some of these shortcomings, we next consider constructions of fractal surfaces over simplices in $\mathbb{R}^2$.

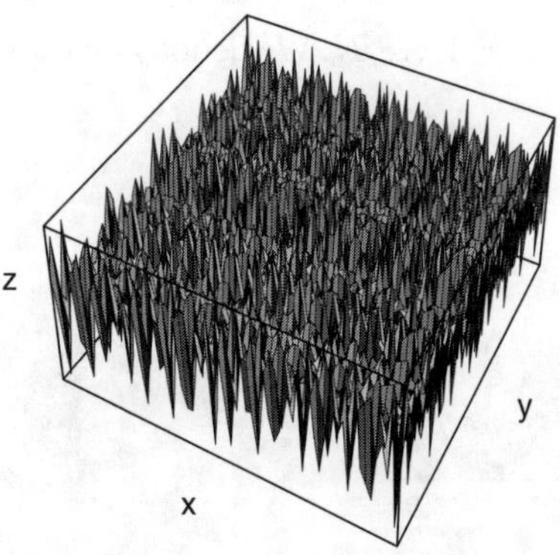

Figure 6.1 An example of an affine tensor product fractal surface.

6.2 Affine Fractal Surfaces in $\mathbb{R}^2$

In this section, we introduce two types of affine fractal surfaces over certain triangular domains in $\mathbb{R}^2$. The constructions given here in $\mathbb{R}^2$ can be generalized to polyhedral domains in $\mathbb{R}^n$ and more general types of fractal surfaces. The interested reader is referred to the literature [71, 73, 74, 83, 110], and [113] for these extensions and generalizations. The presentation given here follows the ideas presented in [71] and [110].

We first need to define what is meant by a regular simplex in $\mathbb{R}^{n+1}$, $n \in \mathbb{N}_0$.

Definition 6.2 A *regular n-simplex* in $\mathbb{R}^{n+1}$ is defined as the set

$$\triangle^n := \left\{ \boldsymbol{x} := (x_1, \ldots, x_{n+1}) \in \mathbb{R}^{n+1} \,\middle|\, \forall i = 1, \ldots, n+1, \, x_i \geq 0 \text{ and } \sum_{i=1}^{n+1} x_i = 1 \right\}$$

$$= \mathrm{conv}\{\boldsymbol{e}^1, \ldots, \boldsymbol{e}^{n+1}\},$$

where $\boldsymbol{e}^i := (0, \ldots, 0, 1, 0, \ldots, 0)^\top$ is the i-th unit vector in $\mathbb{R}^{n+1}$, $i = 1, \ldots, n+1$.

Example 6.3 In Figure 6.2 the simplices $\triangle^1$ and $\triangle^2$ are displayed.

To simplify notation, we set $\triangle := \triangle^1$. Moreover, let

$$C(\triangle) := C(\triangle, \mathbb{R}) := \{f \in \mathrm{Map}(\triangle, \mathbb{R}) \mid f \text{ is continuous on } \triangle\}.$$

Now suppose that $\{\triangle_i \mid i = 1, \ldots, N\}$, $1 < N \in \mathbb{N}$, is a partition of $\triangle$ consisting of nonempty compact sets $\triangle_i$ with the property that

$$\triangle = \bigcup_{i=1}^{N} \triangle_i \quad \text{with} \quad \triangle_i^\circ \cap \triangle_i^\circ = \emptyset, \; \forall i \neq j,$$

and $\triangle_i$ is similar to $\triangle$, $i = 1, \ldots, N$. We denote the class of all such partitions by $\mathscr{P}$.

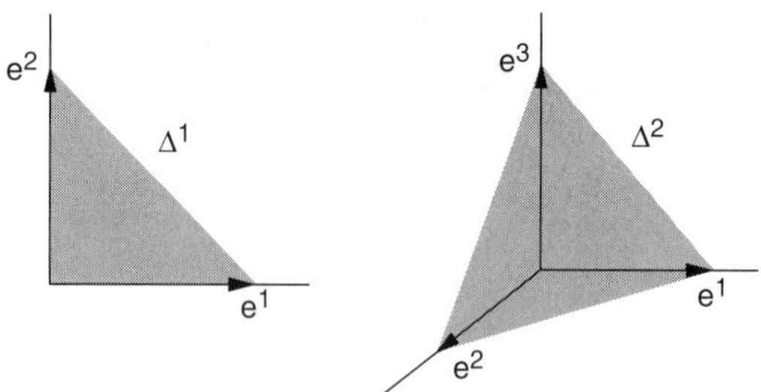

Figure 6.2 The simplex $\triangle^1$ and $\triangle^2$.

The above requirements imply the existence of N unique contractive similitudes $u_i : \triangle \to \triangle_i$ given by

$$u_i = s_i\, O_i + \tau_i, \quad i = 1, \dots, N, \tag{6.2}$$

where $s_i < 1$ is the similarity constant or the similarity ratio for $\triangle_i$ with respect to $\triangle$, O_i is an orthogonal transformation, and τ_i a translation. (Cf. Proposition 4.36.)

Now, suppose that $\{q_i \in \mathrm{Map}(\mathbb{R}^2, \mathbb{R}) \mid i = 1, \dots, N\}$ is a collection of N affine functions and $\{\lambda_i \mid i = 1, \dots, N\}$ a set of N real numbers. As usually, we set $\boldsymbol{q} := (q_1, \dots, q_N)$ and $\boldsymbol{\lambda} := (\lambda_1, \dots, \lambda_N)$. Denoting by $\mathrm{Aff}(\mathbb{R}^2)$ the vector space of all affine mappings $q : \mathbb{R}^2 \to \mathbb{R}$, we define an operator

$$T[\boldsymbol{\lambda}, \boldsymbol{q}] : \mathbb{R}^N \times \overset{N}{\underset{i=1}{\mathsf{X}}} \mathrm{Aff}(\mathbb{R}^2) \times C(\triangle) \to \mathrm{Map}(\triangle, \mathbb{R})$$

by

$$T[\boldsymbol{\lambda}, \boldsymbol{q}]\, f := q_i \circ u_i^{-1} + \lambda_i f \circ u_i^{-1}, \quad \text{on} \quad \triangle_i, \quad i = 1, \dots, N. \tag{6.3}$$

The affine mappings q_i are usually determined via interpolation conditions. To this end, we consider a partition $\{\triangle_i \mid i = 1, \dots, N\} \in \mathscr{P}$, and denote by V the set of all distinct vertices of the subtriangles $\{\triangle_i \mid i = 1, \dots, N\}$, with the set of vertices of $\triangle$ given by $\{v_1, v_2, v_3\}$. Let $n := \mathrm{card}\, V$. We introduce the mapping $\phi : \{1, 2, 3\} \times \{1, \dots, N\} \to \{1, \dots, n\}$ to associate to each $\triangle_i$ its set of three distinct vertices: For each $i = 1, \dots, N$, $\{v_{\phi(j,i)} \mid j = 1, 2, 3\}$ is the set of vertices of subtriangle $\triangle_i$. Note that since we map sets to sets, the order is – for the moment – irrelevant.

Example 6.4   For the partition of $\triangle$ into $N = 4$ subtriangles and the associated set V of six vertices $V = \{v_1, \dots, v_6\}$, we have, for instance, $\{\phi(j, 2) \mid j = 1, 2, 3\} = \{v_2, v_4, v_6\}$. See also Figure 6.3.

Now, consider the interpolation set $Z := \{(v_\nu, z_\nu) \in \triangle \times \mathbb{R} \mid \nu = 1, \dots, n\}$. Then, for all $i = 1, \dots, N$, the affine mapping q_i is uniquely determined by the

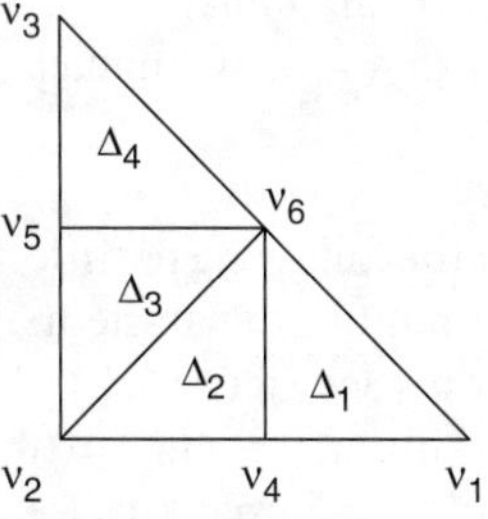

Figure 6.3   An example of vertex assignment.

interpolation conditions

$$q_i(v_j) + \lambda_i z_j = z_{\phi(j,i)}, \quad j = 1, 2, 3.$$

Clearly, $T[\lambda, q] \in C(\triangle_i)$, for all $i = 1, \ldots, N$, and $(T[\lambda, q]f)(v_{\phi(j,i)}) = z_{\phi(j,i)}$, $j = 1, 2, 3$.

To obtain *global* continuity, i.e., continuity on $\triangle$, we need to impose the following conditions. We require that

$$q_i \circ u_i^{-1}(x, y) + \lambda_i f \circ u_i^{-1}(x, y) = q_j \circ u_j^{-1}(x, y) + \lambda_j f \circ u_j^{-1}(x, y), \quad (6.4)$$

for all $(x, y) \in e_{ij} := \triangle_i \cap \triangle_j$, $i, j \in \{1, \ldots, N\}$ with $i \neq j$. The quantity e_{ij} is called a *common edge* of $\triangle_i$ and $\triangle_j$.

We will consider two cases, each of which satisfies the join-up conditions imposed by (6.4).

Case 1: Coplanar Boundary Conditions and Different λ_i.

Let $C_0(\triangle) := \{f \in C(\triangle) \mid f|_{\partial \triangle} \equiv 0\}$ and let $Z_0 := \{(v_v, z_v) \in Z \mid z_v = 0 \text{ whenever } v_v \in \partial \triangle\}$. Then, we have the following result.

Theorem 6.5 Suppose that $q_i \in \mathrm{Aff}(\mathbb{R}^2) \cap C_0(\triangle)$, $i = 1, \ldots, N$ and $\lambda := \max\{|\lambda_i| \mid i = 1, \ldots, N\} < 1$. Then, the operator $T[\lambda, q]$ defined above is a contraction on the Banach space $(C_0(\triangle), \|\ \|_{\infty,\triangle})$.

Proof The main issue to prove is the well-definiteness of $T[\lambda, q]f$. This, however, can be done by considering the values of $T[\lambda, q]f$ along the common edge e_{ij} of two subtriangles $\triangle_i$ and $\triangle_j$, and using the join-up conditions. The details are left to the reader. $\square$

Theorem 6.5 implies that the unique fixed point $f : \triangle \to \mathbb{R}$ of the operator $T[\lambda, q]$ given in (6.3) satisfies the equation

$$f(x, y) = (q_i \circ u_i^{-1})(x, y) + \lambda_i (f \circ u_i^{-1})(x, y), \quad \forall(x, y) \in \triangle_i, \quad i = 1, \ldots, N,$$

and that it interpolates the set Z_0.

The graph of f is called an *affine fractal surface of class* $C_0(\triangle)$. We sometimes, although this is an abuse of language, refer to the function f as a fractal surface.

Example 6.6 We consider the following partition of $\triangle$ into $N = 9$ piece-wise congruent subtriangles and require the affine mappings to pass through the vertices given by this partition. (See Figure 6.4.) All but one of these vertices lie on the boundary of $\triangle$. In addition, we set $\lambda := (0.5, 0.6, -0.3, 0.5, 0.4, -0.6, -0.7, 0.8, 0.6)$. Starting with the initial surface $f_0(x, y) \equiv 0$, we show the iterates f_1, f_2, f_3, and f_4 in Figure 6.5. The resulting fractal surface f vanishes on $\partial \triangle$ and passes through the point $(\frac{1}{3}, \frac{1}{3}, 1)$.

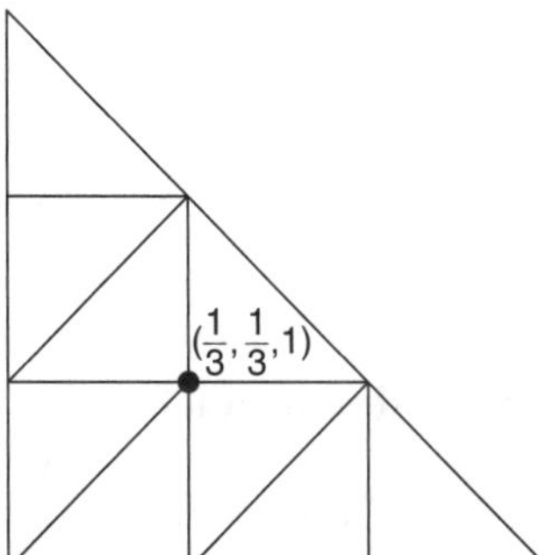

Figure 6.4 The partition of the simplex $\triangle$.

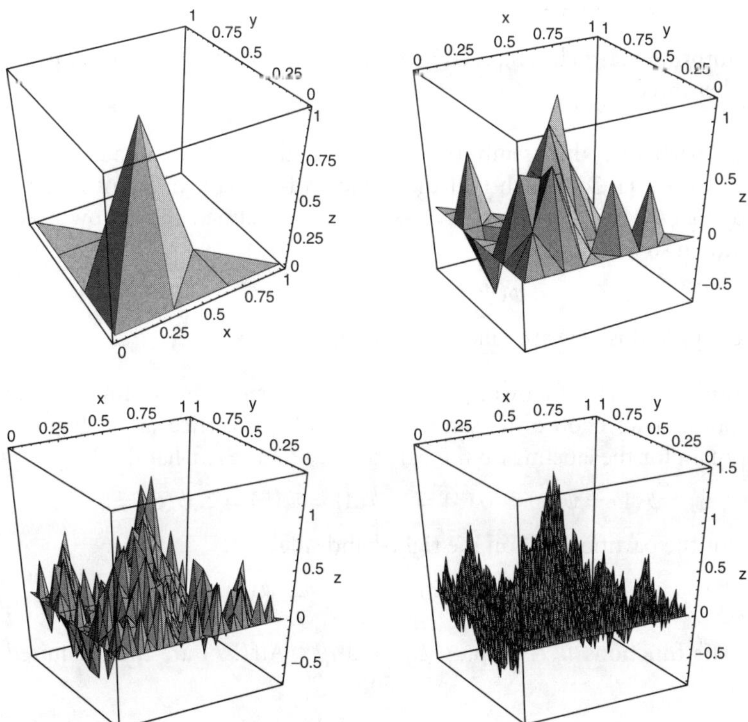

Figure 6.5 The construction of an affine fractal surface of class C_0.

Remark 6.7 Note that by a simple affine transformation one can map the set of zero boundary values to a set of coplanar boundary values. (Derive such an affine transformation!) Hence Theorem 6.5 remains valid in case the interpolation values lie in a plane $P \subset \mathbb{R}^3$.

Although the case above offers the option of choosing different scaling factors λ_i, the requirement of coplanar boundary values is for applications too stringent.

Case 2: General Boundary Conditions and Equal λ_i.

Let $\pi := \{\Delta_i \mid i = 1, \ldots, N\} \in \mathcal{P}$ be a partition of Δ. Note that the collection of all vertices V together with the collection E of all distinct sides of the subtriangles $\Delta_i \in \pi$ generates a graph, which we denote by $G_\pi := (V, E)$. An element $e \in E$ is called an *edge*. Recall that the *chromatic number* of a graph G is the least number of symbols needed to label the vertices of G in such a way that any two vertices that are joined by an edge have distinct symbols.

We impose the following condition on the graph $G_\pi = (V, E)$, i.e., its associated partition π.

Assumption 6.8 The graph G_π associated with the partition π has chromatic number three.

In particular, this assumption guarantees the existence of a mapping $\psi : \{1, \ldots, n\} \to \{1, 2, 3\}$ such that the vertices of Δ_i have different labels. The mapping $\phi : \{1, 2, 3\} \times \{1, \ldots, N\} \to \{1, \ldots, n\}$ defined above now needs to be chosen so that

$$\phi(\psi(j'), i) = j', \quad \forall v_{j'} \in \Delta_i.$$

Labelings of this type are called *admissible*.

Example 6.9 The following partitions of Δ show an example of such labelings. If we choose the same ordering of the vertices as in Figure 6.3, we obtain for the labeling on the partition π^* on the left-hand side

$$\psi(1) = \psi(2) = \psi(3) = 1, \ \psi(4) = \psi(5) = 2, \ \psi(6) = 3,$$

and for the partition π^{**} on the right-hand side

$$\psi(2) = \psi(6) = 1, \ \psi(3) = \psi(4) = 2, \ \psi(1) = \psi(5) = 3.$$

The functions $u_i \in \mathrm{Map}(\Delta, \Delta_i)$ and $q_i \in \mathrm{Aff}(\mathbb{R}^2)$ are then required to satisfy

$$u_i(v_{\psi(j)}) = v_j,$$
$$q_i(v_{\psi(j)}) + \lambda_i z_{\psi(j)} = z_j,$$

(6.5)

for each of the three vertices v_j of Δ_i.

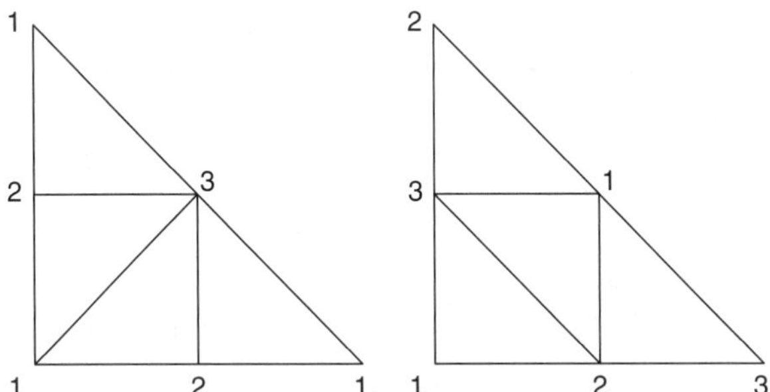

Figure 6.6 Admissible labelings of two partitions π^* and π^{**} of $\triangle$.

We will use partitions of $\triangle$ that allow admissible labelings to construct affine fractal surfaces. For this purpose, we first define, for a given simplex $\triangle$ and a given interpolation set $Z = \{(v_v, z_v) \mid v = 1, \ldots, n\}$, the function space

$$\widetilde{C}(\triangle) := \{f \in C(\triangle) \mid f(v_v) = z_v, \, v = 1, \ldots, n\}.$$

The set $\widetilde{C}(\triangle)$ becomes a complete metric space when endowed with the metric $d_{\infty,\triangle} : C(\triangle) \times C(\triangle) \to \mathbb{R}$, $d_{\infty,\triangle}(f, g) := \|f - g\|_{\infty,\triangle}$, where $\| \ \|_{\infty,\triangle}$ is the Chebyshev norm.

Theorem 6.10 Suppose that π is an admissible partition of $\triangle$. Let u_i and q_i, $i = 1, \ldots, n$, be defined as in (6.5). Furthermore, let $T[\lambda, q]$ be given as in (6.3), but with $\lambda_i = \lambda$, for all $i = 1, \ldots, n$, where $|\lambda| < 1$. Then $T[\lambda, q]$: $\widetilde{C}(\triangle) \to \widetilde{C}(\triangle)$ is well-defined and contractive on $(\widetilde{C}(\triangle, d_{\infty,\triangle}))$. Moreover, $T[\lambda, q] f$ interpolates Z.

Proof Let $i, i' \in \{1, \ldots, N\}$ and assume that $u_i(\triangle)$ is adjacent to $u_{i'}(\triangle)$, and that $e_{jj'}$, $j, j' \in \{1, \ldots, n\}$, is a common edge. By (6.5), we have that $u_i^{-1}(v_j) = u_{i'}^{-1}(v_j) = v_{\psi(j)}$ and $u_i^{-1}(v_{j'}) = u_{i'}(v_{j'}) = v_{\psi(j')}$. Since the mappings u_i are affine, their inverses u_i^{-1} are also affine, implying that $u_i^{-1}(x, y) = u_{i'}^{-1}(x, y), \forall(x, y) \in e_{jj'}$. This, together with the equality of the scaling factors, proves the validity of the join-up conditions (6.4). Thus, the operator $T[\lambda, q]$ is well-defined mapping from $\widetilde{C}(\triangle)$ to itself.

The contractivity of $T[\lambda, q]$ follows directly from the fact that $|\lambda| < 1$ and the interpolation property from (6.3) and (6.5). $\qquad \square$

Theorem 6.10 implies that the unique fixed point $f \in \widetilde{C}(\triangle)$ of the operator $T[\lambda, q]$ satisfies the equation

$$f(x, y) = (q_i \circ u_i^{-1})(x, y) + \lambda(f \circ u_i^{-1})(x, y), \quad \forall(x, y) \in \triangle_i, \quad i = 1, \ldots, N,$$

and that it interpolates Z. The (graph of the) bivariate function $f : \triangle \to \mathbb{R}$ is called an *affine fractal surface of class* $\widetilde{C}(\triangle)$.

Example 6.11   Take the simplex depicted on the left-hand side of Figure 6.6 as the domain $\triangle$ for an affine fractal function. The admissible partition π^* of $\triangle$ induces the four similitudes

$$u_1(x, y) = \frac{1}{2} \begin{pmatrix} 1 & 0 \\ 0 & 1 \end{pmatrix} \begin{pmatrix} x \\ y \end{pmatrix} + \begin{pmatrix} \frac{1}{2} \\ 0 \end{pmatrix},$$

$$u_2(x, y) = \frac{1}{2} \begin{pmatrix} -1 & 0 \\ 0 & 1 \end{pmatrix} \begin{pmatrix} x \\ y \end{pmatrix} + \begin{pmatrix} \frac{1}{2} \\ 0 \end{pmatrix}$$

$$u_3(x, y) = \frac{1}{2} \begin{pmatrix} 1 & 0 \\ 0 & -1 \end{pmatrix} \begin{pmatrix} x \\ y \end{pmatrix} + \begin{pmatrix} 0 \\ \frac{1}{2} \end{pmatrix},$$

$$u_4(x, y) = \frac{1}{2} \begin{pmatrix} 1 & 0 \\ 0 & 1 \end{pmatrix} \begin{pmatrix} x \\ y \end{pmatrix} + \begin{pmatrix} 0 \\ \frac{1}{2} \end{pmatrix}.$$

The similarity ratio σ equals 1/2. Choose $\lambda := 3/5$ and as functions $q_1, \ldots, q_4$:

$$q_1(x, y) := -\frac{1}{5} x + \frac{3}{10} y + \frac{1}{5} =: q_2(x, y),$$

$$q_3(x, y) := \frac{1}{5} x - \frac{3}{10} y + \frac{3}{10} =: q_4(x, y).$$

A short computation shows that this collection of functions $\{q_1, q_2, q_3, q_4\}$ satisfies (6.5) and also the join-up conditions (6.4). The graph of the affine fractal surface generated by these mappings is shown in Figure 6.7.

Following the exposition in Section 5.4, it is natural to ask whether one can prove a result similar to Proposition 5.23 for affine fractal surfaces. To this

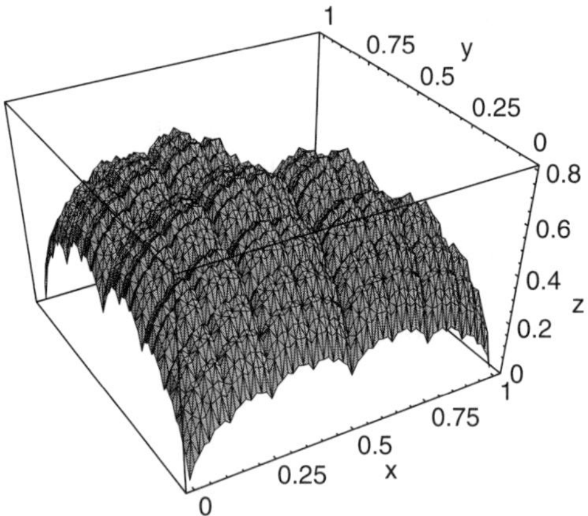

Figure 6.7   An affine fractal surface of class $\widetilde{C}(\triangle)$.

end, note that if we are given a partition π of $\triangle$ from $\mathscr{P}$, functions $q_i \in$ Aff$(\mathbb{R}^2)$, $i = 1, \ldots, N$, and an interpolation set Z, then the space $\mathfrak{A}^2(\triangle; \pi)$ of affine fractal surfaces of class $\widetilde{C}(\triangle)$ over $\triangle$ is $(3N - n)$-dimensional. (3 free parameters for each of the N mappings $q_i \in$ Aff$(\mathbb{R}^2)$ and n interpolation conditions at the vertices of $\triangle$.)

This observation suggests the construction of a two-dimensional system of Lagrange interpolants for each $\mathfrak{f} \in \mathfrak{A}^2(\triangle; \pi)$ of the form

$$\{\mathfrak{e}_k \in \mathfrak{A}^2(\triangle) \mid k = 1, \ldots, n\}, \tag{6.6}$$

where each $\mathfrak{e}_k$ is the unique affine fractal surface interpolating the set $Z_k :=$ $\{(v_j, \delta_{kj}) \mid j = 1, \ldots n\}$, $k = 1, \ldots, n$. We refer to these affine fractal surfaces also as *basis fractal surfaces*. Thus, we have the following result.

Theorem 6.12 Let $\mathfrak{f} \in \mathfrak{A}^2(\triangle; \pi)$ and let $Z := \{(v_j, z_j) \mid j = 1, \ldots, n\}$ be the associated interpolation set. Then there exist n unique affine fractal surfaces of Lagrange type (6.6) so that

$$\mathfrak{f} = \sum_{k=1}^{n} z_k \, \mathfrak{e}_k.$$

Example 6.13 For the affine fractal surface constructed in Example 6.11, we have that $\dim \mathfrak{A}^2(\triangle; \pi^*) = 6$. Three of the six basis fractal surfaces are depicted in Figure 6.8.

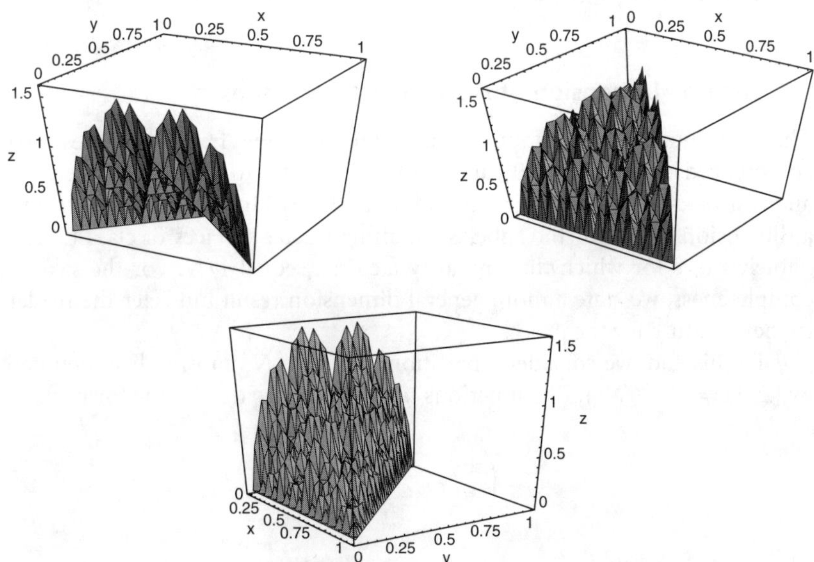

Figure 6.8 Some basis fractal surfaces.

Remark 6.14   The geometric and analytic structure of an affine fractal surface of type C_0 and $\widetilde{C}$ depends on the partition $\pi \in \mathscr{P}$ of $\triangle$. Thus the notation $\mathfrak{A}^2(\triangle; \pi)$.

Remark 6.15   Both constructions considered in this section use affine mappings q_i, $i = 1, \ldots, N$, together with certain boundary conditions (coplanar versus general) and scaling factor requirements (unequal versus equal) to satify conditions (6.4). The first case of coplanar boundary values and unequal scaling factors λ_i, $i = 1, \ldots, N$, generates an affine fractal function whose graph contains many straight-line segments which is an impediment when modeling naturally occurring surfaces. However, as different scaling factors are used, the fractal function has different degrees of "roughness" across its graph.

The second construction uses general boundary values but needs to use equal scaling factors. This type of fractal surface is better suited for modeling naturally occuring surfaces but has the disadvantage of being of equal "roughness" everywhere.

In order to achieve both, namely general boundary values plus different scaling factors, one needs to use so-called *recurrent iterated function systems* [13]. Several constructions based on recurrent IFSs can be found in the literature and we refer the interested reader to [29] and [30], and the references given therein.

In addition, there are constructions of fractal functions $\mathfrak{f} : D \subset \mathbb{R}^m \to \mathbb{R}^n$, where D is a polygonal region in $\mathbb{R}^m$, also using recurrent iterated function systems [83], as well as constructions that yield fractal surfaces of class C^k, $k \in \mathbb{N}_0$, [112, 113], and also [27].

6.3 The Box Dimension of Affine Fractal Surfaces

The computation of the box dimension of affine fractal surfaces first commenced in [71] and [110], and was later extended to more general settings and constructions. (See, for instance, [30] and [83].) In this section, we prove a dimension result for the subclass of affine fractal surfaces of class $C_0(\triangle)$, namely those for which the similarity factors s_i equal $1/N$. For the sake of completeness, we state a more general dimension result but refer the reader to the literature for the proof.

To this end, we consider a partition of $\triangle$ into N^2 subsimplices denoted by $\{\triangle_i \,|\, i = 1, \ldots, N^2\}$. The functions $u_i : \triangle \to \triangle_i$ are then of the form

$$u_i = \begin{pmatrix} \pm\frac{1}{N} & 0 \\ 0 & \pm\frac{1}{N} \end{pmatrix} (\bullet) + \tau_i,$$

where τ_i is a translation vector and $i = 1, \ldots, N^2$. The mappings $q_i(x, y) := a_i x + b_i y + c_i \in \mathrm{Aff}(\mathbb{R}^2) \cap C_0(\triangle)$ and the scaling vector $\boldsymbol{\lambda}$ are as in Section 6.2.

Denote by $\mathfrak{f}^* : \triangle \to \mathbb{R}$ the bivariate affine fractal function generated by these maps and by $\mathfrak{G}^*$ its graph. To proceed, we need two lemmata.

Lemma 6.16 Let B_ε denote a closed ball in $\mathbb{R}^3$ of radius $\varepsilon > 0$. For $\varepsilon > 0$, let Π_ε be the ε-prism $\varepsilon \triangle \times \varepsilon [0, 1]$, where εA denotes the set $\{\varepsilon\, a \,|\, a \in A\}$. Let S be a bounded set in $\mathbb{R}^3$ and let $\mathcal{N}(B_\varepsilon)$ denote the minimum number of ε-balls B_ε necessary to cover S. Then

$$\mathcal{N}(B_\varepsilon) \leq \mathcal{N}(\Pi_\varepsilon) \leq 8 \mathcal{N}(B_\varepsilon),$$

where $\mathcal{N}(\Pi_\varepsilon)$ is the minimum number of ε-prisms needed to cover S.

Proof Every B_ε can be covered by at most eight Π_ε and every Π_ε can be covered by one B_ε. $\square$

Lemma 6.17 If $\displaystyle\sum_{i=1}^{N^2} |\lambda_i| > N$ and if $\displaystyle\bigcup_{i=1}^{N^2} \triangle_i$ is not coplanar, then

$$\lim_{m \to \infty} \frac{N^{2m}}{\mathcal{N}(m)} = 0.$$

Proof The statement of Lemma 6.17 is similar to the statement in Lemma 5.29, as is its proof. We leave the details to the reader. $\square$

Theorem 6.18 Assume that $\displaystyle\sum_{i=1}^{N^2} |\lambda_i| > N$ and that $\displaystyle\bigcup_{i=1}^{N^2} \triangle_i$ is not contained in any hyperplane of $\mathbb{R}^3$. Then, the box dimension of $\mathfrak{G}^* = \operatorname{graph} \mathfrak{f}^*$ is given by

$$\dim_F \mathfrak{G}^* = 1 + \log_N \sum_{i=1}^{N^2} |\lambda_i|;$$

otherwise, $\dim_F \mathfrak{G}^* = 2$. (Here, $\log_N$ denotes the logarithm to the base N.)

Proof Notice, that by Remark 4.15, it suffices to prove the theorem for $\varepsilon = \varepsilon_m := N^{-m}$, $m \in \mathbb{N}$. Denote by $\operatorname{proj}_{z=0}$ the projection $\mathbb{R}^3 \mapsto \{(x, y, z) \,|\, z = 0\}$. Let $m \in \mathbb{N}$ and let $\mathcal{U}_m$ be a collection of covers of $\mathfrak{G}^*$ by ε_m-prisms $\Pi = \Pi_m(p, q)$ of the form $\operatorname{proj}_{z=0} \Pi = \triangle_p$, $p = 1, \ldots, N^{2m}$, $\triangle_p := \overline{w}_{\sigma(p)} \triangle$. Here, $\overline{w}_i : \triangle \times \mathbb{R} \to \mathbb{R}^3$ denotes the set-valued map $\overline{w}_i := (\overline{u}_i, \overline{q}_i)$, $i = 1, \ldots, N^2$, $\sigma(p) \in \Sigma^* := \{1, \ldots, N^2\}^{\mathbb{N}}$ is a finite code of length p, and $q = 1, \ldots, \mathcal{N}(m, p)$, where $\mathcal{N}(m, p) \in \mathbb{N}$ is the number of such ε_m-prisms over $\triangle_p$. Furthermore, it is required that $\Pi_m(p, q)$ and $\Pi_m(p, q + 1)$ intersect along their respective faces.

Now, let $U_m \in \mathcal{U}_m$ be a cover of $\mathfrak{G}^*$ of minimal cardinality $\mathcal{N}(m) \in \mathbb{N}$. Observe that $\mathcal{N}(m) = \sum_{p=1}^{N^{2m}} \mathcal{N}(m, p)$. Let $\mathfrak{G}_p^* := \bigcup_{q=1}^{\mathcal{N}(m,p)} \Pi_m(p, q), p = 1, \ldots, N^{2m}$. It is immediate that $\mathfrak{G}_p^*$ is a compact and connected set.

The image $\mathfrak{G}_{pi}^*$ of $\mathfrak{G}_p^*$ under the map $\overline{w}_i$, $i = 1, \ldots, N^2$, is a compact set in $\triangle_p \times \mathbb{R}$ above the i-th subsimplex $\triangle_{pi}$ of $\triangle_p$. Since $\mathfrak{G}^* = \bigcup_{i=1}^{N^2} \overline{w}_i(\mathfrak{G}^*)$, it follows that $\mathfrak{G}^* \subseteq \bigcup_{i=1}^{N^2} \overline{w}_i \left(\bigcup_{p=1}^{N^{2m}} \mathfrak{G}_p^* \right)$. The compact set $\mathfrak{G}_{pi}^*$ is entirely contained in the prism $\frac{1}{N^{m+1}} \triangle \times (\mathcal{N}(m, p)|\lambda_i| + |a_i| + |b_i|)(N^{-m})[0, 1]$. Hence, if $\mathcal{N}(m+1, p, i)$ denotes the number of ε_{m+1}-prisms above $\triangle_{pi}$, then

$$\mathcal{N}(m+1, p, i) \leq N\left[\mathcal{N}(m, p)|\lambda_i| + |a_i| + |b_i|\right] + 1.$$

Note that $\mathcal{N}(m+1) = \sum_{i=1}^{N^2} \sum_{p=1}^{N^{2m}} \mathcal{N}(m, p, i)$. Summing the preceding inequality over p and i yields

$$\mathcal{N}(m+1) \leq \left[N \sum_{i=1}^{N^2} |\lambda_i| \right] \mathcal{N}(m) + c_1 N^{2m+1},$$

with $c_1 := \sum_{i=1}^{N^2}(|a_i| + |b_i| + N) > 0$. If $\sum_{i=1}^{N^2} |\lambda_i| \leq N$, then induction on m gives

$$\mathcal{N}(m) \leq [\mathcal{N}(1) + c_1 m N^{-1}] N^{2m},$$

and thus,

$$\limsup_{m \to \infty} \frac{\log \mathcal{N}(m)}{\log N^m} \leq 2.$$

Hence, $\dim_F \mathfrak{G}^* = 2$ in this case. Also, if $\bigcup_{i=1}^{N^2} \triangle_i$ is contained in a hyperplane H of $\mathbb{R}^3$, then $\mathfrak{G}^* = H$, and therefore $\dim_F \mathfrak{G}^* = 2$.

If $\gamma := \sum_{i=1}^{N^2} |\lambda_i| > N$ then, again by induction on m,

$$\mathcal{N}(m) \le (N\gamma)\left[\mathcal{N}(1) + \frac{c_1}{\gamma}\left(1 + \frac{N}{\gamma} + \cdots + \left(\frac{N}{\gamma}\right)^{m-1}\right)\right]$$

$$\le (N\gamma)\left[\mathcal{N}(1) + \frac{c_1}{\gamma - N}\right],$$

which implies that

$$\dim_F \mathfrak{G}^* \le 1 + \log_N \gamma.$$

To obtain a lower bound for $\dim_F \mathfrak{G}^*$ when $\gamma > N$ and $\bigcup_{i=1}^{N^2} \Delta_i$ is not contained in any hyperplane of $\mathbb{R}^3$, one proceeds as follows: Choose $i \in \{1, \ldots, N^2\}$ and assume that $\lambda_i \ne 0$. The inverse image of $\mathfrak{G}_p^*$ under $\overline{w}_i^{-1}$ is contained in a prism $(N^{-m+1})\Delta_i \times N^{-m}(\mathcal{N}(m, p)|\lambda_i^{-1}| + |b_i\lambda_i^{-1}|N)[0, 1]$. Hence, in the same notation as before,

$$\mathcal{N}(m - 1, p, i) \le |s_i^{-1}|\mathcal{N}(m, p, i) + (|\lambda_i^{-1}|N)(|a_i| + |b_i|) + 1.$$

Summing over p and i yields

$$\mathcal{N}(m) \ge (N\gamma)\mathcal{N}(m - 1) - c_2 N^{2m-1},$$

with $c_2 := N\sum_{i=1}^{N^2} |\lambda_i^{-1}|(|a_i| + |b_i|) + N > 0$. Note that the preceding inequality holds trivially for $\lambda_i = 0$. Induction on m for all $m_0 \in \{1, \ldots, m\}$ gives

$$\mathcal{N}(m) \ge (N\gamma)^{m-m_0}(\mathcal{N}(m_0) - c_2 N^{2m_0}(\gamma - N)^{-1}).$$

By Lemma 6.17, one can choose m_0 large enough to guarantee that

$$\mathcal{N}(m_0) > c_2 N^{2m_0}(\gamma - N)^{-1}.$$

Let $c_3 := (N\gamma)^{-m_0}(\mathcal{N}(m_0) - c_2 N^{2m_0}(\gamma - N)^{-1}) > 0$. Then

$$\mathcal{N}(m) \ge c_3(N\gamma)^m.$$

Hence,

$$\liminf_{m\to\infty} \frac{\log\mathcal{N}(m)}{\log N^m} \ge 1 + \log_N \gamma.$$

Thus, $\dim_F \mathfrak{G}^* = 1 + \log_N \gamma$ for $\gamma > N$ and $\bigcup_{i=1}^{N^2} \Delta_i$ not contained in any hyperplane of $\mathbb{R}^3$. $\quad\square$

We now state, without a proof, the more general dimension result for affine fractal surfaces of class $C_0(\triangle)$ and class $\widetilde{C}(\triangle)$. The reader is referred to [71] or [113] for more details and complete proofs. We employ the notation used in Theorems 6.5 and 6.10.

Theorem 6.19 Assume that the similitudes $u_i : \triangle \to \triangle_i$ defined in (6.2) have similarity ratios s_i, $i = 1, \ldots, N$. If the interpolation set Z is not coplanar and

$$\sum_{i=1}^{N} |\lambda_i| s_i > 1,$$ then the box dimension of the graph $\mathfrak{G}$ of a fractal function

$f^* : \triangle \to \mathbb{R}$ of either class $C_0(\triangle)$ or $\widetilde{C}(\triangle)$ is the unique positive solution of

$$\sum_{i=1}^{N} |\lambda_i| s_i^{d-1} = 1;$$

otherwise $\dim_B \mathfrak{G} = 2$.

6.4 Hölder Continuity of Affine Fractal Functions

Next it is shown that the bivariate fractal function $f^* : \triangle \to \mathbb{R}$ considered in the previous section is Hölder continuous with exponent $\alpha = 3 - \dim_F \operatorname{graph} f^*$. We use the notation and terminology of Section 6.3.

Theorem 6.20 Let $f^* : \triangle \to \mathbb{R}$ be defined as in the previous section. Then, if $0 \leq h < 1$,

$$|f^*(x + h, y + h) - f^*(x, y)| \leq c\, h^\alpha, \qquad (x, y) \in \triangle,$$

for a positive constant c and $\alpha = 3 - \dim_B \operatorname{graph} f^* = 2 - \log_N \sum_{i=1}^{N^2} |\lambda_i|$.

Proof Let $n \in \mathbb{N}$. For every finite code $\sigma(n) \in \Sigma^* = \{1, \ldots, N^2\}^{\mathbb{N}}$ of length n denote by $\triangle_p$ the simplex $\overline{w}_{\sigma(n)}(\triangle)$, $p = 1, \ldots, N^{2n}$. Let

$$\Pi(m, p) := \bigcup_{k=0}^{\mathcal{N}(m,p)} N^{-m}\triangle \times N^{-m}[k, k + N^{-m}]$$

be the collection of all sets of the above form that cover $\operatorname{graph} f^*|_{\triangle_p}$.

Let $0 \leq h < 1$ be given, and let m be the least integer such that there exists a finite code $\sigma(m) \in \Sigma^*$ of length m so that (x, y) and $(x + h, y + h)$ are both in the interior of $\overline{u}_{\sigma(m)}\triangle$. Then it follows from Theorem 6.18 that

$$|f^*(x + h, y + h) - f^*(x, y)| \leq N^{-m} \max\{\mathcal{N}(m, p) \mid p = 1, \ldots, N^{2m}\}.$$

However, since $\mathcal{N}(m, p) \geq c\, N^{-2m}(N^{-m}N^{-d})$, where $d = \dim_F \operatorname{graph} \mathfrak{f}^*$, we obtain

$$\frac{\log |\mathfrak{f}^*(x + h, y + h) - \mathfrak{f}^*(x, y)|}{\log h} \geq \frac{\log c\, N^{-2m}(N^{-m}N^{-d})}{\log h}$$

$$\geq \frac{\log c}{\log h} + \frac{\log N^{-3m+md}}{\log N^{-m}}$$

$$= \frac{\log c}{\log h} + 3 - d,$$

because $h \leq N^{-m}$. $\square$

The fact that $\operatorname{graph} \mathfrak{f}^*$ has Hölder exponent $\alpha = 3 - \dim_F \operatorname{graph} \mathfrak{f}^*$ has the following consequence.

Corollary 6.21 Suppose that $\mathfrak{f}^*$ is Hölder continuous with exponent $\alpha = 3 - \dim_B \operatorname{graph} \mathfrak{f}^*$. Then $\mathcal{H}^{\dim_B \operatorname{graph} \mathfrak{f}^*}(\operatorname{graph} \mathfrak{f}^*) < +\infty$.

Proof Let $0 \leq h < 1$ and let m be the least integer so that $h \leq N^{-2m}$. Let $\triangle_p$ be any of the N^{2m} subsimplices of $\triangle$. Then $\operatorname{graph} \mathfrak{f}^*|_{\triangle_p}$ can be covered by at most $N^{2m}(ch^\alpha + 1)$ sets of the form $N^{-m}\triangle \times N^{-m}[k, k + N^{-m}]$, $k \in \mathbb{N}$. Thus, setting $\mathfrak{G}^* := \operatorname{graph} \mathfrak{f}^*$, we have for all $m \in \mathbb{N}$

$$\mathcal{H}^{\dim_F \mathfrak{G}^*}_{N^{-2m}}(\mathfrak{G}^*) \leq N^{2m}\left(c\, h^{3-\dim_F \mathfrak{G}^*} + 1\right)\left(N^{-m}\right)^{\dim_B \mathfrak{G}^*}$$

$$\leq N^{2m}\left(c\, N^{(-m)(3-\dim_F \mathfrak{G}^*)} + 1\right)\left(N^{-m}\right)^{\dim_B \mathfrak{G}^*}$$

$$= c\, N^{-m} + \left(N^{-m}\right)^{\dim_B \mathfrak{G}^*-2}$$

$$\leq \frac{c}{N} + \left(\frac{1}{N}\right)^{\dim_B \mathfrak{G}^*-2} < +\infty.$$

Therefore, $\mathcal{H}^{\dim_B \mathfrak{G}^*}(\mathfrak{G}^*) < +\infty$. $\square$

6.5 Bilinear Fractal Surfaces in $\mathbb{R}^2$

In this section, we present a construction of fractal surfaces over rectangular domains of the form $Q := [a, b] \times [c, d]$, $a, b, c, d \in \mathbb{R}$, with nonempty interior. The idea of obtaining fractal surfaces over rectangular domains is similar to that for simplicial domains. To this end, we first consider the following partition of Q into four rectangular subdomains Q_i, $i = 1, 2, 3, 4$, with boundaries parallel to the coordinate axes.

$$Q_1 := \overline{u}_1(Q) := \begin{pmatrix} \alpha & 0 \\ 0 & -\beta \end{pmatrix} Q + \begin{pmatrix} 0 \\ \beta b \end{pmatrix},$$

$$Q_2 := \overline{u}_2(Q) := \begin{pmatrix} \alpha - 1 & 0 \\ 0 & -\beta \end{pmatrix} Q + \begin{pmatrix} a \\ \beta b \end{pmatrix},$$

$$Q_3 := \overline{u}_3(Q) := \begin{pmatrix} \alpha - 1 & 0 \\ 0 & 1 - \beta \end{pmatrix} Q + \begin{pmatrix} a \\ \beta b \end{pmatrix},$$

$$Q_4 := \overline{u}_4(Q) := \begin{pmatrix} \alpha & 0 \\ 0 & 1 - \beta \end{pmatrix} Q + \begin{pmatrix} 0 \\ \beta b \end{pmatrix},$$

$$(6.7)$$

where $\alpha, \beta \in (0, 1)$ are free parameters. The set-valued mappings $\overline{u}_i$, $i = 1, 2, 3, 4$, are defined according to the scheme depicted in Figure 6.9. Label the vertices of Q counterclockwise by **1, 2, 3**, and **4** and the vertices of the subdomains Q_i, $i = 1, 2, 3, 4$, as shown in Figure 6.9. Mapping the vertices of Q onto the corresponding vertices of the subdomains Q_i uniquely defines the four set-valued affine mappings $\overline{u}_i$. If we denote the original vertices by $A := (a, c)$, $B := (b, c)$, $C := (a, d)$, and $D := (b, d)$, and the new vertices by $v_1, \ldots, v_9$, then this particular arrangement of vertices induces a labeling map $\phi : \{1, 2, 3, 4\} \times \{A, B, C, D\} \to \{1, \ldots, 9\}$, $v_{\phi(i, V)} = u_i(V)$.

This procedure can be extended to obtain partitions of Q of higher cardinality. Let $m, n \in \mathbb{N}$ and let $\{x_\mu \mid \mu = 0, 1, \ldots, m\}$ be a partition of the interval $[a, b]$ with $x_0 := a$ and $x_m := b$. Similarly, let $\{y_\nu \mid \nu = 0, 1, \ldots, n\}$ be a partition of the interval $[c, d]$ with $y_0 := c$ and $y_n := d$. The set of points $\{(x_\mu, y_\nu) \mid \mu = 0, 1, \ldots, m; \nu = 0, 1, \ldots, n\}$ divides Q into $N := m \cdot n$ rectangular subdomains. If we denote the vertices of this partition of Q by $v_j := (x_\mu, y_\nu)$ with $j = \mu + 1 + \nu(m + 1)$, and extend the labeling introduced above periodically along both the x- and y-directions, we also extend the labeling map ϕ to a mapping $\{1, \ldots, m \cdot n\} \times \{A, B, C, D\} \mapsto \{1, \ldots, (m + 1) \cdot (n + 1)\}$, and obtain $v_{\phi(i, V)} = u_i(V)$. An example of a partition of cardinality $N = 12 = 4 \cdot 3$ is shown in Figure 6.10.

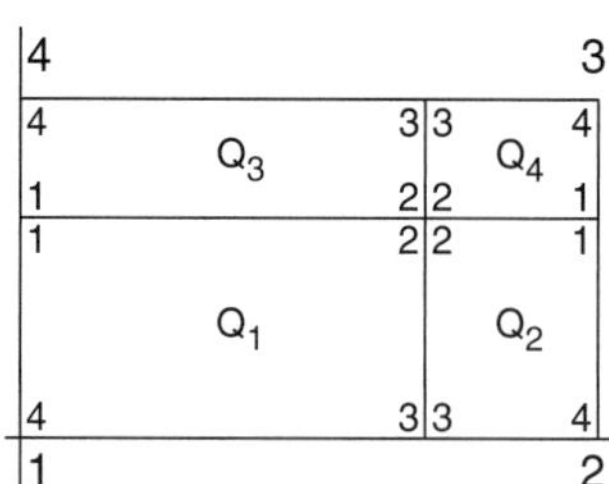

Figure 6.9 The basic partition of Q.

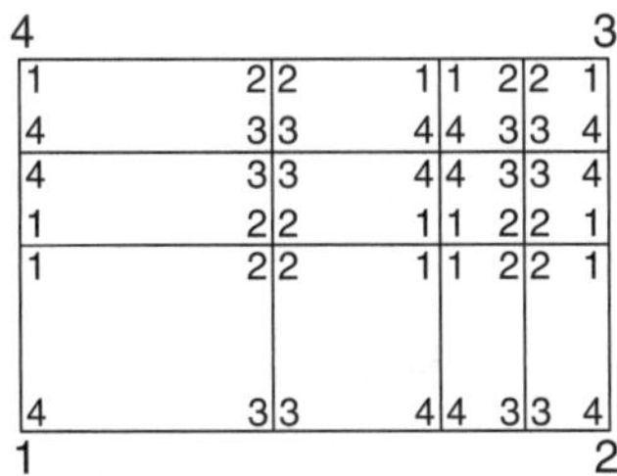

Figure 6.10 A more general partition of Q.

Start labeling the subdomains Q_i, $i = 1, \ldots, N$, beginning with the setting given in Figure 6.9 and continue by matching vertices across edges and preserving the original ordering of vertices. The index i of the subdomain Q_i, $i = 1, \ldots, N$, is determined by the index (μ, ν) of its lower left vertex. We leave it to the reader to show that

$$i = \mu + 1 + \nu m, \quad \mu = 0, 1, \ldots, m; \ \nu = 0, 1, \ldots, n,$$

and that given a subdomain Q_i, the pair (μ, ν) is found via

$$\mu = (i - 1) \bmod m \quad \text{and} \quad \nu = \frac{i - 1 - \mu}{m}, \ i = 1, \ldots, N.$$

We also note that this periodic labeling implies that the set-valued affine maps $\overline{u}_i$, $i = 1, \ldots, N$, are of a specific structure.

The linear parts of all set-valued mappings $\overline{u}_i$, $i = 1, \ldots, N$, are of the form $\begin{pmatrix} -1 & 0 \\ 0 & 1 \end{pmatrix}$ (reflection about the y-axis), $\begin{pmatrix} 1 & 0 \\ 0 & -1 \end{pmatrix}$ (reflection about the x-axis), and $\begin{pmatrix} -1 & 0 \\ 0 & -1 \end{pmatrix}$ (reflection about the origin) multiplying a diagonal matrix $\begin{pmatrix} \alpha & 0 \\ 0 & \beta \end{pmatrix}$ with $\alpha, \beta \in (0, 1)$. Hence, for given parameters α and β, which are determined by the partitions of the intervals $[a, b]$ and $[c, d]$, there are only four different linear parts for the maps $\overline{u}_i$. The class of partitions of Q of the above form will be denoted by $\mathscr{P}_Q$.

To each partition $\pi \in \mathscr{P}_Q$ corresponds a collection of numerical mappings $\{u_i : Q \to Q_i \,|\, i = 1, \ldots, N\}$ with the properties that each u_i is an affine mapping of one of the four forms given in (6.7).

In order to generate a fractal surface over Q given a partition $\pi \in \mathscr{P}_Q$ into subdomains Q_i defined above, we consider a special class of maps q_i, namely so-called *bilinear mappings*.

Definition 6.22 (Bilinear Mapping) Suppose that X, Y, and Z are linear spaces over a field $\mathbb{K}$. A mapping $f : \mathsf{X} \times \mathsf{Y} \to \mathsf{Z}$ is called *bilinear* if the partial mappings $f(x, \bullet) : \mathsf{Y} \to \mathsf{Z}$, $x \in \mathsf{X}$, and $f(\bullet, y) : \mathsf{X} \to \mathsf{Z}$, $y \in \mathsf{Y}$, are linear. In other words,

1. $\forall\, x, x' \in \mathsf{X} \;\forall\, y \in \mathsf{Y} \;\forall\, \alpha \in \mathbb{K}$: $f(\alpha x + x', y) = \alpha f(x, y) + f(x', y)$.

2. $\forall\, y, y' \in \mathsf{Y} \;\forall\, x \in \mathsf{X} \;\forall\, \alpha \in \mathbb{K}$: $f(x, \alpha y + y') = \alpha f(x, y) + f(x, y')$.

In case $\mathsf{Z} = \mathbb{K}$, the mapping f is termed a *bilinear form*.

Example 6.23   Let $\mathsf{X} := \mathbb{R} =: \mathsf{Y}$ and $\mathsf{Z} := \mathbb{R}$. Let $a, b, c, d \in \mathbb{R}$. Then the mapping $f : \mathbb{R}^2 \to \mathbb{R}$, $(x, y) \mapsto (a + bx)(c + dy)$ is bilinear. Also note that f can be written in the form $f(x, y) = (a' + b'x) + (c' + d'x)\, y$, for real numbers a', b', c', and d'.

Now, let $Z := \{(x_\mu, y_\nu, z_{\mu\nu}) \,|\, \mu = 0, 1, \ldots, m;\ \nu = 0, 1, \ldots, n\}$ be a given set of interpolation points. As above, we set

$$v_j := (x_\mu, y_\nu) \quad \text{and} \quad \zeta_j := z_{\mu\nu}, \quad \text{with } j := \mu + 1 + \nu(m+1).$$

In addition, let z_V be the z-value above the original vertex $V \in \{A, B, C, D\}$.

We now define bilinear mappings $q_i : Q \to \mathbb{R}$, $(x, y) \mapsto a_i x + b_i y + c_i xy + d_i$, $a_i, b_i, c_i, d_i \in \mathbb{R}$, by requiring that

$$q_i(V) + \lambda z_V = z_{\phi(i, V)}, \tag{6.8}$$

where $\lambda \in (-1, 1)$ is a free parameter.

As previously, we define for a given rectangular domain $Q = [a, b] \times [c, d]$ with nonempty interior and given interpolation set Z, a class of continuous functions as follows.

$$\widetilde{C}(Q) := \{f \in C(Q) \,|\, f(x_\mu, y_\nu) = z_{\mu\nu},\ \mu = 0, 1, \ldots, m;\ \nu = 0, 1, \ldots, n\}.$$

Endowing $\widetilde{C}(Q)$ with the metric $d_{\infty, Q} : C(Q) \times C(Q) \to \mathbb{R}$, $d_{\infty, Q}(f, g) := \|f - g\|_{\infty, Q}$, where $\|\ \|_{\infty, Q}$ is the Chebyshev norm, creates a complete metric space $(\widetilde{C}(Q), \|\ \|_{\infty, Q})$.

Theorem 6.24   Let Q be a rectangular domain with nonempty interior, and suppose that $\pi \in \mathscr{P}_Q$ generating $N = m \cdot n$ subdomains $\{Q_i \,|\, i = 1, \ldots, N\}$ of Q. Let $u_i : Q \to Q_i$ be the numerical mappings corresponding to the partition π. Furthermore, suppose that Z is an interpolation set over Q, and that $\{q_i \,|\, i = 1, \ldots, N\}$ is a collection of bilinear mappings satisfying the conditions given by (6.8) for a given free parameter $\lambda \in (-1, 1)$. Define an operator $T[\boldsymbol{q}, \lambda] : \widetilde{C}(Q) \to \mathrm{Map}(Q, \mathbb{R})$ by

$$T[\boldsymbol{q}, \lambda]f := \sum_{i=1}^{N} (q_i \circ u_i^{-1} + \lambda f \circ u_i^{-1})\, \chi_{Q_i}.$$

Then, $T[q, \lambda]$ is well-defined, maps $\widetilde{C}(Q)$ into itself, and possesses a unique fixed point $\mathfrak{f} : Q \to \mathbb{R}$. Moreover, the fixed point $\mathfrak{f}$ interpolates Z.

Proof Exercise! $\square$

The graph of the fixed point $\mathfrak{f} : Q \to Q$ is called a *bilinear fractal surface (over Q)*.

Example 6.25 Let $Q := [0, 3] \times [0, 2]$ and suppose that $\pi \in \mathscr{P}_Q$ is such that $Q = Q_1 \cup Q_2 \cup Q_3 \cup Q_4$, where

$$Q_1 := \begin{pmatrix} 0.7 & 0 \\ 0 & -0.65 \end{pmatrix} Q + \begin{pmatrix} 0 \\ 1.3 \end{pmatrix}, \quad Q_2 := \begin{pmatrix} -0.3 & 0 \\ 0 & -0.65 \end{pmatrix} Q + \begin{pmatrix} 3 \\ 1.3 \end{pmatrix},$$

$$Q_3 := \begin{pmatrix} -0.3 & 0 \\ 0 & 0.35 \end{pmatrix} Q + \begin{pmatrix} 3 \\ 1.3 \end{pmatrix}, \quad Q_4 := \begin{pmatrix} 0.7 & 0 \\ 0 & 0.35 \end{pmatrix} Q + \begin{pmatrix} 0 \\ 1.3 \end{pmatrix}.$$

Let u_i, $i = 1, \ldots, 4$, be the numerical mappings corresponding to π, i.e.,

$$u_1(x, y) := \begin{pmatrix} 0.7 & 0 \\ 0 & 0.65 \end{pmatrix} \begin{pmatrix} x \\ y \end{pmatrix} + \begin{pmatrix} 0 \\ 1.3 \end{pmatrix},$$

$$u_2(x, y) := \begin{pmatrix} -0.3 & 0 \\ 0 & -0.65 \end{pmatrix} \begin{pmatrix} x \\ y \end{pmatrix} + \begin{pmatrix} 3 \\ 1.3 \end{pmatrix},$$

$$u_3(x, y) := \begin{pmatrix} -0.3 & 0 \\ 0 & 0.35 \end{pmatrix} \begin{pmatrix} x \\ y \end{pmatrix} + \begin{pmatrix} 3 \\ 1.3 \end{pmatrix},$$

$$u_4(x, y) := \begin{pmatrix} 0.7 & 0 \\ 0 & 0.35 \end{pmatrix} \begin{pmatrix} x \\ y \end{pmatrix} + \begin{pmatrix} 0 \\ 1.3 \end{pmatrix}.$$

Let $\lambda := 0.65$ and suppose that

$$Z := \{(0, 0, 0), (2.1, 0, 0.5), (3, 0, 0), (0, 1.3, -0.5), (0, 2, 0), (2.1, 1.3, 0.25),$$
$$(2.1, 2, 0.5), (3, 1.3, 0.25), (3, 2, 0)\}.$$

The bilinear mappings q_i, $i = 1, \ldots, 4$ are thus given by

$$q_1(x, y) = -\frac{1}{2} + \frac{x}{4} + \frac{y}{4} - \frac{xy}{24}, \qquad q_2(x, y) = \frac{1}{4} - \frac{y}{8} - \frac{xy}{12},$$

$$q_3(x, y) = \frac{1}{4} - \frac{y}{8} - \frac{xy}{12}, \qquad q_4(x, y) = -\frac{1}{2} + \frac{x}{4} + \frac{y}{4} - \frac{xy}{24}.$$

The graph of the resulting bilinear fractal surface is depicted in Figure 6.11.

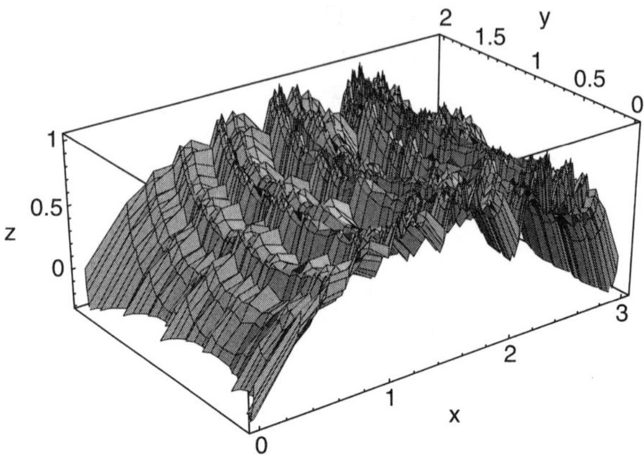

Figure 6.11 A bilinear fractal surface.

6.6 Fractal Surfaces Arising from Quadratic Forms

In this section, a method for constructing fractal surfaces defined by a class of quadratic forms in three variable is presented. We concentrate on defining these fractal surfaces over the unit square $\square \subset \mathbb{R}^2$, and refer to [112] or [113] for additional information and constructions.

To this end, let $\square := [0, 1] \times [0, 1]$, let $e_1 := (1, 0)^\top$, $e_2 = (0, 1)^\top$, and let N be a fixed integer greater than one. Let $\Gamma := \{(m/N)e_1 + (n/N)e_2 : m, n \in \mathbb{Z}\}$ be a lattice in $\mathbb{R}^2$. Suppose that for each lattice point $(x_j, y_j) \in \Gamma \cap Q$ a real number z_{ij}, $i, j \in \{0, 1, \ldots, N\}$ is given. The set $\Xi := \{(x_j, y_j, z_{ij}) \mid i, j = 0, 1, \ldots, N\}$ can be thought of as a given set of data or interpolation points over $\square$. Fractal surfaces containing this interpolating set will be constructed next.

For this purpose, let $u_{ij} : \square \to \square$ be given by

$$u_{ij}(x, y) = \begin{pmatrix} \frac{1}{N} & 0 \\ 0 & \frac{1}{N} \end{pmatrix} \begin{pmatrix} x \\ y \end{pmatrix} + \begin{pmatrix} \frac{i-1}{N} \\ \frac{j-1}{N} \end{pmatrix}, \quad i, j = 1, \ldots, N$$

and let $v_{ij} : \square \times \mathbb{R} \to \mathbb{R}$ be the following quadratic form in three variables:

$$v_{ij}(x, y, z) = A_{ij}\, x^2 + B_{ij}\, y^2 + C_{ij}\, z^2 + D_{ij}\, xy + E_{ij}\, yz + F_{ij}\, zx + G_{ij},$$

for all $i, j \in \{1, \ldots, N\}$. The coefficients $A_{ij}, B_{ij}, C_{ij}, D_{ij}, E_{ij}, F_{ij}$ and G_{ij}, $i, j = 1, \ldots, N$, are real numbers that will be uniquely determined by join-up conditions to ensure continuity, respectively, smoothness across the edges. Initially, we require that for all $i, j = 1, \ldots, N$,

$$\begin{aligned} v_{ij}(0, 0, z_{0,0}) &= z_{i-1,j-1}, & v_{ij}(0, 1, z_{0,N}) &= z_{i-1,j}, \\ v_{ij}(1, 0, z_{N,0}) &= z_{i,j-1}, & v_{ij}(1, 1, z_{N,N}) &= z_{i,j}. \end{aligned}$$

$$(6.9)$$

Furthermore, the following join-up conditions are to be satisfied: For $j = 1, \ldots, N$ and $y \in \left[(j-1)/N, j/N\right]$,

$$
\begin{aligned}
v_{ij}(0, y, \varphi(0, y)) &= v_{i-1,j}(1, y, \varphi(1, y)), & i &= 2, \ldots, N, \\
v_{ij}(1, y, \varphi(1, y)) &= v_{i+1,j}(0, y, \varphi(0, y)), & i &= 1, \ldots, N-1,
\end{aligned}
\tag{6.10}
$$

and, for $i = 1, \ldots, N$ and $x \in \left[(j-1)/N, j/N\right]$,

$$
\begin{aligned}
v_{ij}(x, 0, \varphi(x, 0)) &= v_{i,j-1}(x, 1, \varphi(x, 1)), & j &= 2, \ldots, N, \\
v_{ij}(x, 1, \varphi(x, 1)) &= v_{i,j+1}(x, 0, \varphi(x, 0)), & j &= 1, \ldots, N-1.
\end{aligned}
\tag{6.11}
$$

Here φ denotes any C^0-function interpolating the set Ξ.

Conditions (6.9), (6.10), and (6.11) uniquely determine some of the coefficients $A_{ij}, \ldots, G_{ij}$. For instance, if $z_{0,0} = z_{0,N} = z_{N,0} = z_{N,N} = 0$, then

$$
\begin{aligned}
A_{ij} &= z_{i,j-1} - z_{i-1,j-1}, & B_{ij} &= z_{i-1,j} - z_{i-1,j-1} \\
D_{ij} &= (z_{ij} - z_{i,j-1}) - (z_{i-1,j} - z_{i-1,j-1}), & G_{ij} &= z_{i-1,j-1},
\end{aligned}
\tag{6.12}
$$

for $i, j = 1, \ldots, N$. If $\varphi \equiv 0$ on $\partial\square$, then $A_{ij} = B_{ij} = D_{ij} \equiv 0$, and the join-up conditions are automatically satisfied. If $\varphi(0, \bullet) \equiv \varphi(1, \bullet)$ and $\varphi(\bullet, 0) \equiv \varphi(\bullet, 1)$, then in addition to (6.12) one also needs that

$$
\begin{aligned}
A_{ij} &= A_{i,j-1}, & B_{ij} &= B_{i-1,j}, \\
C_{ij} &= C_{i-1,j} = C_{i,j-1}, & E_{ij} &= F_{ij} = 0,
\end{aligned}
$$

in order for the join-up conditions to hold.

Define the function space

$$
C_\Xi(\square) := \{\varphi \in C(\square, \mathbb{R}) \mid \varphi(x_j, y_j) = z_{ij}, \ i, j = 0, 1, \ldots, N\},
$$

and an RB operator $T : C(\square) \to \mathrm{Map}(\square, \mathbb{R})$ by

$$
(Tf)(x, y) := v_{ij}(u_{ij}^{-1}(x, y), f \circ u_{ij}^{-1}(x, y))), \qquad (x, y) \in u_{ij}(\square). \tag{6.13}
$$

Suppose, without loss of generality, that $\|f\|_\infty \leq 1$ on $\square$ and that

$$
\lambda := 2 \max_{1 \leq i,j \leq N} |C_{ij}| + \max_{1 \leq i,j \leq N} |E_{i,j}| + \max_{1 \leq i,j \leq N} |F_{ij}| < 1.
$$

Theorem 6.26 The RB operator T maps $C_\Xi(\square)$ into itself, is well-defined and contractive in the $\|\ \|_\infty$-norm with contractivity constant λ.

Proof The results follow readily from the definition of T, conditions (6.9), (6.10) and (6.11), and the assumption on λ. Details are left to the reader. $\square$

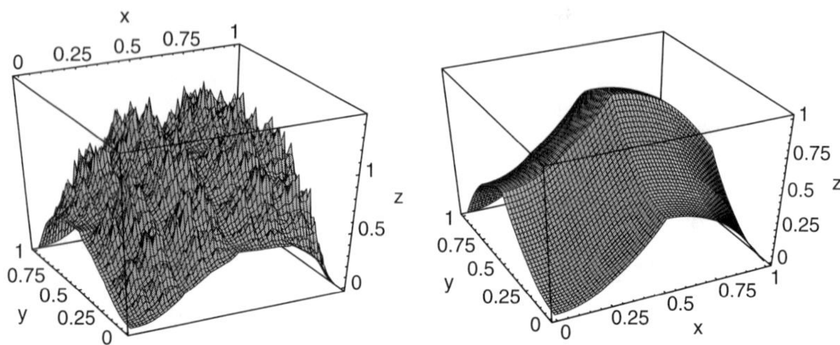

Figure 6.12 A fractal surface of class $C_\Xi(\square)$ with $C_{ij} = \frac{2}{5}$ (left) and $C_{ij} = \frac{1}{20}$ (right), $i, j = 1, 2$.

The unique fixed point of the RB operator T is the graph of a C^0-function $\mathfrak{f} : \square \to \mathbb{R}$ that interpolates Ξ. The graph of $\mathfrak{f}$ is called a fractal surface of class $C_\Xi(\square)$.

Example 6.27 Let $z_{00} = z_{02} = z_{20} = z_{22} = 0, z_{01} = z_{10} = z_{12} = z_{21} = \frac{1}{2}$, and $z_{11} = 1$. Suppose that φ is the piecewise linear function interpolating $\{(\frac{i}{2}, \frac{j}{2}, z_{ij}) \mid i, j = 0, 1, 2\}$. Then, one obtains for the coefficients $A_{ij}, \ldots, G_{ij}$ the following values:

$$A_{11} = A_{12} = -A_{21} = -A_{22} = \frac{1}{2}, \qquad B_{11} = -B_{12} = B_{21} = -B_{22} = \frac{1}{2},$$

$$D_{11} = D_{12} = D_{21} = D_{22} = 0, \qquad G_{11} = 0, \, G_{12} = G_{21} = \frac{1}{2}, \, G_{22} = 1.$$

The coefficients E_{ij} and F_{ij} are zero, for all $i, j = 0, 1, 2$. Figure 6.12 displays the graphs of two fractal surfaces generated by these coefficients for different values of $C_{ij}, i, j = 1, 2$.

6.7 Smooth Fractal Surfaces via Indefinite Integrals

In this section, we construct smooth fractal surfaces using indefinite two-dimensional integrals. This construction employs the same ideas as those that were developed in Section 5.12. The fractal surfaces that are obtained in this way provide one means of constructing smooth fractal surfaces.

To this end, we use the same terminology and notation as in the previous section. In particular, let $\square := [0, 1] \times [0, 1]$ and let $u_{ij}, i, j = 1, \ldots, N$, be chosen as in section 6.6. The mappings v_{ij}, however, are to be of the following form:

(i) $v_{ij}(\cdot, \cdot, z) - v_{ij}(0, 0, z_0)$ is a symmetric quadratic form for all $z, z_0 \in \mathbb{R}$;

(ii) $v_{ij}(x, y, \cdot) - v_{ij}(x, y, z_0)$ is a bilinear form for all $(x, y) \in \square$ and $z_0 \in \mathbb{R}$.

Furthermore, it is required that conditions (6.9), (6.10), and (6.11) also hold for this particular choice of v_{ij}. In the special case where $z_{0,0} = z_{N,0} = z_{0,N} = z_{N,N} = 0$, the same expressions for A_{ij}, B_{ij}, D_i, and G_{ij} are obtained as above, if we select

$$v_{ij}(x, y, z) = A_{ij}x^2 + B_{ij}y^2 + C_{ij}z + D_{ij}xy + G_{ij}.$$

Note that (6.10) and (6.11) follow whether $\varphi \equiv 0$ on $\partial\square$ or $\varphi\big|_{[0,1]\times\{0\}} \equiv \varphi\big|_{[0,1]\times\{1\}}$ and $\varphi\big|_{\{0\}\times[0,1]} \equiv \varphi\big|_{\{1\}\times[0,1]}$, $\varphi \in C_\Xi(\square)$.

Defining an operator T as in (6.13) and assuming that $\lambda := \max\{|C_{ij}| \mid i, j = 1, \ldots, N\} < 1$, the following theorem is obtained, whose straight-forward proof is left for the reader.

Theorem 6.28   The unique fixed point of the RB operator T is a C^0-function $f : \square \to \mathbb{R}$ such that $f(x_j, y_j) = z_{ij}$, for all $i, j = 0, 1, \ldots, N$.

The reason for choosing this particular form of the v_{ij} will shortly become clear. Let

$$\widetilde{f}(x, y) := \widetilde{z}_{0,0} + \int_0^x \int_0^y f(s, t)\, dt\, ds, \quad \text{for some } \widetilde{z}_{0,0} \in \mathbb{R},$$

be the two-dimensional indefinite integral of f over $\square$. Denote the operator $\int_0^x \int_0^y (\bullet)\, dt\, ds$ by $\mathscr{I}_{(0,0)}^{(x,y)}(\bullet)$, and set $u_{ij}(x, y) := (\xi_i(x), \eta_j(y))$, where $\xi_i(x) := \frac{x+i-1}{N}$ and $\eta_j(y) := \frac{y+j-1}{N}$, $i, j = 0, 1, \ldots, N$. Then, using the fixed point equation for f, we have

$$\widetilde{f} \circ u_{ij}(x, y) = \widetilde{z}_{0,0} + \mathscr{I}_{(0,0)}^{(x_{i-1}, y_{j-1})}(f) + \mathscr{I}_{(x_{i-1},0)}^{(\xi_i(x), y_{j-1})}(f)$$

$$+ \mathscr{I}_{(0, y_{i-1})}^{(x_{i-1}, \eta_j(y))}(f)\, \mathscr{I}_{(x_{i-1}, y_{j-1})}^{(\xi_i(x), \eta_j(y))}(f)$$

$$= \widetilde{z}_{0,0} + \mathscr{I}_{(0,0)}^{(x_{i-1}, y_{j-1})}(f) + \mathscr{I}_{(x_{i-1},0)}^{(\xi_i(x), y_{j-1})}(f) + \mathscr{I}_{(0, y_{j-1})}^{(x_{i-1}, \eta_j(y))}(f)$$

$$+ \frac{1}{N^2}\, \mathscr{I}_{(0,0)}^{(x_{i-1}, y_{j-1})}(f \circ u_{ij}).$$

Since $f \circ u_{ij} = v_{ij}(\bullet, f)$, this produces

$$\widetilde{f} \circ u_{ij}(x, y) = \left[\widetilde{z}_{0,0} + \mathscr{I}_{(0,0)}^{(x_{i-1}, y_{j-1})}(f) + \mathscr{I}_{(x_{i-1},0)}^{(\xi_i(x), y_{j-1})}(f) + \mathscr{I}_{(0, y_{j-1})}^{(x_{i-1}, \eta_j(y))}(f)\right]$$

$$+ \frac{C_{ij}}{N^2}\, \mathscr{I}_{(0,0)}^{(x,y)}(f) + \frac{1}{N^2}\, \mathscr{I}_{(0,0)}^{(x,y)}(v_{ij}\big|_{z=0})$$

$$=: \frac{C_{ij}}{N^2}\, \widetilde{f}(x, y) + R_{ij}(x, y).$$

Hence, $\widetilde{f}$ is the unique fixed point of the operator $\Psi : C^1(\square, \mathbb{R}) \to C^1(\square, \mathbb{R})$,

$$\Psi\varphi := \widetilde{v}_{ij}(u_{ij}^{-1}(\bullet, \bullet), \varphi \circ u_{ij}^{-1}(\bullet, \bullet)),$$

where $\widetilde{v}_{ij}(x, y, z) := R_{ij}(x, y, z) + \frac{C_{ij}}{N^2} z$, or equivalently, graph f is the unique attractor of the IFS $(\Box \times \mathbb{R}, \widetilde{W})$ with $\widetilde{W} := \{\widetilde{w}_{ij} : \Box \times \mathbb{R} \to \Box \times \mathbb{R} \mid \widetilde{w}_{ij} = (u_{ij}, \widetilde{v}_{ij}),\ i, j = 1, \dots, N\}$. Since the operator $\mathscr{I}_{(0,0)}^{(x,y)}(\bullet)$ is continuous, $\widetilde{\mathsf{f}}$ is continuous at its interpolating set $\widetilde{\Xi} := \{(x_j, y_j, \widetilde{z}_{ij}) :\ i, j = 0, 1, \dots, N\}$. To determine the values of the $\widetilde{z}_{ij}$, notice that $u_{ij}(0, 0) = (x_{i-1}, y_{j-1}) = z_{i-1,j-1}$, and thus $\widetilde{\mathsf{f}} \circ u_{ij}(0, 0) = \widetilde{z}_{0,0} + \mathscr{I}_{(0,0)}^{(x_{i-1}, y_{j-1})}(\mathsf{f}) =: \widetilde{z}_{i-1,j-1}$. Therefore,

$$\widetilde{z}_{ij} = \widetilde{z}_{i-1,j-1} + \frac{C_{ij}}{N^2} \mathscr{I}_{(0,0)}^{(1,1)}(\mathsf{f}) + \frac{1}{N^2} \mathscr{I}_{(0,0)}^{(1,1)}(v_{ij}\mid_{z=0}) + \mathscr{I}_{(x_{i-1},0)}^{(x_i, y_{j-1})}(\mathsf{f})$$

$$+ \mathscr{I}_{(0, y_{j-1})}^{(x_{i-1}, y_j)}(\mathsf{f})$$

$$= \frac{C_{ij}}{N^2}(\widetilde{z}_{N,N} - \widetilde{z}_{0,0}) + \frac{1}{N^2} \mathscr{I}_{(0,0)}^{(1,1)}(v_{ij}\mid_{z=0}) + (\widetilde{z}_{i-1,j} - \widetilde{z}_{i-1,j-1}) + \widetilde{z}_{i,j-1}.$$

Hence the $\widetilde{z}_{ij}$, $i, j = 0, 1, \dots, N$, can be expressed in terms of $\widetilde{z}_{0,0}$, C_{ij}, and $\mathscr{I}_{(0,0)}^{(1,1)}(v_{ij}\mid_{z=0})$, $(i, j) \neq (0, 0)$. These results are now summarized in a theorem.

Theorem 6.29   Let graph f be a fractal surface generated by the IFS $(\Box \times \mathbb{R}, \mathcal{W})$, where $w_{ij} := (u_{ij}, v_{ij})$, $i, j = 1, \dots, N$, with

$$u_{ij}(x, y) = \begin{pmatrix} \frac{1}{N} & 0 \\ 0 & \frac{1}{N} \end{pmatrix} \begin{pmatrix} x \\ y \end{pmatrix} + \begin{pmatrix} \frac{i-1}{N} \\ \frac{j-1}{N} \end{pmatrix},$$

and

$$v_{ij}(x, y, z) = A_{ij} x^2 + B_{ij} y^2 + C_{ij} z + D_{ij} xy + G_{ij},$$

such that $\max\{|C_{ij}| \mid i, j = 1, \dots, N\} < 1$. Let

$$\widetilde{\mathsf{f}}(x, y) := \widetilde{z}_{0,0} + \int_0^x \int_0^y \mathsf{f}(s, t)\, dt\, ds, \quad \text{for some } \widetilde{z}_{0,0} \in \mathbb{R}.$$

Then graph f is the attractor of the IFS $(\Box \times \mathbb{R}, \widetilde{\mathcal{W}})$ with $\widetilde{w}_{ij} = (u_{ij}, \widetilde{v}_{ij})$, where

$$\widetilde{v}_{ij}(x, y, z) = \widetilde{z}_{i-1,j-1} + \frac{C_{ij}}{N^2} z + \frac{1}{N^2} \int_0^x \int_0^y v_{ij}(s, t, 0)\, dt\, ds, \quad i, j = 1, \dots N.$$

Furthermore, the $\widetilde{z}_{ij}$, $(i, j) \neq (0, 0)$, are recursively and uniquely determined by $\widetilde{z}_{0,0}$ which is a free parameter, C_{ij}, and $\int_0^x \int_0^y v_{ij}(s, t, 0)\, dt\, ds$. Also, $\nabla \widetilde{\mathsf{f}}(x, y) = (\mathfrak{g}_y(x), \mathfrak{h}_x(y))$, where

$$\mathfrak{g}_y(x) = \int_0^y \mathsf{f}(x, t)\, dt \ \text{ and } \ \mathfrak{h}_x(y) = \int_0^x \mathsf{f}(s, y)\, ds, \quad x, y \in [0, 1].$$

Moreover, $\frac{\partial}{\partial x} \frac{\partial}{\partial y} \widetilde{\mathsf{f}} = \frac{\partial}{\partial y} \frac{\partial}{\partial x} \widetilde{\mathsf{f}} = \mathsf{f}$.

Proof The last part of the theorem follows from calculus. $\square$

It should now be clear how one can construct C^n-interpolating fractal surfaces for any $n \in \mathbb{N}$: the foregoing procedure can be iterated an arbitrary number of times.

Exercises

1. Bilinear mappings can be classified in terms of the tensor product in the following way. Suppose that X, Y, and Z are linear spaces over a field $\mathbb{K}$. For every bilinear mapping $f : \mathsf{X} \times \mathsf{Y} \to \mathsf{Z}$, there exists a uniquely determined linear mapping $\theta : \mathsf{X} \otimes \mathsf{Y} \to \mathsf{Z}$, $x \otimes y \overset{\theta}{\mapsto} f(x, y)$. Conversely, for every linear mapping $\psi : \mathsf{X} \otimes \mathsf{Y} \to \mathsf{Z}$, there exists a bilinear mapping $g : \mathsf{X} \times \mathsf{Y} \to \mathsf{Z}$, $(x, y) \overset{g}{\mapsto} \psi(x \otimes y)$. Show that these two correspondences define a bijection between the space of bilinear mappings $\mathsf{X} \times \mathsf{Y} \to \mathsf{Z}$ and the space of linear mappings $\mathsf{X} \otimes \mathsf{Y} \to \mathsf{Z}$.

2. Complete the proof of Theorem 6.5.

3. Prove Lemma 6.17.

4. Given a general partition of Q, derive an explicit formula for the labeling map $\phi : \{1, \ldots, m \cdot n\} \times \{A, B, C, D\} \to \{1, \ldots, (m+1) \cdot (n+1)\}$ given in Section 6.5 in terms of the indices of the mappings u_i and the original vertices.

5. Prove Theorem 6.24.

6. Verify the join-up conditions mentioned in Section 6.6, in particular Equations (6.9) to (6.12).

7. Give the proof of Theorem 6.28.

8. Complete the proof of Theorem 6.29.

9. Set $\widetilde{z}_{0,0} := 0$ and compute $\mathscr{I}_{(0,0)}^{(x,y)} f$, where f is either one of the fractal functions constructed in Example 6.27. $\square$

7

Superfractals

The concept of a *superfractal* is a recent addition to fractal geometry and the focus of current research. Superfractals are, like standard fractals, constructed via IFSs. The difference, however, between a fractal and a superfractal lies in the type of IFS used to generate it and the spaces onto which this IFS acts. A superfractal can be viewed as the attractor of a superIFS, i.e., an IFS whose mappings are IFSs themselves. Thus, the functions in a superIFS do not map the space $\mathbf{X}$ into itself, but are rather contractions on the space $\mathcal{H}(\mathbf{X})$, the hyperspace of all nonempty compact subsets of $\mathbf{X}$. The points of a superfractal are so-called V-variable fractal sets, where $V \in \mathbb{N} \cup \{\infty\}$, is a parameter describing the variability of shapes or forms encountered in the geometric structure of these fractals.

Loosely speaking, a V-variable fractal is characterized by the fact that it possesses at most V distinct local patterns at each level of magnification, where the class of patterns depends on the level. The fractals considered so far correspond to $V = 1$ with one IFS, since the pattern that describes them at each level is determined by one IFS only. Such fractals are called *homogeneous random fractals*. The larger the parameter V, the more random the fractal becomes, and in the limiting case $V = \infty$, one obtains *standard random fractals*. In other words, the parameter V describes a ladder of fractal families ranging from homogeneous fractals to completely random fractals which have no correlations between the patterns of shapes at each level. Therefore, V-variable fractals have the potential to be employed for modeling purposes and for geometric applications that require random fractals with a controlled degree of strict self-similarity at each scale.

We first describe the $V = 1$ case elaborating on the geometric and code space structure of 1-variable fractal sets and also 1-variable fractal measures.

Using the superfractal version of the chaos game, we then present the general case as well as the limiting case $V = \infty$. The material upon which this chapter is built can be found in [9] and the original papers on the subject: [10, 16, 17], and [18]. For applications of superfractals and V-variable fractals to natural images, computer graphics, bioinformatics, and numerous others, we refer the interested reader to [9].

There also exist formulae for the Hausdorff and box dimension of superfractal sets. Again, we refer the interested reader to the literature, most notably [9, 17, 18] and the references given therein.

7.1 1-Variable Fractal Sets

In this section, we introduce V-variable fractal sets, where V is a positive integer. For this purpose, we recall the definition of $\mathcal{H}(\mathbf{X})$, the class of all nonempty compact subsets of a set $\mathbf{X}$, and the content of Proposition 4.21. In Section 4.2 we showed that the complete metric space $(\mathcal{H}(\mathbf{X}), d_{\mathcal{H}})$ has a rich geometric structure and is the space in which fractals reside.

Instead of requiring $(\mathbf{X}, d)$ to be complete, we impose the stronger condition of compactness on $(\mathbf{X}, d)$. Considering Proposition 4.30, this requirement is not too much of a restriction. Exercise 1 shows that under the assumption that $(\mathbf{X}, d)$ is a compact metric space, the hyperspace $(\mathcal{H}(\mathbf{X}), d_{\mathcal{H}})$ is also a compact metric space. Starting then with a compact metric space $(\mathbf{X}, d)$, one can construct a sequence of spaces $\{(\mathcal{H}^k(\mathbf{X}), d_{\mathcal{H}}^k) \mid k \in \mathbb{N}\}$ as follows:

$$\mathcal{H}^1(\mathbf{X}) := \mathcal{H}(\mathbf{X}) \quad \text{and} \quad d_{\mathcal{H}}^1 := d_{\mathcal{H}},$$

$$\mathcal{H}^k(\mathbf{X}) := \mathcal{H}(\mathcal{H}^{k-1}(\mathbf{X})) \quad \text{and} \quad d_{\mathcal{H}}^k := d_{\mathcal{H}}(d_{\mathcal{H}}^{k-1}), \quad 1 < k \in \mathbb{N},$$

where

$$d_{\mathcal{H}}^k : \mathcal{H}^k(\mathbf{X}) \times \mathcal{H}^k(\mathbf{X}) \to \mathbb{R},$$

$$(\alpha, \beta) \longmapsto \max \left\{ \max_{A \in \alpha} \min_{B \in \beta} d_{\mathcal{H}}^{k-1}(A, B), \max_{B \in \beta} \min_{A \in \alpha} d_{\mathcal{H}}^{k-1}(A, B) \right\},$$

with $A, B \in \mathcal{H}^{k-1}(\mathbf{X})$. The compactness of $\mathbf{X}$ and the fact that $d_{\mathcal{H}}^k$ is a complete metric on the set $\mathcal{H}^k(\mathbf{X})$ imply that all spaces $(\mathcal{H}^k(\mathbf{X}), d_{\mathcal{H}}^k)$, $k \in \mathbb{N}$, are compact metric spaces. Note that this in particular means that a sequence $\{\alpha_v \mid v \in \mathbb{N}\} \subset \mathcal{H}^k(\mathbf{X})$ converges to $\alpha \in \mathcal{H}^k(\mathbf{X})$ iff for each $A \in \alpha$ there exists a sequence $\{A_v \mid v \in \mathbb{N}\} \subset \mathcal{H}^{k-1}(\mathbf{X})$ of sets $A_v \in \alpha_v$ with the property that $\{A_v \mid v \in \mathbb{N}\}$ converges to A.

Note that since $\mathbf{X}$ is a compact metric space, $\mathbf{X} \in \mathcal{H}(\mathbf{X})$ and $\{\mathbf{X}\} \in \mathcal{H}^2(\mathbf{X})$.

Example 7.1 Let $\mathbf{X}$ be a nonempty compact subset of $\mathbb{R}^2$ endowed with the usual Euclidean metric. Then a point of $\mathcal{H}^2(\mathbf{X}) = \mathcal{H}(\mathcal{H}(\mathbf{X}))$ is a compact subset of $\mathcal{H}(\mathbf{X})$. Hence, the collection of compact sets in Figure 7.1 is a point of $\mathcal{H}^2(\mathbf{X})$.

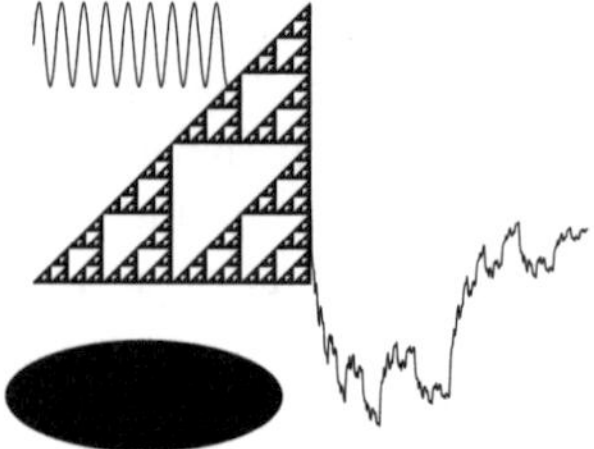

Figure 7.1 A point of $\mathcal{H}^2(\mathbf{X})$, where $\mathbf{X}$ is a compact subset of $\mathbb{R}^2$.

Let $(\mathbf{X}, d)$ be a compact metric space and let $1 < N \in \mathbb{N}$. We denote an IFS on $\mathbf{X}$ by $\mathcal{F} := (\mathbf{X}; F)$, where $F := \{f_i : \mathbf{X} \to \mathbf{X} \,|\, i = 1, \ldots, N\}$ is a collection of contractions on $\mathbf{X}$. Suppose we are given a collection of IFSs $\{\mathcal{F}_\alpha := (\mathbf{X}; F_\alpha) \,|\, \alpha = 1, \ldots, A\}$ on $\mathbf{X}$, where $A \in \mathbb{N}$. The elements of F_α will be denoted by $\{f_i^\alpha \,|\, i = 1, \ldots, N_\alpha\}$, where $1 < N_\alpha \in \mathbb{N}$ is fixed and $\alpha \in \{1, \ldots, A\}$.

Note: *In order to ease notation, we will from now on use the same letter for a numerical function and its induced set-valued mapping.*

Definition 7.2 (SuperIFS)   Let $(\mathbf{X}, d)$ be a compact metric space and let $\{\mathcal{F}_\alpha \,|\, \alpha = 1, \ldots, A\}$ be a finite family of IFSs on $\mathbf{X}$. The pair $(\mathbf{X}; \{\mathcal{F}_1, \ldots, \mathcal{F}_A\})$ is called a *superIFS on* $\mathbf{X}$.

It is important to note that a superIFS is *not* an IFS: The IFSs $\mathcal{F}_\alpha$, $\alpha \in \{1, \ldots, A\}$, act on the spaces $\mathcal{H}(\mathbf{X})$ and not on $\mathbf{X}$. However, to each superIFS one can associate an IFS by setting

$$\mathcal{F}^{(1)} := (\mathcal{H}(\mathbf{X}); \{\mathcal{F}_\alpha \,|\, \alpha \in \{1, \ldots, A\}\}).$$

It is easily verified that $\mathcal{F}^{(1)}$ is an IFS on $(\mathcal{H}(\mathbf{X}))$: The compactness of $(\mathbf{X}, d)$ implies the compactness of the metric space $(\mathcal{H}(\mathbf{X}), d_{\mathcal{H}})$ and each IFS $\mathcal{F}_\alpha$ acts on $\mathcal{H}(\mathbf{X})$ via

$$\mathcal{F}_\alpha : \mathcal{H}(\mathbf{X}) \to \mathcal{H}(\mathbf{X})$$

$$\mathcal{F}_\alpha(S) = \bigcup_{i=1}^{N_\alpha} f_i^\alpha(S). \tag{7.1}$$

Each mapping $\mathcal{F}_\alpha$ is, by Lemma 4.22, a contraction on $\mathcal{H}(\mathbf{X})$ with contractivity constant $\lambda_\alpha := \max\{\mathrm{Lip}(f_i^\alpha) \,|\, i = 1, \ldots, N_\alpha\} < 1$, for all $\alpha \in \{1, \ldots, A\}$.

Definition 7.3 (1-Variable IFS)   The pair $\mathcal{F}^{(1)} := (\mathcal{H}(\mathbf{X}); \{\mathcal{F}_\alpha \,|\, \alpha \in \{1, \ldots, A\}\})$, where the IFS $\mathcal{F}_\alpha$, $\alpha \in \{1, \ldots, A\}$, is given by (7.1), is called a *1-variable IFS* (*on* $\mathcal{H}(\mathbf{X})$).

Since $\mathcal{F}^{(1)}$ is an IFS, there exists a unique element $\mathfrak{A}^{(1)} \in \mathcal{H}^2(\mathbf{X}) = \mathcal{H}(\mathcal{H}(\mathbf{X}))$ so that

$$\mathfrak{A}^{(1)} = \bigcup_{\alpha \in \{1,\dots,A\}} \mathcal{F}_\alpha(\mathfrak{A}^{(1)}).$$

The element $\mathfrak{A}^{(1)}$ is called a 1-*variable superfractal (set)* and is a collection of fractal sets. Its structure can be best understood by looking at the code space associated with the IFS $\mathcal{F}^{(1)}$.

To this end, we recall Theorem 4.44 and the mapping $\gamma_{\mathcal{F}}$ from code space to the attractor. In the current setting, we have $\gamma_{\mathcal{F}^{(1)}} : \Sigma_A \to \mathcal{H}(\mathbf{X})$ with $\gamma_{\mathcal{F}^{(1)}}(\Sigma_A) = \mathfrak{A}^{(1)}$, where $\Sigma_A := \{1,\dots,A\}^{\mathbb{N}} = \mathrm{Map}(\mathbb{N},\{1,\dots,A\})$ is the code space associated with the IFS $\mathcal{F}^{(1)}$. According to the definition of $\gamma_{\mathcal{F}^{(1)}}$ (cf. Theorem 4.44), the 1-variable superfractal $\mathfrak{A}^{(1)}$ is the collection of all sets

$$\mathfrak{A}_\sigma := \gamma_{\mathcal{F}^{(1)}}(\sigma) = \lim_{n\to\infty} \mathcal{F}_{\sigma_1} \circ \cdots \circ \mathcal{F}_{\sigma_n}(\mathbf{X}) \in \mathcal{H}(\mathbf{X}), \tag{7.2}$$

for $\sigma := (\sigma_1,\dots,\sigma_n,\dots) \in \Sigma_A$:

$$\mathfrak{A}^{(1)} = \{\mathfrak{A}_\sigma \mid \sigma \in \Sigma_A\}.$$

Here, we defined $\mathcal{F}_1 \circ \cdots \circ \mathcal{F}_\alpha$ to be the IFS

$$(\mathbf{X}; \{f_{i_1}^1 \circ \cdots \circ f_{i_\alpha}^\alpha \mid i_1 = 1,\dots,N_1; \dots; i_\alpha = 1,\dots,N_\alpha\}), \quad \alpha \in \{1,\dots,A\}. \tag{7.3}$$

Definition 7.4 (1-Variable Fractal Set)   The sets $\mathfrak{A}_\sigma \in \mathcal{H}(\mathbf{X})$, $\sigma \in \Sigma_A$, are called 1-*variable fractal sets* (arising from the superfractal $\mathfrak{A}^{(1)}$).

Denoting by $\overline{\alpha} := (\alpha,\dots,\alpha,\dots)$ a constant code in Σ_A, it easily follows from (7.2) that $\mathfrak{A}_{\overline{\alpha}}$ is the set attractor of the IFS $\mathcal{F}_\alpha$, $\alpha \in \{1,\dots,A\}$.

Example 7.5   Consider $\mathbf{X} := \square$ and the two IFSs $\mathcal{F}_1 := (\square; \{f_1^1, f_2^1, f_3^1\})$, where

$$f_1^1(x,y) := \left(\frac{2x}{5}, \frac{2y}{5}\right),$$

$$f_2^1(x,y) := \left(\frac{2x+3}{5}, \frac{2y}{5}\right),$$

$$f_3^1(x,y) := \left(\frac{4x+3}{10}, \frac{2y+3}{5}\right),$$

and $\mathcal{F}_2 := (\square; \{f_1^2, f_2^2\})$, with

$$f_1^2(x,y) := \left(\frac{x}{2}, x+\frac{y}{2}\right),$$

$$f_2^2(x,y) := \left(\frac{x+1}{2}, 1-x+\frac{y}{2}\right).$$

The set attractors $\mathfrak{A}_{\overline{1}}$ and $\mathfrak{A}_{\overline{2}}$ are displayed in Figure 7.2 (top). The 1-variable fractal set $\mathfrak{A}_{\overline{12}}$ is shown on the bottom part of Figure 7.2.

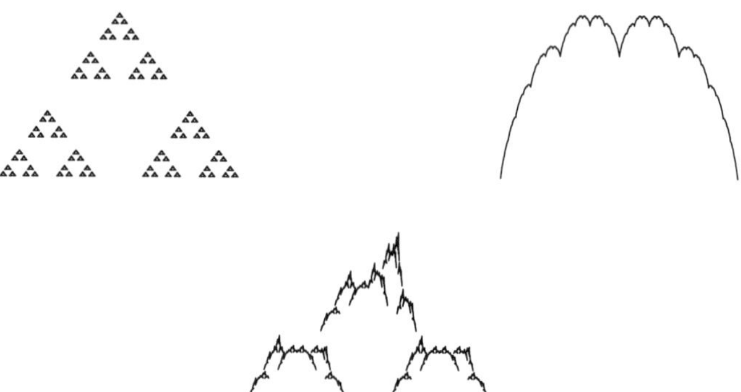

Figure 7.2 Elements of a 1-variable superfractal set.

With each 1-variable IFS $\mathcal{F}^{(1)}$ one can associate an *underlying IFS*, denoted by $\mathcal{F}_\sqcup$ and defined by

$$\mathcal{F}_\sqcup := (\mathbf{X}; \{f_1^1, \ldots, f_{N_1}^1, f_1^2, \ldots, f_{N_2}^2, \ldots, f_1^A, \ldots, f_{N_A}^A\}).$$

Setting $N := \sum_{\alpha=1}^{A} N_\alpha$, $v_0 := 0$, $v_\alpha := N_1 + \cdots + N_\alpha$, $\alpha = 1, \ldots, A$, and letting $f_{v_{\alpha-1}+i} := f_i^\alpha$, $i = 1, \ldots, N_\alpha$, $\alpha = 1, \ldots, A$, we may express the underlying IFS more succinctly in the form

$$\mathcal{F}_\sqcup = (\mathbf{X}; \{f_j \,|\, j = 1, \ldots, N\}).$$

The attractor of the underlying IFS $\mathcal{F}_\sqcup$, which is an element of $\mathcal{H}(\mathbf{X})$, will be denoted by $\mathfrak{A}_\sqcup$. We further set

$$\Upsilon_\alpha := \{v_{\alpha-1} + 1, \ldots, v_\alpha\}, \quad \alpha \in \{1, \ldots, A\},$$

and, for $\sigma \in \Sigma_A$, define

$$\Sigma_\sigma := \mathop{\mathsf{X}}_{n=1}^{\infty} \Upsilon_{\sigma_n}.$$

Then, clearly, $\Sigma_\sigma \subset \Sigma_N$. Moreover, it is not difficult to show that

$$\mathfrak{A}_\sigma = \gamma_{\mathcal{F}_\sqcup}(\Sigma_\sigma). \tag{7.4}$$

Example 7.6 We continue with Example 7.5, and consider the underlying IFS

$$\mathcal{F}_\sqcup = (\square; \{f_j \,|\, j = 1, \ldots, 5\}).$$

The contractions in $\mathcal{F}_\sqcup$ are given by $f_1 = f_1^1$, $f_2 = f_2^1$, $f_3 = f_3^1$, $f_4 = f_1^2$, and $f_5 = f_2^2$, and the attractor $\mathfrak{A}_\sqcup$ of this IFS is shown in Figure 7.3.

Our next goal is the introduction of the analogue of the tops function and the tops code space for 1-variable IFSs. For this purpose, define the *restricted*

Figure 7.3 The underlying attractor $\mathfrak{A}_{\sqcup}$ in Example 7.6.

tops function by

$$\tau_\sigma : \mathfrak{A}_\sigma \to \Sigma_\sigma,$$

$$\tau_\sigma(\mathfrak{a}) = \max\{\omega \in \Sigma_\sigma \mid \gamma_{\mathcal{F}_{\sqcup}}(\omega) = \mathfrak{a}\}.$$

Note that the maximum is taken with respect to the code space ordering $\prec$ introduced in Equation (4.8). The collection

$$\{\tau_\sigma \mid \sigma \in \Sigma_A\}$$

is called the *set of 1-variable tops of the IFS $\mathcal{F}^{(1)}$*.

Now, consider the underlying IFS $(\mathsf{X}; \mathcal{F}_{\sqcup})$ and the associated code space mapping $\gamma_{\mathcal{F}_{\sqcup}} : \Sigma_N \to \mathfrak{A}_{\sqcup}$. The mapping

$$\gamma_\sigma : \overline{\Sigma}_\sigma \to \mathfrak{A}_\sigma$$

is the restriction of $\gamma_{\mathcal{F}_{\sqcup}}$ to the closure of Σ_σ in Σ_N. We call the collection of sets

$$\mathcal{Q}_\sigma := \left\{ \gamma_\sigma^{-1}(\mathfrak{a}) \mid \mathfrak{a} \in \mathfrak{A}_\sigma \right\}$$

the *restricted address structure of the 1-variable fractal set $\mathfrak{A}_\sigma$*, $\sigma \in \Sigma_N$. A theorem, similar to Theorem 4.62, holds for restricted address structures and 1-variable fractal sets. To state this theorem, we need the following definition.

Definition 7.7 (Homeomorphic Restricted Address Structures)   Two restricted address structures $\mathcal{Q}_\sigma$ and $\mathcal{Q}_\omega$ are called *homeomorphic*, in symbol $\mathcal{Q}_\sigma \cong \mathcal{Q}_\omega$, provided that there exists a homeomorphism $\eta : \overline{\Sigma}_\sigma \to \overline{\Sigma}_\omega$ such that $q \in \mathcal{Q}_\sigma$ iff $\eta(q) \in \mathcal{Q}_\omega$.

Now, we can present the analogue of Theorem 4.62 for 1-variable fractal sets.

Theorem 7.8   Suppose that $\mathcal{F}^{(1)} = (\mathcal{H}(\mathsf{X}); \{\mathcal{F}_\alpha \mid \alpha \in A\})$ is a 1-variable IFS with set attractor $\mathfrak{A}^{(1)} = \{\mathfrak{A}_\sigma \mid \sigma \in \Sigma_A\}$. Further, suppose that $\{\tau_\sigma \mid \sigma \in \Sigma_A\}$ is the associated set of 1-variable tops. Let $\sigma, \omega \in \Sigma_A$. If $\mathcal{Q}_\sigma \cong \mathcal{Q}_\omega$, then there exists a homeomorphism $\mathfrak{h} : \mathfrak{A}_\sigma \to \mathfrak{A}_\omega$ so that $\mathfrak{h}(\mathfrak{A}_\sigma) = \mathfrak{A}_\omega$. If, in addition, $\eta(\overline{\Sigma}_\sigma) = \overline{\Sigma}_\omega$, then

$$\tau_\omega \circ \mathfrak{h} = \eta \circ \tau_\sigma.$$

Proof The proof follows the same lines as the proofs of Theorems 4.59 and 4.62, and is therefore left to the reader. $\square$

7.2 1-Variable Fractal Measures

Here, we extend the concept of fractal measure encountered in Section 4.7 to the superfractal setting. For this purpose, suppose that $(X; \{\mathcal{F}_\alpha, \mathcal{P}_\alpha \,|\, \alpha = 1, \ldots, A\})$, $A \in \mathbb{N}$, is a collection of IFSs with probabilities on a compact metric space (X, d). As before, we denote the elements of the IFS $\mathcal{F}_\alpha$ by f_i^α and the probabilities in $\mathcal{P}_\alpha$ by p_i^α, $i = 1, \ldots, N_\alpha$, $1 < N_\alpha \in \mathbb{N}$. Further suppose that $\mathcal{P} := \{P_1, \ldots, P_A\}$ is a given set of probabilities and $\mathsf{P}(X)$ the set of normalized Borel measures on X.

Proposition 7.9 The triple $\mathcal{F}^{\sharp\,(1)} := \big(\mathsf{P}(X); \{\mathcal{F}_\alpha^\sharp \,|\, \alpha = 1, \ldots, A\}, \mathcal{P}\big)$ is an IFS on $\mathsf{P}(X)$, called a 1-*variable measure IFS*.

Proof By Theorem 4.83 we know that $(\mathsf{P}(X), d_\mathsf{P})$ is a compact metric space. The mappings $\mathcal{F}_\alpha^\sharp : \mathsf{P}(X) \to \mathsf{P}(X)$ are of the form

$$\mathcal{F}_\alpha^\sharp(\mu) = \sum_{i=1}^{N_\alpha} p_i^\alpha \, (f_i^\alpha)^\sharp(\mu), \quad \alpha = 1, \ldots, A, \tag{7.5}$$

and are, by Theorem 4.86, contractive in the Monge–Kantorovich metric. $\square$

The unique *set* attractor $\mathfrak{A}^{\sharp\,(1)}$ (see Theorem 4.91) of the IFS $\mathcal{F}^{\sharp\,(1)}$ must therefore be an element of $\mathcal{H}(\mathsf{P}(X))$, viz. a nonempty set, compact with respect to the Monge–Kantorovich metric d_P, whose elements are normalized Borel measures on X. Employing the code space mapping $\gamma_{\mathcal{F}^{\sharp(1)}} : \Sigma_A \to \mathsf{P}(X)$ that is associated with the IFS $\mathcal{F}^{\sharp\,(1)}$, we see that this attractor must be given by

$$\mathfrak{A}^{\sharp\,(1)} = \gamma_{\mathcal{F}^{\sharp(1)}}(\Sigma_A).$$

Thus, the elements of $\mathfrak{A}^{\sharp\,(1)}$ are exactly those measures in $\mathsf{P}(X)$ that can be written in the form

$$\mu_\sigma = \gamma_{\mathcal{F}^{\sharp(1)}}(\sigma) = \lim_{n \to \infty} \mathcal{F}_{\sigma_1}^\sharp \circ \cdots \circ \mathcal{F}_{\sigma_n}^\sharp(\mu_0),$$

for some arbitrary $\mu_0 \in \mathsf{P}(X)$. Here, any composition of the form $\mathcal{F}_1^\sharp \circ \cdots \circ \mathcal{F}_\alpha^\sharp$ is defined as in (7.3) adjusted to the current setting, namely to be the IFS with probabilities

$$\big(\mathsf{P}(X); \{(f_{i_1}^1)^\sharp \circ \cdots \circ (f_{i_\alpha}^\alpha)^\sharp ; \, p_{i_1}^1 \cdot \ldots \cdot p_{i_\alpha}^\alpha \,|\, i_1 = 1, \ldots, N_1; \, \ldots; \, i_\alpha = 1, \ldots, N_\alpha\}\big)$$

for all $\alpha \in \{1, \ldots, A\}$. The measure μ_σ is called a 1-*variable fractal measure* and the set

$$\mathfrak{A}^{\sharp\,(1)} = \{\mu_\sigma \,|\, \sigma \in \Sigma_A\}$$

Figure 7.4   The measure attractors for Example 7.10.

a *superfractal of 1-variable fractal measures* associated with the superIFS $(\mathsf{X}; \{\mathcal{F}_\alpha \mid \alpha = 1, \ldots, A\}, \mathcal{P})$.

Example 7.10   We may reconsider Example 7.5 with the current setting in mind. Elements of the superfractal $\mathfrak{A}^{\sharp(1)}$ of 1-variable fractal measures are $\mu_{\overline{1}}$, $\mu_{\overline{2}}$, and $\mu_{\overline{12}}$. In Figure 7.4, these fractal measure attractors are shown. The sets $\mathcal{P}_1 := \{\frac{1}{2}, \frac{1}{4}, \frac{1}{4}\}$ and $\mathcal{P}_2 := \{\frac{2}{3}, \frac{1}{3}\}$ were chosen as probability sets for the two IFSs $\mathcal{F}_1$ and $\mathcal{F}_2$. The probabilities associated with the IFS $\mathcal{F}_1 \circ \mathcal{F}_2$ are then given by $\mathcal{P}_1 \circ \mathcal{P}_2 := \{\frac{2}{5}, \frac{1}{6}, \frac{1}{10}, \frac{1}{5}, \frac{1}{12}, \frac{1}{20}\}$. Note that the support of the measures are the set attractors but the distribution of points as reflected by the gray level is different.

As $\mathcal{F}^{\sharp(1)}$ is an IFS, it possesses a unique measure attractor, which we denote by $\mathrm{m}^{(1)}$. This measure attractor is an element of $\mathsf{P}^2(\mathsf{X}) := \mathsf{P}(\mathsf{P}(\mathsf{X}))$, the space of all normalized Borel measures on $\mathsf{P}(\mathsf{X})$. Theorem 4.92 that gave a random iteration algorithm, the so-called chaos game, for the construction of fractal sets can now be extended to the superfractal setting as follows.

Theorem 7.11   Let $\mathcal{F}^{\sharp(1)} := \big(\mathsf{P}(\mathsf{X}); \{\mathcal{F}_\alpha^\sharp, P_\alpha \mid \alpha = 1, \ldots, A\}\big)$ be a 1-variable measure IFS and let $\mathrm{m}^{(1)} \in \mathsf{P}^2(\mathsf{X})$ denote its unique measure attractor. For an arbitrary $\mu_0 \in \mathsf{P}(\mathsf{X})$ define iterates $\mu_k := \mathcal{F}_\alpha^\sharp \mu_{k-1}$, $k \in \mathbb{N}$, where the IFS $\mathcal{F}_\alpha^\sharp$ is chosen with probability p_α. Then, for almost all random sequences $\{\mu_k \mid k \in \mathbb{N}\}$, we have

$$\mathrm{m}^{(1)}(A) = \lim_{k \to \infty} \frac{\mathcal{N}(A \cap \{\mu_0, \mu_1, \ldots, \mu_k\})}{k+1},$$

for all $A \in \mathcal{B}(\mathsf{P}(\mathsf{X}))$ with $\mathrm{m}^{(1)}(\partial A) = 0$. Here, $\mathcal{N}(A)$ denotes the number of points in set A.

The interpretation of Theorem 7.11 is essentially the same as for Theorem 4.92. For applications, we again refer the reader to [9].

7.3 *V*-Variable Fractal Sets and Fractal Measures

Here, we generalize the results from the previous section for $V = 1$ to general values $V \in \mathbb{N}$, which provides the basis for the limiting case $V = \infty$ described in the next section. We begin the presentation (see also [9, 18]) by elaborating on the case $V = 2$, and then provide the details for general V.

7.3.1 *V*-Variable Fractal Sets

The structure and construction of $(V = 2)$-variable fractal sets is best illustrated by hand of the following procedure originally presented in [9] and [18]. To this end, suppose that $(\Box_1, \Box_2)$ and $(\Box_{1'}, \Box_{2'})$ are two pairs of digital buffers. We refer to $(\Box_1, \Box_2)$ as the input screens and to $(\Box_{1'}, \Box_{2'})$ as the output screens. Furthermore, suppose that $\mathcal{F}_1 := \{f_1^1, f_2^1\}$ and $\mathcal{F}_2 := \{f_1^2, f_2^2\}$ are two IFSs with probabilities P_1, respectively, P_2, on $\mathbf{X}$. Suppose that $A_1 \subseteq \Box_1$ and $A_2 \subseteq \Box_2$ are elements from $\mathcal{H}(\mathbf{X})$. We call A_1 and A_2 initial images on the input screens. Further, suppose that the output screens are clear, i.e., that they contain no initial image. We successively apply the following algorithm to generate images on the output screens.

1. Choose randomly one of the two IFSs $\mathcal{F}_1$ and $\mathcal{F}_2$, say $\mathcal{F}_{\alpha_1}$. Now apply $f_1^{\alpha_1}$ to either $A_1 \subseteq \Box_1$ or $A_2 \subseteq \Box_2$, selected randomly, to generate an image $f_1^{\alpha_1}(A_{\beta_1})$ on the output screen $\Box_{1'}$. After that, apply the mapping $f_2^{\alpha_1}$ to either $A_1 \subseteq \Box_1$ or $A_2 \subseteq \Box_2$, again chosen randomly, to obtain an image $f_2^{\alpha_1}(A_{\beta_2})$ on $\Box_{1'}$. Then, form the union of these two images: $f_1^{\alpha_1}(A_{\beta_1}) \cup f_2^{\alpha_1}(A_{\beta_2}) \subseteq \Box_{1'}$.

2. Again select randomly one of the two IFSs $\mathcal{F}_1$ and $\mathcal{F}_2$, say $\mathcal{F}_{\alpha_2}$. Apply $f_1^{\alpha_2}$ to either $A_1 \subseteq \Box_1$ or $A_2 \subseteq \Box_2$, chosen randomly, to generate an image $f_1^{\alpha_2}(A_{\gamma_1})$ on the output screen $\Box_{2'}$. Then apply the mapping $f_2^{\alpha_2}$ to either $A_1 \subseteq \Box_1$ or $A_2 \subseteq \Box_2$, again selected randomly, to obtain an image $f_2^{\alpha_2}(A_{\gamma_2})$ on $\Box_{2'}$. Form the union $f_1^{\alpha_2}(A_{\gamma_1}) \cup f_2^{\alpha_2}(A_{\gamma_2}) \subseteq \Box_{2'}$.

3. Switch input and output screens, $\Box_i \leftrightarrow \Box_{i'}$, $i = 1, 2$, and apply steps 1. and 2. again.

4. Repeat this process indefinitely.

The above algorithm can be expressed mathematically as follows. The two input, respectively, output screens derive from $V = 2$. In the notation introduced above, we have $N_1 = N_2 = 2$ and $A = 2$. Let $v \in \{1, 2\}$ $(V = 2)$ and let $\alpha_v \in \{1, 2\}$ $(A = 2)$. Furthermore, introduce indices $v_{v,i} \in \{1, 2\}$ $(V = 2)$. Set $\boldsymbol{\alpha} := (\alpha_1, \alpha_2) \in \{1, 2\}^2$ and $\boldsymbol{v} := (v_{1,1}, v_{1,2}, v_{2,1}, v_{2,2}) \in \{1, 2\}^{2+2}$. Define the index set $\mathcal{J}$ to consist of all indices $\mathbf{j} := (\boldsymbol{\alpha}, \boldsymbol{v})$, where $\boldsymbol{\alpha}$ and $\boldsymbol{v}$ are given as above. For $\mathbf{j} \in \mathcal{J}$, define mappings $f_{\mathbf{j}} : \mathcal{H}(\mathbf{X}) \times \mathcal{H}(\mathbf{X}) \to \mathcal{H}(\mathbf{X}) \times \mathcal{H}(\mathbf{X})$ by

$$f_{\mathbf{j}}(A_1, A_2) := \left(\bigcup_{i=1}^{N_{\alpha_1}} f_i^{\alpha_1}(A_{v_1, i}), \bigcup_{i=1}^{N_{\alpha_2}} f_i^{\alpha_2}(A_{v_2, i}) \right),$$

and denote their associated set of probabilities by $\{p_{\mathbf{j}} \mid \mathbf{j} \in \mathcal{J}\}$ where $p_{\mathbf{j}} := \dfrac{P_{\alpha_1} P_{\alpha_2}}{2^{N_{\alpha_1}+N_{\alpha_2}}}$. The index of $A_{v_v,i}$ indicates which map (indexed by i) is applied to which one of the two ($v = 1, 2$) sets $A_v \subseteq \square_v$.

The generalization to $V \in \mathbb{N}$ is now at hand. To this end, consider the superIFS

$$(\mathbf{X}; \{\mathcal{F}_\alpha \mid \alpha = 1, \ldots, A\}, \mathcal{P}). \tag{7.6}$$

For any $V \in \mathbb{N}$, define

$$\mathcal{J} := \Big\{ \mathbf{j} = (\boldsymbol{\alpha}, \boldsymbol{v}) \mid \boldsymbol{\alpha} := (\alpha_1, \ldots, \alpha_V), \, \boldsymbol{v} := (v_{1,1}, \ldots, v_{1,N_{\alpha_1}}, \ldots, v_{V,1}, \ldots, v_{V,N_{\alpha_V}}),$$

$$v \in \{1, \ldots, V\}, \alpha_v \in \{1, \ldots, A\}, v_{v,i} \in \{1, \ldots, V\}, i = 1, \ldots, N_{\alpha_v} \Big\} \tag{7.7}$$

$$= \{1, \ldots, A\}^V \times \{1, \ldots, V\}^{\sum_{v=1}^{V} N_{\alpha_v}}$$

and, for each $\mathbf{j} \in \mathcal{J}$, a mapping $f_{\mathbf{j}} : \mathcal{H}(\mathbf{X})^V \to \mathcal{H}(\mathbf{X})^V$ by

$$f_{\mathbf{j}}(A_1, \ldots, A_V) := \left(\bigcup_{i=1}^{N_{\alpha_1}} f_i^{\alpha_1}(A_{v_1,i}), \ldots, \bigcup_{i=1}^{N_{\alpha_V}} f_i^{\alpha_V}(A_{v_V,i}) \right). \tag{7.8}$$

Here, we set $\mathcal{H}(\mathbf{X})^V := \overset{V}{\underset{i=1}{\times}} \mathcal{H}(\mathbf{X})$. The Hausdorff metric $d_{\mathcal{H}}$ on $\mathcal{H}(\mathbf{X})$ induces a metric $d_{\mathcal{H}^V}$ on the cartesian product $\mathcal{H}(\mathbf{X})^V$, which is given by

$$d_{\mathcal{H}^V}(A, B) := \max\{d_{\mathcal{H}}(A_v, B_v) \mid v = 1, \ldots, V\}.$$

Note that since $(\mathbf{X}, d)$ is a compact metric space, so it $(\mathcal{H}(\mathbf{X})^V, d_{\mathcal{H}^V})$. (The reader is encouraged to verify this statement.)

Using the set of mappings $\{f_{\mathbf{j}} \mid \mathbf{j} \in \mathcal{J}\}$ given in (7.8), we define the system

$$\mathcal{F}^{(V)} := \big(\mathcal{H}(\mathbf{X})^V; \{f_{\mathbf{j}}, p_{\mathbf{j}} \mid \mathbf{j} \in \mathcal{J}\}\big), \tag{7.9}$$

where the probabilities $p_{\mathbf{j}}$ are given by

$$p_{\mathbf{j}} := \frac{P_{\alpha_1} \cdot \ldots \cdot P_{\alpha_V}}{V^{N_{\alpha_1} + \cdots + N_{\alpha_V}}}. \tag{7.10}$$

Theorem 7.12 The system $\mathcal{F}^{(V)} := \big(\mathcal{H}(\mathbf{X})^V; \{f_{\mathbf{j}}, p_{\mathbf{j}} \mid \mathbf{j} \in \mathcal{J}\}\big)$ is an IFS for all $V \in \mathbb{N}$.

Proof Let $\lambda := \max\{\|\mathrm{Lip}(f_i^\alpha)\| \mid i = 1, \ldots, N_\alpha; \alpha = 1, \ldots, A\}$. Since each $\mathcal{F}_\alpha$ is an IFS, $\lambda \in [0, 1)$. Using Lemma 4.23 and, in particular, Exercise 8 in Chapter 4, we obtain the following chain of inequalities

$$d_{\mathcal{H}^V}\left(\bigcup_{i=1}^{N_{\alpha_v}} f_i^{\alpha_v}(A_i), \bigcup_{i=1}^{N_{\alpha_v}} f_i^{\alpha_v}(B_i) \right) \leq \max\{d_{\mathcal{H}}(f_i^{\alpha_v}(A_i), f_i^{\alpha_v}(B_i)) \mid i = 1, \ldots, N_{\alpha_v}\}$$

$$\leq \max\{\lambda \, d_{\mathcal{H}}(A_i, B_i) \mid i = 1, \ldots, N_{\alpha_v}\}$$

$$= \lambda \, d_{\mathcal{H}^{N_{\alpha_v}}}(A, C),$$

for all $A := (A_1, \ldots, A_{N_{\alpha_v}})$ and $B := (B_1, \ldots, B_{N_{\alpha_v}})$ in $\mathcal{H}(\mathbf{X})^{N_{\alpha_v}}$, and all $v = 1, \ldots, V$. Here, we assumed that the mappings in $\mathcal{F}^{(V)}$ are given in a fixed order. Therefore,

$$d_{\mathcal{H}^V}\left(f_{\mathbf{j}}(A_1, \ldots, A_V), f_{\mathbf{j}}(B_1, \ldots, B_V)\right)$$

$$= \max_{v=1,\ldots,V}\left\{d_{\mathcal{H}}\left(\bigcup_{i=1}^{N_{\alpha_v}} f_i^{N_{\alpha_v}}(A_{v_v,i}), \bigcup_{i=1}^{N_{\alpha_v}} f_i^{N_{\alpha_v}}(B_{v_v,i})\right)\right\}$$

$$\leq \max_{v=1,\ldots,V}\left\{\lambda\, d_{\mathcal{H}^{N_{\alpha_v}}}\left((A_{v_v,1}, \ldots, A_{v_v,N_{\alpha_v}}), (B_{v_v,1}, \ldots, B_{v_v,N_{\alpha_v}})\right)\right\}$$

$$\leq \lambda\, d_{\mathcal{H}^V}\left((A_1, \ldots, A_V), (B_1, \ldots, B_V)\right),$$

for all $(A_1, \ldots, A_V), (B_1, \ldots, B_V) \in \mathcal{H}(\mathbf{X})^V$. $\qquad\square$

Definition 7.13 (V-variable IFS with probabilities) The system $\mathcal{F}^{(V)}$ is called a V-*variable IFS with probabilities* $\{p_{\mathbf{j}} \mid \mathbf{j} \in \mathcal{J}\}$.

Theorems 4.91 and 7.12 now imply that $\mathcal{F}^{(V)} = \left(\mathcal{H}(\mathbf{X})^V; \{f_{\mathbf{j}}, p_{\mathbf{j}} \mid \mathbf{j} \in \mathcal{J}\}\right)$ has a unique set attractor

$$\mathfrak{A}^{(V)} \in \mathcal{H}(\mathcal{H}(\mathbf{X})^V),$$

as well as a unique measure attractor

$$\widetilde{\mathfrak{m}}^{(V)} \in \mathsf{P}(\mathcal{H}(\mathbf{X})^V).$$

Definition 7.14 (V-variable Superfractal and V-variable Fractal Set) The set

$$\mathfrak{A}^{(V)} := \left\{A \in \mathcal{H}(\mathbf{X}) \mid A \text{ is a component of a point in } \mathfrak{A}^{(V)}\right\}$$

is called the V-*variable superfractal associated with the IFS* $\mathcal{F}^{(V)}$. The elements of $\mathfrak{A}^{(V)}$ are referred to as V-*variable fractal sets*.

We remark that $\mathfrak{A}^{(1)} = \mathfrak{A}^{(1)}$ and that the following sequence of inclusions holds.

$$\mathfrak{A}^{(1)} \subset \mathfrak{A}^{(2)} \subset \cdots \subset \mathfrak{A}^{(V)} \subset \mathcal{H}(\mathcal{H}(\mathbf{X})), \quad V \in \mathbb{N}.$$

7.3.2 *V*-Variable Fractal Measures

We begin again with the superIFS (7.6) but take as the base space $\mathsf{P}(\mathbf{X})^V := \overset{V}{\underset{i=1}{\mathsf{X}}} \mathsf{P}(\mathbf{X})$ instead of $\mathcal{H}(\mathbf{X})^V$. It is not difficult to show that $(\mathsf{P}(\mathbf{X})^V, d_{\mathsf{P}^V})$, where for $A := (A_1, \ldots, A_V)$, $B := (B_1, \ldots, B_V)$

$$d_{\mathsf{P}^V}(A, B) := \max\{d_{\mathsf{P}}(A_v, B_v) \mid v = 1, \ldots, V\},$$

is a compact metric space. (The reader is encouraged to verify this statement.) Using the indexing introduced in the previous section, we now define measure-valued mappings $f_j^\sharp : \mathsf{P(X)}^V \to \mathsf{P(X)}^V$ by

$$f_j^\sharp(\mu_1, \ldots, \mu_V) := \left(\sum_{i=1}^{N_{\alpha_1}} p_i^{\alpha_1} (f_i^{\alpha_1})^\sharp (\mu_{v_1, i}), \ldots, \sum_{i=1}^{N_{\alpha_V}} p_i^{\alpha_V} (f_i^{\alpha_V})^\sharp (\mu_{v_V, i}) \right),$$

for any $(\mu_1, \ldots, \mu_V) \in \mathsf{P(X)}^V$.

The analoguous result to Theorem 7.12 now reads.

Theorem 7.15   The system $\mathcal{F}^{\sharp(V)} := \left(\mathsf{P(X)}^V; \{(f_j)^\sharp, p_j \,|\, \mathbf{j} \in \mathcal{J}\} \right)$, where the probabilities p_j are given by (7.10), is an IFS for all $V \in \mathbb{N}$.

Proof   The arguments are similar to those in the proof of Theorem 7.12. The details are left to the reader. (See also [18].) $\square$

Hence, the IFS $\mathcal{F}^{\sharp(V)}$ has a unique fixed point, i.e., measure attractor,

$$\mathbf{m}^{(V)} \in \mathsf{P}(\mathsf{P(X)}^V).$$

Thus, the measure attractor $\mathbf{m}^{(V)}$ is a normalized Borel measure on a set of V-tuples of normalized Borel measures, each of which is a fractal measure. By Theorem 4.91, there also exists a unique *set* attractor of the IFS $\mathcal{F}^{\sharp(V)}$, denoted by $\mathfrak{A}^{\sharp(V)}$, which is an element of $\mathcal{H}(\mathsf{P(X)}^V) \subset \mathsf{P(X)}^V$.

Definition 7.16 (V-variable Superfractal Set of Measures and V-variable Fractal Measure)   The set

$$\mathfrak{A}^{\sharp(V)} := \left\{ \mu \in \mathcal{H}(\mathsf{P(X)}^V) \,\big|\, \mu \text{ is a component of a point in } \mathfrak{A}^{\sharp(V)} \right\}$$

is called the *V-variable superfractal or measures associated with the IFS* $\mathcal{F}^{\sharp(V)}$. The elements of $\mathfrak{A}^{\sharp(V)}$ are referred to as *V-variable fractal measures*.

Remark 7.17   It is shown in [18], that $\mathfrak{A}^{(V)} = \mathfrak{A}_v^{(V)}$, for all $v = 1, \ldots, V$, where

$$\mathfrak{A}_v^{(V)} := \{A_v \in \mathcal{H}(\mathsf{X}) \,|\, A = (A_1, \ldots, A_v, \ldots, A_V) \in \mathfrak{A}^{(V)}\}$$

is the set consisting of the v-components of elements of $\mathfrak{A}^{(V)}$.

Finally, we state the equivalent of the random iteration algorithm or "chaos game" (Theorem 4.92), in the V-variable setting. For proofs, we need to refer the reader to [18].

Using Remark 4.94, we proceed as follows.

- Choose an arbitrary point $A^0 := (A_1^0, \ldots, A_V^0) \in \mathcal{H}(\mathsf{X})^V$.

- Define a sequence of V-tuples of set $\{A^k \in \mathcal{H}(\mathsf{X})^V \,|\, k \in \mathbb{N}\}$ in $\mathcal{H}(\mathsf{X})^V$ by

$$A^{k+1} := (A_1^k, \ldots, A_V^k) := f_{j(k)}(A^k),$$

where $\mathbf{j}(k) = \mathbf{j} \in \mathcal{J}$ is chosen with probability p_j given by (7.10).

- Define a sequence $\{\Delta_k^{(V)} \mid k \in \mathbb{N}\}$ of averaged Dirac measures on $\mathsf{P}(\mathcal{H}(\mathbf{X}))$ by

$$\Delta_k^{(V)} := \frac{\delta_{A_1^0} + \delta_{A_1^1} + \ldots + \delta_{A_1^k}}{k+1}, \quad k \in \mathbb{N}.$$

- Then

$$\lim_{k \to \infty} \Delta_k^{(V)}(B) = \widetilde{\mathfrak{m}}^{(V)}(B),$$

for all $B \in \mathcal{B}(\mathcal{H}(\mathbf{X}))$ with $\widetilde{\mathfrak{m}}^{(V)}(\partial B) = 0$. Here, $\widetilde{\mathfrak{m}}^{(V)} \in \mathsf{P}(\mathcal{H}(\mathbf{X}))$ is given by

$$\widetilde{\mathfrak{m}}^{(V)}(B) = \widetilde{\mathfrak{m}}^{(V)}(B, \mathcal{H}(\mathbf{X}), \ldots, \mathcal{H}(\mathbf{X})), \quad B \in \mathcal{B}(\mathcal{H}(\mathbf{X})).$$

Note that in this algorithm we made use of the result stated in Remark 7.17 by employing only the first component of the sets $A^k \in \mathcal{H}(\mathbf{X})^V$.

7.4 Graph Theory and Random Fractals

The construction of superfractals and V-variable fractals can also be described in terms of some graph-theoretical notions. Indeed, several aspects achieve more lucidity through the graph-theoretical representation, which also allows the generation of random fractals and random fractals measures as specified in this section. Additionally, the description of the limit case $V = \infty$, which is one of the set goals of this section, is based on these notions. At this point, we refer the reader back to Section 4.5 for cross-reference and similarities.

7.4.1 Code Trees

To this end, we let $1 < N \in \mathbb{N}$ and consider the *alphabet* $\mathcal{N} := \{1, \ldots, N\}$ together with the set $\mathrm{Map}(\mathbb{N}, \mathcal{N})$ of all sequences with values in $\mathcal{N}$. Each sequence $\sigma \in \mathrm{Map}(\mathbb{N}, \mathcal{N})$ is written in the form $\sigma := \sigma_1 \cdots \sigma_n \cdots$, where $\sigma_n := \sigma(n) \in \mathcal{N}$, $n \in \mathbb{N}$. We call a sequence $\sigma \in \mathrm{Map}(\mathbb{N}, \mathcal{N})$ *finite*, if there exists an $n \in \mathbb{N}$ so that for all $m > n$, $\sigma_m = 0$. We identify a finite sequence $\sigma = \sigma_1 \cdots \sigma_n 00 \cdots$ with $\sigma = \sigma_1 \cdots \sigma_n$, i.e., with the function $\sigma \in \mathrm{Map}(\mathbb{N}_n, \mathcal{N})$, where $\mathbb{N}_n := \{1, \ldots, n\}$, $n \in \mathbb{N}$. The empty function will be identified with the empty sequence $\sigma = \emptyset$.

Definition 7.18 (Tree)   The set of all finite sequences with range $\mathcal{N}$ is called a *tree* $\mathsf{T} := \mathsf{T}(\mathcal{N})$ and the elements $\sigma \in \mathsf{T}$ the *nodes of the tree* T. The *length* of a finite sequence $\sigma = \sigma_1 \cdots \sigma_n$ is defined to be the function $|\,| : \mathsf{T} \to \mathbb{N}_0$, $|\sigma| := n$, with the understanding that $|\emptyset| = 0$. We refer to the length n of a node σ also as the *level* or *height* of σ in T.

Note that we implicitly require that $\emptyset \in \mathsf{T}$. The node $\emptyset$ is called the *root* of the tree T. It has length or level 0. The *concatenation* of two nodes $\sigma, \sigma' \in \mathsf{T}$ is defined as the operation $c : \mathsf{T} \times \mathsf{T} \to \mathsf{T}$, $c(\sigma, \sigma') := \sigma_1 \cdots \sigma_n \sigma_1' \ldots \sigma_{n'}'$, where n

and n' is the length of σ and σ', respectively. For short, we write $\sigma\sigma'$ instead of $c(\sigma, \sigma')$.

The main idea behind the concept of a tree is to obtain a graph-theoretical representation of the codes associated with a superIFS and to define *random fractal sets* and *random fractal measures* in a way specified below. In this context, we like to associate the nodes σ of a tree $\mathsf{T}(\mathcal{N})$ with the IFSs $\mathcal{F}_\alpha$, $\alpha \in \mathcal{A} := \{1, \ldots, A\}$, in the superIFS $\mathcal{F}$. This association is obtained via a *code tree*.

Definition 7.19 (Code Tree)   A function $\omega : \mathsf{T} \to \mathcal{A} := \{1, \ldots, A\}$, where $A \in \mathbb{N}$, is called a *code tree*. A *finite code tree of level n* or *height n* is a function $\omega_n : \{\sigma \in \mathsf{T} \mid |\sigma| \le n\} \to \mathcal{A}$, $n \in \mathbb{N}_0$.

Example 7.20   Consider the binary alphabet $\mathcal{N} := \{1, 2\}$ and its tree $\mathsf{T} = \mathsf{T}(\mathcal{N})$. A code tree ω creates a labeling of T by the elements of $\mathcal{A} := \{1, 2\}$. See Figure 7.5.

Define the set

$$\Omega := \{\omega \mid \omega : \mathsf{T} \to \mathcal{A}\}$$

of all code trees and a function $d_\Omega : \Omega \times \Omega \to \mathbb{R}$ by

$$d_\Omega(\omega, \omega') := \begin{cases} \frac{1}{N^n}, & \omega \ne \omega' \\ 0, & \omega = \omega', \end{cases}$$

where n is the least integer for which $\omega_\sigma \ne \omega'_\sigma$ for some $\sigma \in \mathsf{T}$ with $|\sigma| = n$. The pair (Ω, d_Ω) is a compact metric space. (The reader is encouraged to verify this statement.)

Each code tree ω possesses an *initial segment*, namely the finite code tree $\omega \lfloor n$ defined by $(\omega \lfloor n)(\sigma) := \omega(\sigma)$, for $|\sigma| \le n$, $n \in \mathbb{N}_0$. Graphically one obtains the finite code tree $\omega \lfloor n$ by truncating ω at height or level n. We refer to the code tree $\omega \lfloor n$ as a *finite code subtree of height n of* ω. Also note that each finite code tree ω_n is the initial segment of an infinite code tree ω: $\omega_n = \omega \lfloor n$, $n \in \mathbb{N}_0$.

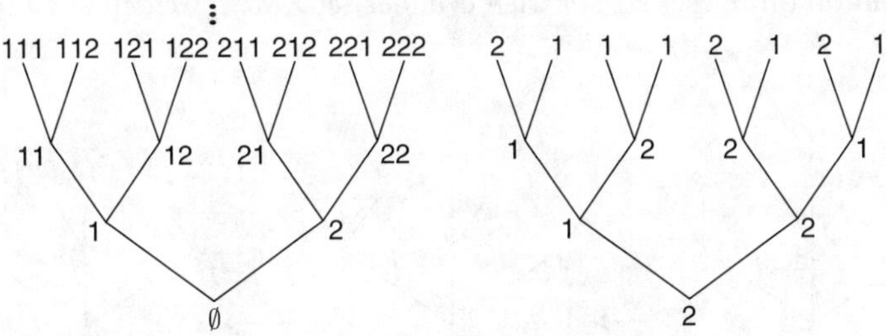

Figure 7.5   A binary tree T (left) and the initial segment of a code tree (right).

Given an $n \in \mathbb{N}_0$, we associate with each finite code tree ω_n of height n the cylinder set

$$Z(\omega_n) := \{\omega' \in \Omega \,|\, d_\Omega(\omega_n, \omega') = N^{-n}\}.$$

Notice that $\mathrm{diam}(Z(\omega_n)) \le N^{-(n+1)}$, and that $Z(\omega_n)$ is the collection of all code trees ω whose initial segment of length n is $\omega_n = \omega \lfloor n$.

Now we connect the nodes in a code tree to the mappings in a superIFS. For this purpose, consider the superIFS

$$\mathcal{F} := (\mathsf{X}; \{\mathcal{F}_\alpha \,|\, \alpha = 1, \ldots, A\}, \mathcal{P}), \tag{7.11}$$

where each $\mathcal{F}_\alpha := \{f_i^\alpha; p_i^\alpha \,|\, i = 1, \ldots, N\}$, $1 < N \in \mathbb{N}$, is an IFS with probabilities defined on a compact metric space (X, d), and $\mathcal{P} := \{P_\alpha \,|\, \alpha = 1, \ldots, A\}$ a given set of probabilities, i.e.,

$$\sum_{\alpha=1}^{A} P_\alpha = 1, \quad P_\alpha \ge 0, \quad \alpha = 1, \ldots, A.$$

Let $\sigma \in \mathsf{T}$ be a node and $\omega \in \Omega$ a code tree. With the node σ we associate the IFS $\mathcal{F}_{\omega(\sigma)} \in \{\mathcal{F}_\alpha \,|\, \alpha = 1, \ldots, A\}$. We denote this association $\sigma \mapsto \mathcal{F}_{\omega(\sigma)}$ again by the letter ω regarding it now as a mapping from T to $\{\mathcal{F}_\alpha \,|\, \alpha = 1, \ldots, A\}$. Moreover, $\omega : \mathsf{T} \to \{\mathcal{F}_\alpha \,|\, \alpha = 1, \ldots, A\}$ also associates to each nonempty node $\sigma = \sigma' i$, where $i \in \{1, \ldots, N\}$, a contraction $f_i^{\omega(\sigma')} \in \mathcal{F}_{\omega(\sigma')}$. Writing again ω for this latter association, we have

$$\mathsf{T} \ni \sigma = \sigma' i \quad \overset{\omega}{\longmapsto} \quad f_i^{\omega(\sigma')} \in \mathcal{F}_{\omega(\sigma')}. \tag{7.12}$$

Example 7.21 Let $N := 3$ and let $A := 2$. Consider the tree $\mathsf{T} = \mathsf{T}(\mathcal{N})$ and the code tree $\omega : \mathsf{T} \to \mathcal{A}$ as shown in Figure 7.6, where $\mathcal{F}_1 := \{f_1^1, f_2^1, f_3^1\}$ and $\mathcal{F}_2 := \{f_1^2, f_2^2, f_3^2\}$. For $\sigma = 1 = \emptyset 1$, we have $\sigma' = \emptyset$ and $i = 1$. Thus, since $\omega(\emptyset) = 2$, the map f_1^2 is associated with the node $\sigma = 1$. Similarly, for $\sigma = 22$, one has $\sigma' = 2$ and $i = 2$. Since $\omega(2) = 1$, the associated map is f_2^1.

The probabilities P_α, $\alpha = 1, \ldots, A$, induce in a natural way a probability distribution on the set Ω: For each cylinder set $Z(\omega_n)$, we define a function

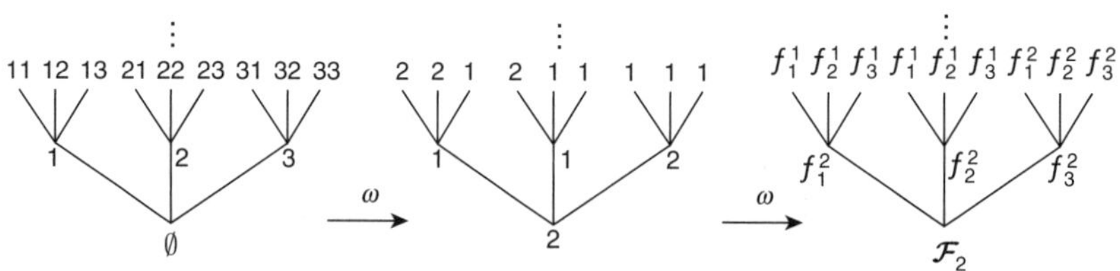

Figure 7.6 The association between a tree T and the mappings in a superIFS.

ϱ by setting

$$\varrho(Z(\omega_n)) := \prod_{1 \leq |\sigma| \leq n} P_{\omega_n(\sigma)}.$$

Via Exercise 45 in Section 4.7.2, we extend ϱ to the Borel σ-algebra generated by the cylinder sets to obtain a probability measure, again denoted by ϱ, on Ω. Having defined a probability measure ϱ on Ω, it now makes sense to speak of code trees with probabilities and of selecting elements of $\mathcal{A}$ and IFSs from $\{\mathcal{F}_\alpha \,|\, \alpha = 1, \ldots, A\}$ according to ϱ. We refer to the measure $\varrho \in$ $\mathsf{P}(\Omega)$ as the *code tree measure*. In reference to Example 7.20, we see that the assignment of an IFS $\mathcal{F}_\alpha$ occurs according to the probability distribution $\varrho \in \mathsf{P}(\Omega)$ induced by $\mathcal{P}$.

7.4.2 Random Fractal Sets and Random Fractal Measures

Assume that $\mathcal{F}$ is a superIFS as given in (7.11), and that $\mathsf{T} = \mathsf{T}(\mathcal{A})$ is a tree associated with this superIFS. For each $n \in \mathbb{N}$, we define a mapping

$$\mathcal{F}(n) : \Omega \times \mathcal{H}(\mathbf{X}) \to \mathcal{H}(\mathbf{X})$$

by

$$\mathcal{F}(n)(\omega)(K_0) := \bigcup_{\sigma = \sigma_1 \cdots \sigma_n \in T} f_{\sigma_1}^{\omega(\emptyset)} \circ f_{\sigma_2}^{\omega(\sigma_1)} \circ \cdots \circ f_{\sigma_n}^{\omega(\sigma_1 \sigma_2 \cdots \sigma_{n-1})}(K_0), \qquad (7.13)$$

for all code trees $\omega \in \Omega$ and all nonempty compact subsets K_0 of $\mathbf{X}$.

Example 7.22 We refer back to Example 7.20 and the binary tree $\mathsf{T} = \mathsf{T}(\mathcal{N})$ on the left-hand panel in Figure 7.5, to illustrate Equation (7.13).

In the case $n := 1$, (7.13) reads

$$\mathcal{F}(1)(\omega)(K_0) = \bigcup_{\sigma = \sigma_1 \in T} f_{\sigma_1}^{\omega(\emptyset)}(K_0) = f_1^{\omega(\emptyset)}(K_0) \cup f_2^{\omega(\emptyset)}(K_0),$$

and for $n := 2$, the expression becomes

$$\mathcal{F}(2)(\omega)(K_0) = \bigcup_{\sigma = \sigma_1 \sigma_2 \in T} f_{\sigma_1}^{\omega(\emptyset)} \circ f_{\sigma_2}^{\omega(\sigma_1)}(K_0)$$

$$= (f_1^{\omega(\emptyset)} \circ f_1^{\omega(1)})(K_0) \cup (f_1^{\omega(\emptyset)} \circ f_2^{\omega(1)})(K_0)$$

$$\cup (f_2^{\omega(\emptyset)} \circ f_1^{\omega(2)})(K_0) \cup (f_2^{\omega(\emptyset)} \circ f_2^{\omega(2)})(K_0)$$

$$= f_1^{\omega(\emptyset)} \left(f_1^{\omega(1)}(K_0) \cup f_2^{\omega(1)}(K_0) \right) \cup f_2^{\omega(\emptyset)} \left(f_1^{\omega(2)}(K_0) \cup f_2^{\omega(2)}(K_0) \right),$$

for all $\omega \in \Omega$ and all $K_0 \in \mathcal{H}(\mathbf{X})$. A graphical representation of $K_1(\omega) :=$ $\mathcal{F}(1)(\omega)(K_0)$ and $K_2(\omega) := \mathcal{F}(2)(\omega)(K_0)$ is shown in Figures 7.7 and 7.8, respectively.

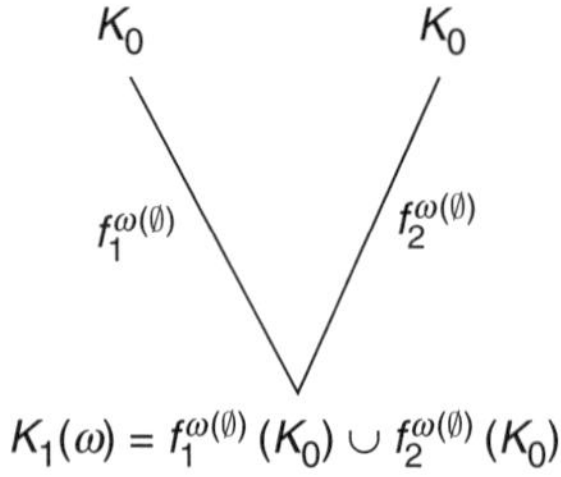

Figure 7.7 Construction of the set $K_1(\omega) = f_1^{\omega(\emptyset)}(K_0) \cup f_2^{\omega(\emptyset)}(K_0)$.

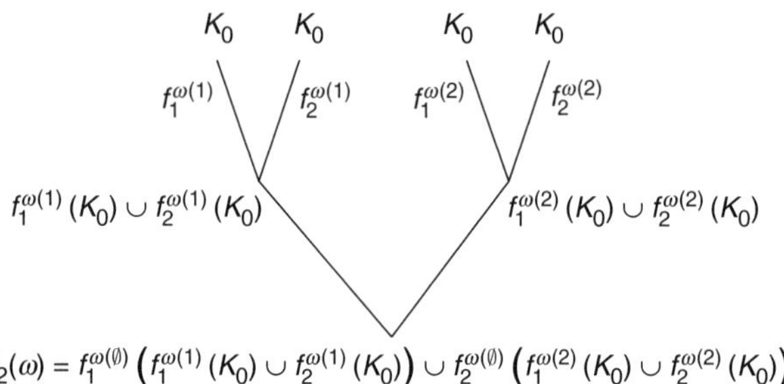

Figure 7.8 Construction of the set $K_2(\omega)$.

Definition 7.23 (Random Prefractal) Let $K_0 \in \mathcal{H}(X)$ be arbitrary and let $\omega \in \Omega$ be a code tree. Furthermore, let $n \in \mathbb{N}$. The set $K_n(\omega) := \mathcal{F}(n)(\omega)(K_0)$ is called a *random prefractal of rank n associated with the tree* $\mathsf{T} = \mathsf{T}(\mathcal{A})$.

The next result gives information about the limit of random prefractals when $n \to \infty$ in the Hausdorff metric $d_{\mathcal{H}}$.

Theorem 7.24 The sequence of random prefractals $\{K_n(\omega) \,|\, n \in \mathbb{N}\}$ has a limit

$$\mathcal{K}_\infty(\omega) := \mathcal{F}(\omega)(K_0) := \lim_{n \to \infty} \mathcal{F}(n)(\omega)(K_0)$$

taken in the Hausdorff metric. The convergence is uniform in K_0 and ω, and the limit is independent of the initially chosen compact set K_0. Moreover, the resulting mapping $\mathcal{F} : \Omega \to \mathcal{H}(X)$ is continuous.

Proof The mappings $\mathcal{F}(n)(\omega)(K_0)$, $n \in \mathbb{N}$, are compositions of contractions and the statements in the theorem follow from standard arguments based on the Banach Fixed Point Theorem 1.35 in compact metric spaces. The details are left to the reader. $\square$

Definition 7.25 (Standard Random Fractal Set)   The set $\mathfrak{K}_\infty(\omega)$ is called a *standard random fractal associated with the code tree ω*.

Remark 7.26   Introducing the notation $(\omega \rfloor \sigma)(\sigma') := \omega(\sigma\sigma')$ for a code tree ω and nodes $\sigma, \sigma' \in \mathsf{T}$, we can write the standard random fractal set $\mathfrak{K}_\infty(\omega)$ in the following way:

$$\mathfrak{K}_\infty(\omega) = \bigcup_{\sigma=\sigma_1\cdots\sigma_n \in T} \mathfrak{K}_\infty^\sigma(\omega),$$

where

$$\mathfrak{K}_\infty^\sigma(\omega) := f_{\sigma_1}^{\omega(\emptyset)} \circ f_{\sigma_2}^{\omega(\sigma_1)} \circ \cdots \circ f_{\sigma_n}^{\omega(\sigma_1\sigma_2\cdots\sigma_{n-1})}(\mathfrak{K}_\infty(\omega\rfloor\sigma)).$$

Notice that $\omega\rfloor\sigma$ is the code tree obtained from ω by starting at node σ instead of $\emptyset$. We call $\omega\rfloor\sigma$ a *code subtree of ω with base node σ*. In [16], the N^n sets $\mathfrak{K}_\infty^\sigma(\omega)$ are called *subfractals of $\mathfrak{K}_\infty(\omega)$ at level n*.

In a completely analogous fashion, we define a measure-valued mapping

$$\mathcal{F}^\sharp(n) : \Omega \times \mathsf{P}(\mathsf{X}) \to \mathsf{P}(\mathsf{X})$$

by

$$\mathcal{F}^\sharp(n)(\omega)(\mu_0) := \sum_{\sigma=\sigma_1\cdots\sigma_n \in T} \left[p_{\sigma_1}^{\omega(\emptyset)} p_{\sigma_2}^{\omega(\sigma_1)} \cdot \ldots \cdot p_{\sigma_n}^{\omega(\sigma_1\sigma_2\cdots\sigma_{n-1})} \right] \tag{7.14}$$

$$\cdot (f_{\sigma_1}^{\omega(\emptyset)})^\sharp \circ (f_{\sigma_2}^{\omega(\sigma_1)})^\sharp \circ \cdots \circ (f_{\sigma_n}^{\omega(\sigma_1\sigma_2\cdots\sigma_{n-1})})^\sharp(\mu_0),$$

for all code trees $\omega \in \Omega$ and all normalized Borel measures $\mu_0 \in \mathsf{P}(\mathsf{X})$. It is easy to verify that $\mathcal{F}^\sharp(n)(\omega)(\mu_0) \in \mathsf{P}(\mathsf{X})$ since

$$\sum_{i=1}^N p_i^\alpha = 1, \quad \alpha = 1, \ldots, A.$$

Definition 7.27 (Random Prefractal Measure)   Let $n \in \mathbb{N}$, $\mu_0 \in \mathsf{P}(\mathsf{X})$, and let $\omega \in \Omega$. The normalized Borel measures $\mu_n(\omega) := \mathcal{F}^\sharp(n)(\omega)(\mu_0)$ are called *random prefractal measures of rank n associated with the tree $\mathsf{T} = \mathsf{T}(\mathcal{A})$*.

We also obtain the analogous result to Theorem 7.24.

Theorem 7.28   The sequence of random prefractal measures $\{\mu_n(\omega) \mid n \in \mathbb{N}\}$ converges in the Monge–Kantorovich metric d_P to a normalized Borel measure

$$\mathfrak{m}_\infty(\omega) := \mathcal{F}^\sharp(\omega)(\mu_0) := \lim_{n\to\infty} \mathcal{F}^\sharp(n)(\omega)(\mu_0).$$

The convergence is uniform in μ_0 and ω, and the limit is independent of the initially chosen Borel measure μ_0. Moreover, the resulting mapping $\mathcal{F}^\sharp : \Omega \to \mathsf{P}(\mathsf{X})$ is continuous.

Proof   Exercise! $\square$

Definition 7.29 (Random Fractal Measure)   The normalized Borel measure $\mathfrak{m}_\infty(\omega)$ is called a *random fractal measure associated with the code tree ω.*

Theorems 7.24 and 7.28 show that the mappings $\mathcal{F} : \Omega \to \mathcal{H}(\mathbf{X})$ and $\mathcal{F}^\sharp : \Omega \to \mathsf{P}(\mathbf{X})$ are continuous. Under an additional assumption on the IFSs $\mathcal{F}_\alpha$, $\alpha \in \{1, \ldots, A\}$, namely that the family $\{\mathcal{F}_\alpha \,|\, \alpha \in \{1, \ldots, A\}$ is *uniformly bounded* on $(\mathbf{X}, d)$, i.e.,

$$\beta := \max\left\{ d(f_i^\alpha(\xi), \xi) \,|\, i \in \{1, \ldots, N\},\ \alpha \in \{1, \ldots, A\} \right\} < \infty,$$

for some $\xi \in \mathbf{X}$, more can be shown.

Proposition 7.30   Suppose that the family of IFSs $\{\mathcal{F}_\alpha \,|\, \alpha \in \{1, \ldots, A\}$ is uniformly bounded by $\beta > 0$ and that

$$\lambda := \max\left\{ \|\mathrm{Lip}(f_i^\alpha)\| \,|\, i \in \{1, \ldots, N\},\ \alpha \in \{1, \ldots, A\} \right\} < \infty.$$

Then the mappings $\mathcal{F} : \Omega \to \mathcal{H}(\mathbf{X})$, $\omega \mapsto \mathfrak{K}_\infty(\omega)$, and $\mathcal{F}^\sharp : \Omega \to \mathsf{P}(\mathbf{X})$, $\omega \mapsto \mathfrak{m}_\infty(\omega)$, are in $\mathrm{Lip}^\alpha(\Omega, \mathcal{H}(\mathbf{X}))$ and $\mathrm{Lip}^\alpha(\Omega, \mathsf{P}(\mathbf{X}))$, respectively, where the Hölder exponent is given by

$$\alpha := -\frac{\log \lambda}{\log N}.$$

More precisely,

$$d_{\mathcal{H}}(\mathfrak{K}_\infty(\omega), \mathfrak{K}_\infty(\omega')) \leq \frac{2\beta}{1 - \lambda}\, d_\Omega(\omega, \omega')^\alpha$$

and

$$d_{\mathsf{P}}(\mathfrak{m}_\infty(\omega), \mathfrak{m}_\infty(\omega')) \leq \frac{2\beta}{1 - \lambda}\, d_\Omega(\omega, \omega')^\alpha.$$

Proof   We prove the first of the two inequalities and leave the proof of the second one to the reader. To this end, set $K_0 := \{\xi\}$ in (7.13) and use the triangle inequality repeatedly to obtain

$$d_{\mathcal{H}}(\mathfrak{K}_\infty(\omega), \{\xi\}) \leq (1 + \lambda + \cdots + \lambda^n + \cdots)\beta = \frac{\beta}{1 - \lambda}.$$

(See also the proof of the Banach Fixed Point Theorem, Exercise 19, in Chapter 1!) Thus,

$$d_{\mathcal{H}}(\mathfrak{K}_\infty(\omega), \mathfrak{K}_\infty(\omega')) \leq \frac{2\beta}{1 - \lambda},$$

for any $\omega, \omega' \in \Omega$. Now suppose that $\omega, \omega' \in \Omega$ are such that $d_\Omega(\omega, \omega') = N^{-n}$, some $n \in \mathbb{N}$. Then $\omega(\sigma) = \omega'(\sigma)$ for all $\sigma \in T$ with $|\sigma| < n$. By Remark 7.26

and the notation introduced therein, we have that

$$d_{\mathcal{H}}(\mathfrak{K}_{\infty}(\omega\rfloor\sigma), \mathfrak{K}_{\infty}(\omega'\rfloor\sigma)) \leq \frac{2\beta}{1-\lambda},$$

and, therefore,

$$d_{\mathcal{H}}(\mathfrak{K}_{\infty}(\omega), \mathfrak{K}_{\infty}(\omega')) \leq \frac{2\beta\lambda^n}{1-\lambda}.$$

Note that $\lambda^n = N^{-n\alpha} = d_{\Omega}(\omega, \omega')^{\alpha}$. Hence, the result follows. $\square$

Definition 7.31 (Set of Standard Random Fractals)   The set

$$\mathfrak{F} := \mathfrak{F}(\Omega) := \{\mathcal{F}(\omega) \in \mathcal{H}(\mathbf{X}) \,|\, \omega \in \Omega\}$$

is called the *set of standard random fractals distributed according to the probability distribution* $\varrho \circ \mathcal{F}^{-1} \in \mathsf{P}(\mathfrak{F}(\Omega))$.

Analogously, we define the *set of standard random fractal measures.*

Definition 7.32 (Set of Standard Random Fractal Measures)   The set

$$\mathfrak{M} := \mathfrak{M}(\Omega) := \{\mathcal{F}^{\sharp}(\omega) \in \mathsf{P}(\mathbf{X}) \,|\, \omega \in \Omega\}$$

is called the *set of standard random fractals distributed according to the probability distribution* $\varrho \circ (\mathcal{F}^{\sharp})^{-1} \in \mathsf{P}(\mathfrak{M}(\Omega))$.

The sets $\mathfrak{F}(\Omega)$ and $\mathfrak{M}(\Omega)$ have a highly complex structure and are hard to compute. For instance, it is remarked in [18] that there do not appear to exist simple algorithms, i.e., a "chaos game," for the practical computation of their elements, even in two dimensions.

Now we are in a position to connect V-variable IFSs $\mathcal{F}^{(V)}$ given by Equation (7.9) to code trees. For this purpose, we make the following definition.

Definition 7.33 (V-Variable Code Tree)   A code tree $\omega \in \Omega$ is called V-*variable* provided that for all $n \in \mathbb{N}$ there are at most V *distinct* code subtrees of the form $\omega\rfloor\sigma$ with $|\sigma| = n$. In case that ω is a finite code tree of height m, it is called V-*variable* if for all $\mathbb{N} \ni n < m$ there exist at most V code subtrees $\omega\rfloor\sigma$ with $|\sigma| = n$.

Remark 7.34   Let $V := 1$. A code tree ω is 1-variable iff ($|\sigma| = |\sigma'| \Rightarrow \omega(\sigma) = \omega(\sigma')$) iff $\forall\, n \in \mathbb{N}$ the values of $\omega(\sigma)$ at level n are equal. For $V > 1$, if a code tree ω is V-variable then for each $n \in \mathbb{N}$ there are at most V distinct values of $\omega(\sigma)$ at level $n = |\sigma|$.

Example 7.35   In Figure 7.9, the initial segment of a 2-variable but not a 1-variable code tree and of a 3-variable but not a 2-variable code tree is displayed.

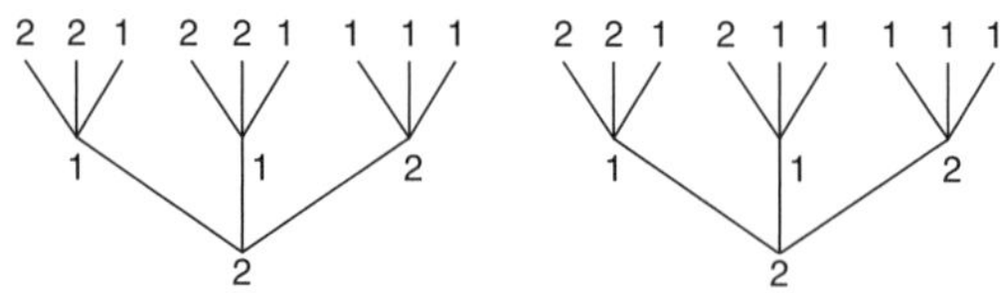

Figure 7.9 Initial segments of a 2-variable code tree (left) and a 3-variable code tree (right).

It follows from Definition 7.33 that a V-variable IFS is characterized by a V-variable code tree with the V distinct values corresponding to the V different buffers. If $V = 1$, there is at most one value for $\omega(\sigma)$ at all levels $n \in \mathbb{N}$ and thus only one shape or pattern that defines the resulting fractal. The elements in the class of 1-variable fractal sets are therefore called *homogeneous fractals*.

The next result tells what happens when $V \to \infty$. For the proof of this result, we refer the reader to [18].

Theorem 7.36 Assume that $V \in \mathbb{N}$ and $1 < N \in \mathbb{N}$. Let $\mathcal{F}$ be the superIFS defined in (7.11), and $\mathcal{F}^{(V)}$ the V-variable IFS with probabilities associated with $\mathcal{F}$ given by (7.9), where the probabilities have the values stated in (7.13) with $N_{\alpha_v} = N$, $v = 1, \ldots, V$. Then the set of V-variable fractal sets $\mathfrak{A}^{(V)}$ converges in the metric of $\mathcal{H}(\mathcal{H}(\mathbf{X}))$ to the set of standard random fractals $\mathfrak{F}$:

$$\lim_{V \to \infty} \mathfrak{A}^{(V)} = \mathfrak{F}.$$

Moreover, the normalized Borel measure $\widetilde{\mathrm{m}}^{(V)}$ converges in the metric of $\mathsf{P}(\mathcal{H}(\mathbf{X}))$ to the measure $\mathcal{F}^{\sharp}\varrho = \varrho \circ \mathcal{F}^{-1}$,

$$\lim_{V \to \infty} \widetilde{\mathrm{m}}^{(V)} = \varrho \circ \mathcal{F}^{-1},$$

where ϱ is the code tree measure.

Note that Theorem 7.36 gives the precise specifications of the statements made in the introduction to this chapter. In the case that $V = \infty$, at each level $n \in \mathbb{N}$ in the fractal construction, there is no correlation between the shapes of patterns used to generate the standard random fractal set. However, the existence of the probability distribution $\varrho \circ \mathcal{F}^{-1}$ allows the resulting fractal set to be interpreted as a statistical self-referential set.

We end this section by looking at a simple example which nevertheless captures the main points in the construction of homogeneous and standard random fractals.

Example 7.37 Let $(\mathbf{X}, d)$ be the compact metric space $(\square, d_E)$, where d_E denotes the Euclidean metric on $\mathbb{R}^2$ restricted to $\square$. Consider the two IFSs $\mathcal{F}_1 := \{f_1^1, f_2^1, f_3^1\}$ and $\mathcal{F}_2 := \{f_1^2, f_2^2, f_3^2\}$, where the functions $f_i^\alpha : \square \to \square$,

Figure 7.10 A random prefractal set.

$\alpha = 1, 2$, $i = 1, 2, 3$, are given by

$$f_1^1(x, y) := \left(\frac{x}{2}, \frac{y}{2}\right),$$

$$f_2^1(x, y) := \left(\frac{x+1}{2}, \frac{y}{2}\right),$$

$$f_3^1(x, y) := \left(\frac{2x+1}{4}, \frac{y+1}{2}\right),$$

and

$$f_1^2(x, y) := \left(\frac{x}{3}, \frac{y}{3}\right),$$

$$f_2^2(x, y) := \left(\frac{x+2}{3}, \frac{y}{3}\right),$$

$$f_3^2(x, y) := \left(\frac{x+1}{3}, \frac{y+2}{3}\right),$$

respectively. Using the finite code tree $\omega \lfloor 2$ from Example 7.21 together with the two IFSs above, we can compute the prefractal $\mathcal{F}(2)(\omega)(\blacktriangle)$. The result is shown in Figure 7.10.

Exercises

1. Recall that a metric space $(\mathbf{X}, d)$ is *compact* provided every sequence $\{x_\nu \mid \nu \in \mathbb{N}\} \subset \mathbf{X}$ has a convergent subsequence $\{x_{\mu_\nu} \mid \nu \in \mathbb{N}\}$. Show that if $(\mathbf{X}, d)$ is a compact metric space, then $(\mathcal{H}^V(\mathbf{X}), h^V)$, $V \in \mathbb{N}$, is a compact metric space.

2. Let Σ_A^f denote the set of all finite codes from the alphabet $\{1, \ldots, A\}$, i.e., codes of the form $\sigma = (\sigma_1, \ldots, \sigma_n)$, $n \in \mathbb{N}$, inclusive the empty code $\emptyset$. The length of a code $\emptyset \neq \sigma \in \Sigma_A^f$ is denoted by $|\sigma|$, and the length of the

empty code $|\emptyset|$ is defined to be zero. Define the IFS $\mathcal{F}_\sigma$ on $\mathbf{X}$ by

$$\mathcal{F}_\sigma := \mathcal{F}_{\sigma_1} \circ \cdots \circ \mathcal{F}_{\sigma_{|\sigma|}}, \quad \sigma \in \Sigma_A^f \setminus \{\emptyset\}.$$

Denote by $\mathfrak{A}_\sigma \in \mathcal{H}(\mathbf{X})$ the set attractor of the IFS $\mathcal{F}_\sigma$, $\sigma \in \Sigma_A^f \setminus \{\emptyset\}$. Show that

$$\mathfrak{A}^{(1)} \supset \widetilde{\mathfrak{A}}^{(1)} := \left\{ \mathcal{F}_\sigma(\mathfrak{A}_\rho) \mid \rho, \sigma \in \Sigma_A^f;\ \rho \neq \{\emptyset\} \right\},$$

and that the closure of $\widetilde{\mathfrak{A}}^{(1)}$ with respect to the Hausdorff metric h is $\mathfrak{A}^{(1)}$.

3. Verify Equation (7.4).

4. Prove Theorem 7.8.

5. Show that $(\mathcal{H}(\mathbf{X})^V, d_{\mathcal{H}^V})$ is a compact metric space whenever $(\mathbf{X}, d)$ is one.

6. Show that $(\mathbf{P}(\mathbf{X})^V, d_{\mathbf{P}^V})$ is a compact metric space whenever $(\mathbf{X}, d)$ is one.

7. Let $\mu, \nu \in \mathbf{P}(\mathbf{X})^k$, $k \in \mathbb{N}$. Further let $\lambda := \max\{|\mathrm{Lip}(f_i^\alpha)| \mid i = 1, \ldots, N_\alpha;\ \alpha = 1, \ldots, A\}$, where the mappings f_i^α are as in the superIFS (7.6). Show that

$$d_{\mathbf{P}}\left(\sum_{i=1}^k p_i^\alpha\, (f_i^\alpha)^\sharp(\mu_i), \sum_{i=1}^k p_i^\alpha\, (f_i^\alpha)^\sharp(\nu_i) \right) \le \lambda\, d_{\mathbf{P}^k}(\mu, \nu).$$

8. Use Exercise 7 to establish Theorem 7.15.

9. Show that (Ω, d_Ω) is a compact metric space and that $\mathrm{diam}(\Omega) = 1$.

10. Give a proof of Theorem 7.24.

11. Prove Theorem 7.28.

12. Fill in all the details in the proof of Proposition 7.30 and prove the second inequality. (*Hint: Set* $\mu_0 := \delta_\xi$)

13. Show that in the case that $V > 1$, the condition given in Remark 7.34 is not sufficient to imply V-variability for a code tree.

14. Prove that a code tree ω is V-variable iff for all $n \in \mathbb{N}$ the finite code trees $\omega\lfloor n$ are V-variable. $\quad\square$

8

Superfractal Functions

This short chapter deals with the new concept of superfractal functions and V-variable fractal functions. As the graphs of fractal functions are special cases of fractal sets, so are the graphs of superfractal functions special cases of superfractals. By specializing to a subclass of IFSs, namely *interpolating* IFSs, one can extend the notions introduced in Chapter 5 to the setting of superfractals that was presented in Chapter 7. Superfractal functions are constructed by joining together pieces of individual fractal functions coming from a finite family of RB operators defined on appropriate function spaces. The notions of homogeneity and standard randomness encountered in the theory of superfractal sets carry over to superfractal functions. Indeed, there exists an entire family of such fractal functions indexed by the variability parameter $V \in \mathbb{N} \cup \{\infty\}$. The possibility of employing different shapes and patterns at successive levels of construction allows for better and more realistic use in interpolation and approximation theory.

The theory of superfractal functions is still in its infancy. It is currently a topic of ongoing research with many open problems yet to be addressed. This chapter is to give a brief overview of some of the fundamental notions and ideas from the theory, and is based on the most recent original publications in this area, such as [9], [18], and [136].

The outline of this chapter is as follows, After some preliminaries regarding the set-up of the construction, including a slightly more general construction of fractal interpolation functions, we generate V-variable fractal interpolation functions using the ideas and methods developed in Chapter 7. Then we introduce fractal functions constructed via finite collections of RB operators and code trees. Several examples illustrate the general concepts.

8.1 Preliminaries

We briefly review the construction of fractal interpolation functions via IFSs and via the RB operator, and explicitly describe the relationship between the two approaches. (See Section 5.2 for reference!) The constructions here are slightly more general than those presented earlier.

To this end, let $N \in \mathbb{N}$, and let $X := \{x_\nu \mid \nu = 0, 1, \ldots, N\}$ be a set of ordered knots. Without loss of generality, we may assume that $0 =: x_0 < x_1 < \cdots < x_N := 1$. Let $I := [0, 1]$. Further assume that $Y := \{y_\nu \mid \nu = 0, 1, \ldots, N\}$ is a set of associated interpolation values for X. Setting $a := y_0$ and $b := y_N$, we recall the definition of the function spaces $C_{a,b}(I)$ given in (5.12). In addition, we define the set of functions

$$C_{X \times Y}(I) := \{f \in C(I) \mid f(x_\nu) = y_\nu, \ \nu = 0, 1, \ldots, N\}.$$

Notice that the set $C_{a,b}(I)$ is a special case of the above defined set of functions; indeed $C_{a,b}(I) = C_{\{0,1\} \times \{a,b\}}(I)$. The sets of functions $C_{a,b}(I)$ and $C_{X \times Y}(I)$ become complete metric spaces when endowed with the metric $d_{\infty, I}$. (See (5.13)!)

Moreover, for all $i \in \{1, \ldots, N\}$, let

$$u_i \in \mathrm{Map}(I, I),$$
$$u_i(0) = x_{i-1} \quad \text{and} \quad u_i(1) = x_i, \tag{8.1}$$

be a contractive homeomorphism, and

$$v_i \in C(I \times \mathbb{R}, \mathbb{R}),$$
$$\forall x \in I: \ v_i(x, \bullet) \in \mathrm{Lip}^1(I \times \mathbb{R}, \mathbb{R}) \text{ with } \lambda_i := \mathrm{Lip}(v_i(x, \bullet)) < 1 \tag{8.2}$$
$$v_i(0, a) = y_{i-1} \quad \text{and} \quad v_i(1, b) = y_i.$$

Setting $\mathsf{X} := I \times \mathbb{R}$ and noting that X is a complete metric space, we consider the IFS $(\mathsf{X}, \mathcal{F})$ with $\mathcal{F} := \{f_i \mid i = 1, \ldots, N\}$ and

$$f_i : \mathsf{X} \to \mathsf{X}$$
$$f_i(x, y) := (u_i(x), v_i(x, y)). \tag{8.3}$$

Remark 8.1 In Chapter 5, we considered the case when $v_i(x, y) := p_i(x) + \lambda_i y$, for continuous functions p_i and real numbers $|\lambda_i| < 1$, $i = 1, \ldots, N$.

Denoting the above IFS by $\mathcal{F} := (\mathsf{X}, \{f_i \mid i = 1, \ldots, N\})$, it is straightforward to establish the existence of a unique set $\mathfrak{G} \in \mathcal{H}(\mathsf{X})$ with the property that

$$\mathfrak{G} = \mathrm{graph}\, \mathfrak{f},$$

where $\mathfrak{f} \in C(I, \mathbb{R})$ satisfies $\mathfrak{f}|_X = Y$, i.e., $\mathfrak{f}(x_\nu) = y_\nu$, $\nu = 0, 1, \ldots, N$. (The reader is encouraged to verify this statement.)

Definition 8.2 (Interpolating IFS) An IFS $\mathcal{F} := (\mathbf{X}, \{f_i \,|\, i = 1, \ldots, N\})$ where the f_i are defined as in (8.3), i.e., with the property that its unique attractor is the graph of a continuous function that interpolates a given set of values Y on a given set of ordered knots X, is called an *interpolating IFS* and is denoted by $\mathcal{F}_{X \times Y}$.

Notice that each interpolating IFS $\mathcal{F}_{X \times Y}$ induces an RB operator

$$T = T_{\mathcal{F}_{X \times Y}} : C_{a,b}(I) \to C_{X \times Y}(I) \subset C_{a,b}(I)$$

via

$$(Tf)(x) = v_i\left(u_i^{-1}(x), (f \circ u_i^{-1})(x)\right), \quad \forall\, x \in [x_{i-1}, x_i], \tag{8.4}$$

and all $i = 1, \ldots, N$. The operator T is contractive in the metric $d_{\infty, I}$ induced by the Chebychev norm and its unique attractor is a continuous function $\mathfrak{f} : I \to \mathbb{R}$ with the property that $\mathfrak{f}|_X = Y$. Moreover, the graph of the unique fixed point $\mathfrak{f}$ of T is the unique set-valued attractor $\mathfrak{G}$ of $\mathcal{F}_{X \times Y}$. (The reader is encouraged to verify this statement.)

On the other hand, given an RB operator $T : C_{a,b}(I) \to C_{X \times Y}(I)$ of the form (8.4) with functions u_i and v_i satisfying (8.1), respectively, (8.2) for a given set of ordered knots X and interpolation values Y, there exists an associated interpolating IFS $\mathcal{F} = \mathcal{F}_T$ with contractions of the form (8.3) whose unique set-valued attractor $\mathfrak{G}$ is the graph of the unique fixed point $\mathfrak{f}$ of T. (The reader is encouraged to verify this statement.)

Hence, we arrive at the following result.

Proposition 8.3 Let X be a given set of ordered knots and Y the associated set of interpolation values. Suppose the elements in the sets of functions $\{u_i \in \mathrm{Map}(I, I) \,|\, i = 1, \ldots, N\}$ and $\{v_i \in \mathrm{Map}(I \times \mathbb{R}, \mathbb{R})\}$ are defined as in (8.1) and (8.2), respectively. Denote by $\mathbf{IFS}_{X \times Y}$ the class of all interpolating IFSs $\mathcal{F}_{X \times Y}$ and by $\mathbf{RB}_{X \times Y}$ the class of all RB operators $T : C_{a,b}(I) \to C_{X \times Y}(I)$. Then there exists a bijection $\Theta : \mathbf{IFS}_{X \times Y} \to \mathbf{RB}_{X \times Y}$.

Remark 8.4 The above interpolation problem for fractal functions, namely to find an $\mathfrak{f} : I \to \mathbb{R}$ so that $\mathfrak{f}|_X = Y$, for a given set of ordered knots X and associated set Y of interpolation values, can be extended to the more general setting of a type of *Birkhoff*[1] *interpolation*.

To this end, let $v_i \in C(I \times \mathbb{R}, \mathbb{R})$ be given by $v_i(x, y) := p_i(x) + \lambda_i y$, where $p_i \in C(I, \mathbb{R})$ and $\lambda_i \in (-1, 1)$, $i = 1, \ldots, N$. Let $k \in \mathbb{N}_0$ and let $\mathbf{y}_v :=$ $(y_v^0, y_v^1, \ldots, y_v^k)^\top \in \mathbb{R}^{k+1}$ be a *vector* of real values associated with each knot x_v, $v = 0, 1, \ldots, N$. We require a function $\mathfrak{f} \in C^k(I)$ so that

$$(D^\varkappa \mathfrak{f})(x_v) = y_v^\varkappa, \quad \varkappa = 0, 1, \ldots, k, \quad v = 0, 1, \ldots, N.$$

---

1 Garrett Birkhoff, 19 January 1911–22 November 1996. American mathematician whose main contributions are the establishment of abstract algebra as a new discipline of mathematics and the development of lattice theory as a new mathematical area.

For this purpose, set $Y := \{y_\nu \,|\, \nu = 0, 1, \ldots, N\}$ and define function spaces

$$C^k_{a,b}(I) = \left\{f \in C^k(I) \,|\, (D^\varkappa f)(0) = y_0^\varkappa \text{ and } (D^\varkappa f)(1) = y_N^\varkappa, \forall \varkappa \in \{0, 1, \ldots, k\}\right\},$$

where $a := y_0$ and $b := y_N$, and

$$C^k_{X \times Y}(I) := \left\{f \in C^k(I) \,|\, (D^\varkappa f)(x_\nu) = y_\nu^\varkappa, \forall \varkappa \in \{0, 1, \ldots, k\}; \forall \nu \in \{0, 1, \ldots, N\}\right\}.$$

Endowed with the metric $d_k : C^k(I) \to C^k(I)$, $d_k(f, g) := \|f - g\|_{C^k, I}$ (see (5.31)) the spaces $C^k_{a,b}(I)$ and $(C^k_{X \times Y}(I), d_k)$ become complete metric spaces. (The reader is encouraged to verify this statement.) Let mappings $u_i : I \to I$, $i = 1, \ldots, N$, be defined as in (8.1) and let $T : C^k_{a,b}(I) \to \mathrm{Map}(I, \mathbb{R})$ be given by

$$(Tf)(x) := \sum_{i=1}^{N} \left[p_i \circ u_i^{-1}(x) + \lambda_i (f \circ u_i^{-1})(x)\right] \chi_{[x_{i-1}, x_i]}(x), \quad x \in [0, 1].$$

Moreover, we require that the action of T on an element $f \in C^k_{a,b}(I)$ satisfies the following join-up conditions.

$$(D^\varkappa T_i f)(x_i) = (D^\varkappa T_{i+1} f)(x_i), \quad \forall i \in \{1, \ldots, N-1\}; \forall \varkappa \in \{0, 1, \ldots, k\};$$

$$(D^\varkappa T_1 f)(0) = y_0^\varkappa \quad \text{and} \quad (D^\varkappa T_N f)(1) = y_N^\varkappa, \quad \forall \varkappa \in \{0, 1, \ldots, k\},$$

where we set

$$T_i := T|_{[x_{i-1}, x_i]} = T|_{u_i(I)}, \quad i = 1, \ldots, N. \tag{8.5}$$

In addition, we impose the condition

$$|\lambda_i| < (\mathrm{Lip}\, u_i)^k,$$

on the scaling factors λ_i, $i = 1, \ldots, N$. Under these conditions, one can easily show that T actually maps $C^k_{a,b}(I)$ into itself and is contractive on $(C^k_{a,b}(I), d_k)$. (The reader is encouraged to verify this statement.) Hence the unique fixed point $\mathfrak{f}$ of T is an element of $C^k_{a,b}(I)$. Moreover, $\mathfrak{f} \in C^k_{X \times Y}(I)$. (The reader is encouraged to show this.)

8.2 *V*-Variable Fractal Interpolation

We use Proposition 8.3 to define V-variable fractal interpolation and V-variable fractal interpolation functions. For this purpose, let $X := \{x_\nu \,|\, \nu = 0, 1, \ldots, N\}$ be a fixed set of ordered knots with $x_0 := 0$ and $x_N := 1$, and let $\mathbf{X} := I \times \mathbb{R}$. Furthermore, let $A \in \mathbb{N}$ be fixed and let $\{\mathcal{F}^\alpha_{X \times Y_\alpha} := (\mathbf{X}; \mathcal{F}^\alpha_{X \times Y_\alpha}) \,|\, \alpha = 1, \ldots, A\}$ be a collection of interpolating IFSs on $\mathbf{X}$ where each $\mathcal{F}^\alpha_{X \times Y_\alpha}$ consists of functions of the form (8.3), and where $Y_\alpha := \{y_\nu^\alpha \,|\, \nu = 0, 1, \ldots, N\}$ is a set of associated interpolation values for X with $a := y_0^\alpha$ and $b := y_N^\alpha$, for all $\alpha \in \{1, \ldots, A\}$ and fixed $a, b \in \mathbb{R}$. Then each interpolating IFS $\mathcal{F}^\alpha_{X \times Y_\alpha}$ consists of exactly N contractions and has a continuous function $\mathfrak{f}^\alpha : I \to \mathbb{R}$ as its unique fixed point that satisfies

$$\mathfrak{f}^\alpha(0) = a, \quad \mathfrak{f}^\alpha(1) = b, \quad \text{and} \quad \mathfrak{f}^\alpha(x_\nu) = y_\nu^\alpha, \ \forall \nu = 1, \ldots, N-1.$$

Using the finite family of interpolating IFSs $\{\mathcal{F}^\alpha_{X \times Y_\alpha} := (\mathbf{X}; \mathcal{F}^\alpha_{X \times Y_\alpha}) \,|\, \alpha = 1, \ldots, A\}$, we define the associated *interpolating superIFS* by

$$\mathcal{F}_{X \times \{Y_\alpha\}} := (\mathbf{X}; \{\mathcal{F}^\alpha_{X \times Y_\alpha} \,|\, \alpha = 1, \ldots, A\}). \tag{8.6}$$

As the unique attractors of the elements in $\mathcal{F}_{X \times \{Y_\alpha\}}$ are the graphs of continuous interpolation functions, this superIFS is also called a *fractal interpolation function superIFS*.

Employing the result in Proposition 8.3 and denoting $\Theta(\mathcal{F}^\alpha_{X \times Y_\alpha})$ by T^α, we refer to the set

$$\mathbf{T}^\alpha := \{(C_{a,b}(I), T^\alpha) \,|\, \alpha = 1, \ldots, A\} \tag{8.7}$$

as the *interpolating superRB operator*. To the interpolating superIFS (8.6) or, equivalently, the interpolating superRB operator (8.7), we can apply all the machinery developed in Chapters 5 and 7.

Example 8.5 Using the definitions in Chapter 7 together with the interpolating superIFS (8.6), we can construct, for instance, the 1-variable superfractal of affine fractal interpolation functions

$$\mathfrak{G}^{(1)} = \{\mathfrak{G}_\sigma \,|\, \sigma \in \Sigma_2\}$$

where

$$\mathfrak{G}_\sigma = \lim_{n \to \infty} \mathcal{F}^{\sigma_1}_{X \times Y_{\sigma_1}} \circ \cdots \circ \mathcal{F}^{\sigma_n}_{X \times Y_{\sigma_n}}(\mathbf{X}),$$

and $\mathcal{F}^1_{X \times Y_1}$ and $\mathcal{F}^2_{X \times Y_2}$ are the interpolating IFS given by the ordered knot set $X := \{0, \frac{1}{3}, 1\}$, the interpolation values

$$Y_1 := \left\{\frac{1}{2}, 1, \frac{1}{3}\right\}, \quad \text{respectively,} \quad Y_2 := \left\{\frac{1}{2}, \frac{1}{4}, \frac{1}{3}\right\},$$

the mappings

$$u_1(x) = \frac{1}{3}x \quad \text{and} \quad u_2(x) = \frac{2}{3}x + \frac{1}{3},$$

and

$$v_1^1(x, y) = \frac{5}{8}x + \frac{1}{8} + \frac{3}{4}y, \qquad v_2^1(x, y) = -\frac{13}{18}x + \frac{7}{6} + \left(-\frac{1}{3}\right)y,$$

$$v_1^2(x, y) = -\frac{13}{36}x + \frac{5}{6} + \left(-\frac{2}{3}\right)y, \quad v_2^2(x, y) = \frac{1}{6}x + \frac{1}{2}y.$$

Note that the scaling vectors for these IFSs are given by $\lambda^1 = (\frac{3}{4}, -\frac{1}{3})$ and $\lambda^2 = (-\frac{2}{3}, \frac{1}{2})$, respectively.

In Figure 8.1, the graphs of $\mathfrak{G}_{\overline{1}}$ (upper left) and $\mathfrak{G}_{\overline{2}}$ (upper right) are displayed, as well as $\mathfrak{G}_{12\overline{1}} = \mathcal{F}^1_{X \times Y_1}(\mathfrak{G}_{2\overline{1}})$.

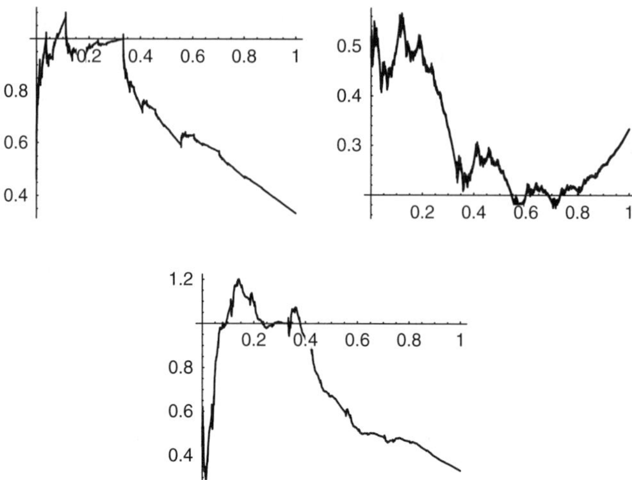

Figure 8.1 Elements of a 1-variable superfractal of affine fractal interpolation functions.

The generalization to the V-variable setting is now at hand. Using interpolating IFSs instead of regular IFSs, we can construct V-variable fractal interpolation functions, more precisely, their graphs, by utilizing the procedures and algorithms presented and discussed in Section 7.3. The elements of the superfractal of V-variable fractal interpolation functions are then better suited for approximation and interpolation purposes as they provide more flexibility in their geometric shape and structure.

To this end, let $\mathcal{J}$ be the index set defined in (7.7), let the set of functions $\{f_j \,|\, j \in \mathcal{J}\}$ in (7.8) be given by the interpolating superIFS $\mathcal{F}_{X \times \{Y_\alpha\}}$, and define a V-variable interpolating IFS $\mathcal{F}^{(V)}_{X \times \{Y_\alpha\}}$ as in (7.9), where the probabilities are given by (7.10). Denote the unique set attractor of $\mathcal{F}^{(V)}_{X \times \{Y_\alpha\}}$ by $\mathfrak{G}^{(V)}$. Then the set

$$\mathfrak{G}^{(V)} := \left\{ G \in \mathcal{H}(\mathbf{X}) \,\middle|\, G \text{ is a component of a point in } \mathfrak{G}^{(V)} \right\}$$

is called the V-variable superfractal (of graphs) of fractal interpolation functions. Its elements are referred to as V-variable (graphs of) fractal interpolation functions. An element $\mathfrak{g}^{(V)} \in \mathfrak{G}^{(V)}$ is thus the graph of a continuous function $\mathfrak{f}^{(V)} \in C_{ab}(I)$.

Remark 8.6 Note that each element $\mathfrak{g}^{(V)} \in \mathfrak{G}^{(V)}$ depends on the set of scaling factors $\boldsymbol{\Lambda} := \{\lambda^\alpha \,|\, \alpha = 1, \ldots, A\}$ and that we should write $\mathfrak{g}^{(V)}[\boldsymbol{\Lambda}]$, $\mathfrak{G}^{(V)}[\boldsymbol{\Lambda}]$, etc. to express this dependence. For the sake of notational simplicity, however, we will suppress this dependence. As before, the elements $\mathfrak{g}^{(V)}$ can be thought of as the graphs of fractal interpolation functions parametrized by the free parameters λ^α whose components are real numbers between -1 and 1.

Remark 8.7 Let $\mathfrak{G}^{(1)}$ be the 1-variable superfractal (of graphs) of fractal interpolation functions and suppose that X is an ordered knot set whose elements are equally spaced, i.e., $x_i - x_{i-1} = \frac{1}{N}$, $i = 1, \ldots, N$. Suppose further that all scaling factors are equal, i.e., $\lambda_i^\alpha := \lambda \in (-1, 1)$, for all $i = 1, \ldots, N$ and all $\alpha = 1, \ldots, A$. Then, for almost all codes $\sigma \in \Sigma_A$,

$$\dim_B \mathfrak{g}^{(1)} = 2 + \frac{\log |\lambda|}{\log N}.$$

(See formula (5.25) for this special case!) For the general case, $V > 1$, we refer the interested reader to [136].

To present an example of 2-variable fractal interpolation functions, we recall the construction in Section 7.3 for the case that $N_1 := N_2 := 2$ and $A := 2$. Then, given an initial pair $A^0 := (A_1^0, A_2^0)$, the general sequence of iterates $\{A^k \mid k \in \mathbb{N}_0\}$ is of the form

$$(A_1^k, A_2^k) \xmapsto{f_{pqrsuv}} \left(f_1^p(A_r^k) \cup f_2^p(A_s^k), f_1^q(A_u^k) \cup f_2^q(A_v^k) \right),$$

where $p, q, r, s, u, v \in \{1, 2\}$. (Here, the index **j** was written out.)

Example 8.8 Suppose that $A := 2$ and $N_1 := N_2 := 2$. Let $X := \{0, \frac{1}{2}, 1\}$ and $Y_1 := \{0, \frac{1}{2}, 0\}$ and $Y_2 := \{0, 1, 0\}$. Moreover, let $\lambda^1 := (\frac{2}{3}, \frac{1}{2})$ and $\lambda^2 := (\frac{1}{2}, -\frac{3}{4})$. Define functions

$$u_i : I \to I,$$

$$x \mapsto \frac{1}{2}(x + i - 1), \quad i = 1, 2,$$

and

$$v_i^\alpha : I \times \mathbb{R} \to \mathbb{R}, \quad i = 1, 2; \ \alpha = 1, 2$$

$$v_1^1(x, y) := \frac{1}{2}x + \frac{2}{3}y, \qquad v_2^1(x, y) := -\frac{1}{2}(x - 1) + \frac{1}{2}y,$$

$$v_1^2(x, y) := \frac{1}{4}\sin\frac{\pi x}{2} + \frac{1}{2}y, \qquad v_2^2(x, y) := \frac{e}{1 - e}(e^{x-1} + 1) - \frac{3}{4}y.$$

Furthermore, let $\mathbf{X} := I \times \mathbb{R}$ and set

$$f_i^\alpha : \mathbf{X} \to \mathbf{X},$$

$$(x, y) \mapsto f_i^\alpha(x, y) := (u_i(x), v_i^\alpha(x, y)).$$

Hence, we have two interpolating IFSs $\mathcal{F}_{X \times Y_\alpha}^\alpha := (\mathbf{X}; \{f_1^\alpha, f_2^\alpha\}), \alpha = 1, 2$, whose attractors are displayed in Figure 8.2.

Setting $A_1^0 := A_2^0 := \{(x, y) \in \mathbb{R}^2 \mid x \in [0, 1] \wedge y = 0\}$ and employing the algorithm in Section 7.3, we can construct successive pairs of iterates of which the first few are shown in Figure 8.3. The reader is encouraged to construct the two components of the attractor $\mathfrak{G}^{(2)}$.

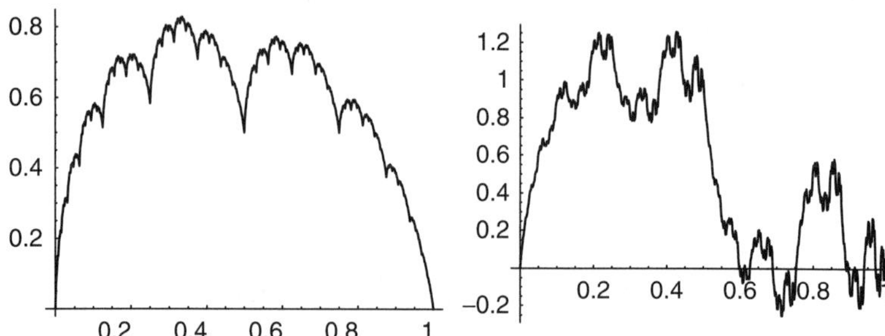

Figure 8.2 The attractors for the two interpolation IFSs.

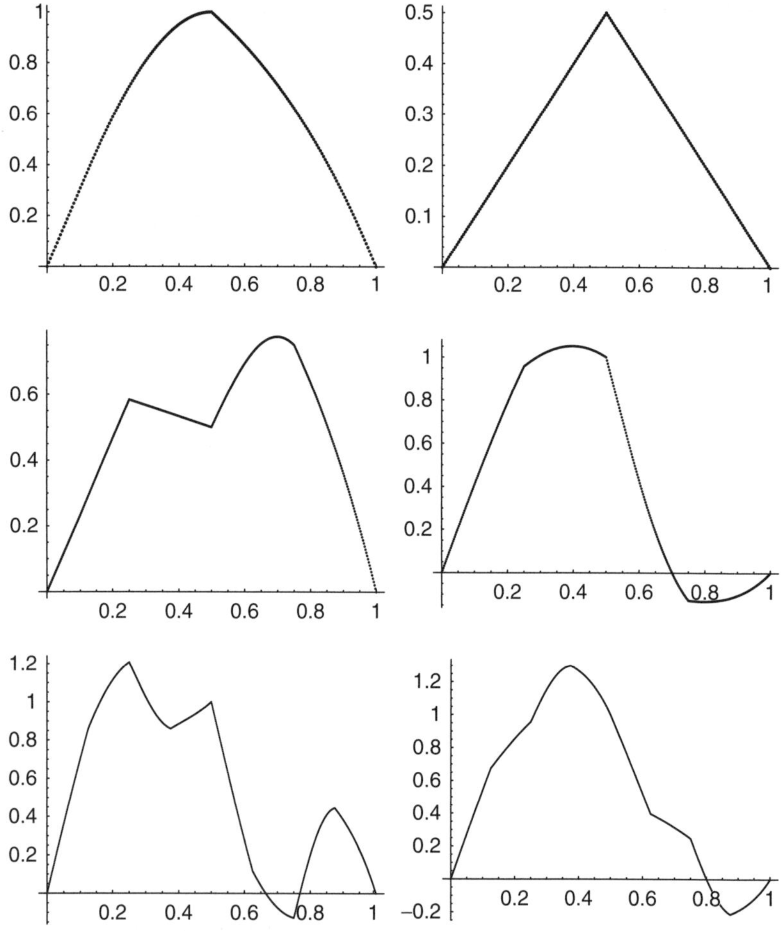

Figure 8.3 Pairs of successive iterates in example 8.8.

The construction of the set-valued V-variable IFS $\mathcal{F}^{(V)}$ using interpolating IFSs as described above and its associated set-valued attractor $\mathfrak{G}^{(V)}$ induces, by Proposition 8.3, a corresponding construction of a function-valued V-variable RB operator $T^{(V)}$ and its associated function-valued fixed point $\mathfrak{f}^{(V)}$.

To this end, let $\mathcal{J}$ be the index set given by Equation (7.7) and let T^{α} be the interpolating superRB operator. Set $C_{a,b}(I)^V := \underset{v=1}{\overset{V}{\times}} C_{a,b}(I)$, $V \in \mathbb{N}$, and introduce the metric

$$d_{\infty,I}^V : C_{a,b}(I)^V \times C_{a,b}(I)^V \to \mathbb{R}$$

$$d_{\infty,I}^V(g,h) := \max\{d_{\infty,I}(f_v,g_v) \,|\, v=1,\dots,V\}.$$

Then $(C_{a,b}(I)^V, d_{\infty,I}^V)$ is a complete metric space. We define operators $T_{\mathbf{j}} : C_{a,b}(I)^V \to C_{a,b}(I)^V$ by

$$T_{\mathbf{j}}(g_1,\dots,g_V) := \left(\sum_{i=1}^{N} T_i^{\alpha_1}(g_{v_1,i}), \dots, \sum_{i=1}^{N} T_i^{\alpha_V}(g_{v_V,i}) \right),$$

where the subscripts on the functions $g_{v_v,i}$ have the same meaning as for the compact sets $A_{v_v,i}$. (See Section 7.3 and, in particular, Equation (7.8)!) Note that the union over subsets now corresponds to a sum over subintervals.

Analogous to (7.9) we set

$$T^{(V)} := (C_{a,b}(I)^V; \{T_{\mathbf{j}} \,|\, \mathbf{j} \in \mathcal{J}\}) \tag{8.8}$$

and refer to it as a *V-variable superRB operator*. The system $T^{(V)}$ is an RB operator that possesses a unique fixed point $\mathfrak{f}^{(V)} \in C_{a,b}^V$. We refer to $\mathfrak{f}^{(V)}$ as a *superfractal function*. The components $\mathfrak{f}_v$, $v \in \{1,\dots,V\}$, of the fixed point $\mathfrak{f}^{(V)}$ are called *V-variable fixed points*. They are fractal functions $\mathfrak{f}_v : I \to \mathbb{R}$ with the property that $\mathfrak{f}(0) = a$ and $\mathfrak{f}(1) = b$.

Remark 8.9 Notice that the above definition of a V-variable superRB operator reduces to that of an RB operator in the case $V = 1$. (It is customary to call the pair (X, T) an operator, where X is the domain of the mapping T.)

Example 8.10 Suppose that $A := 3$ and $N := 2$. Let $X := \{0, \frac{1}{2}, 1\}$ and $Y_1 := \{0, \frac{1}{2}, 0\}$, $Y_2 := \{0, \frac{2}{3}, 0\}$, and $Y_3 := \{0, -\frac{1}{4}, 0\}$. Moreover, let $\boldsymbol{\lambda}^1 := (\frac{2}{3}, \frac{1}{2})$, $\boldsymbol{\lambda}^2 := (\frac{1}{2}, -\frac{3}{4})$, and $\boldsymbol{\lambda}^3 := (-\frac{1}{3}, -\frac{3}{4})$. Define functions

$$u_i : I \to I,$$

$$x \mapsto \frac{1}{2}(x+i-1), \quad i = 1, 2,$$

and

$$v_i^{\alpha} : I \times \mathbb{R} \to \mathbb{R},$$

$$v_i^{\alpha}(x,y) := a_i^{\alpha} x + b_i^{\alpha} + \lambda_i^{\alpha} y, \quad i = 1, 2; \ \alpha = 1, 2, 3.$$

Furthermore, let $\mathsf{X} := I \times \mathbb{R}$ and set

$$f_i^\alpha : \mathsf{X} \to \mathsf{X},$$

$$(x, y) \mapsto f_i^\alpha(x, y) := (u_i(x), v_i^\alpha(x, y)).$$

Hence, we have defined three interpolating IFSs $\mathcal{F}_{\mathsf{X} \times Y_\alpha}^\alpha := (\mathsf{X}; \{f_1^\alpha, f_2^\alpha\})$, $\alpha = 1, 2, 3$. Thus, the restriction of the RB operator associated with the interpolating IFS $\mathcal{F}_{\mathsf{X} \times Y_\alpha}^\alpha$ to the interval $[x_{i-1}, x_i]$ is given by

$$T_i^\alpha(f)(x) = a_i^\alpha x + b_i^\alpha + \lambda_i^\alpha f(x), \quad i = 1, 2; \ \alpha = 1, 2, 3.$$

Example 8.11 In Example 8.8, we actually displayed successive pairs of iterates of the initially chosen function pair $g^0 := (g_1^0, g_2^0)$, where $g_\nu^0(x) := 0$, $x \in I, \nu = 1, 2$, under the action of a certain 2-variable RB operator. Determine this RB operator!

The above arguments show that one can construct V-variable fractal functions by starting with an appropriate metric space of functions $(\mathsf{F}, d_\mathsf{F})$ together with a collection of RB operators that map the function space F into itself and defining an V-variable superRB operator of the form (8.8). This extends the notions considered earlier in Chapter 5.

We can also construct random fractal interpolation functions in the spirit of Section 7.4. To this end, suppose $\omega \in \Omega$ is a V-variable code tree (see Definition 7.33) and define $\mathcal{F}(\omega)(\bullet)$ as in (7.13) but now with the underlying superIFS given by (8.6). By Proposition 8.3, we can state the definition of $\mathcal{F}(\omega)(\bullet)$ in terms of functions rather than compact sets, namely,

$$\mathcal{F}(\omega)(f) := \lim_{n \to \infty} \mathcal{F}(n)(\omega)(f)$$

where

$$\mathcal{F}(n)(\omega)(f)|_{u_{\sigma_1 \cdots \sigma_n}(I)} = T_{\sigma_1}^{\omega(\emptyset)} \circ \cdots \circ T_{\sigma_n}^{\omega(\sigma_1 \cdots \sigma_{n-1})}(f), \quad n \in \mathbb{N}, \tag{8.9}$$

with $u_{\sigma_1 \cdots \sigma_n} := u_{\sigma_1} \circ \cdots \circ u_{\sigma_n}$ and arbitrary $f \in C_{a,b}(I)$. Here $\sigma = \sigma_1 \cdots \sigma_n \in \mathsf{T}$ denotes as usually a node of the tree $\mathsf{T} = \mathsf{T}(\mathcal{N})$ and ω a code tree. The quantity $T_{\sigma_i}^{\omega(\sigma_i \cdots \sigma_{i-1})}$ is the RB operator that corresponds to the mapping $f_{\sigma_i}^{\omega(\sigma_i \cdots \sigma_{i-1})}$ in the associated set-valued IFS. (See Equation (8.5) for the notation regarding the RB operator $T_{\sigma_i}^{\omega(\sigma_i \cdots \sigma_{i-1})}$.)

By Theorem 7.24 and the fact that the RB operator associated with each interpolating IFS $(\mathsf{X}; \mathcal{F}_{\mathsf{X} \times Y_\alpha}^\alpha)$ is mapping $C_{a,b}(I)$ into itself, we have the following result.

Proposition 8.12 Let $\omega \in \Omega$ be a V-variable code tree. Then the *function* $\mathcal{F}(\omega)(f), f \in C_{a,b}(I)$ arbitrary, is an element of $C_{a,b}(I)$.

Remark 8.13 Since the function $\mathcal{F}(\omega)(f)$ does not depend on the particular choice of $f \in C_{a,b}(I)$, we simply write $\mathcal{F}(\omega)$ for the limit.

Definition 8.14 (*V*-Variable Fractal Interpolation Function)   Let $\omega \in \Omega$ be a *V*-variable code tree. The function $\mathcal{F}(\omega)$ is called a *V-variable fractal interpolation function*. The functions $\mathcal{F}(n)(\omega)$, $n \in \mathbb{N}$, are referred to as *iterates* (of an arbitrary starting point $f \in C_{a,b}(I)$) or as the *prefractal functions of rank n*.

In analogy to the notions introduced in Chapter 7, we call a 1-variable fractal (interpolation) function a *homogeneous fractal (interpolation) function* and the *V*-variable (interpolation) function where $V = \infty$ a *standard random fractal (interpolation) function*. The interpretation of these terms and their meaning is the same as for superfractal sets.

Example 8.15   This is a continuation of Example 8.10. Suppose that ω is a code tree whose initial segment is shown in Figure 8.4. Computing explicit expressions for the affine functions v_i^{α}, $i = 1, 2$, $\alpha = 1, 2, 3$, and employing Equation (8.9), generates a prefractal of high rank representing a 2-variable fractal interpolation function $\mathfrak{f}$. The graph of $\mathfrak{f}$ is displayed in Figure 8.6

The constructions presented in this chapter can be extended and generalized to interpolation problems of the form considered in Remark 8.4 and Exercise 7 at the end of this chapter. Indeed, one can construct fractal functions of class C^k, $k \in \mathbb{N}$, using the ideas discussed here. Moreover, it is possible to generate *V*-variable vector-valued fractal functions and also

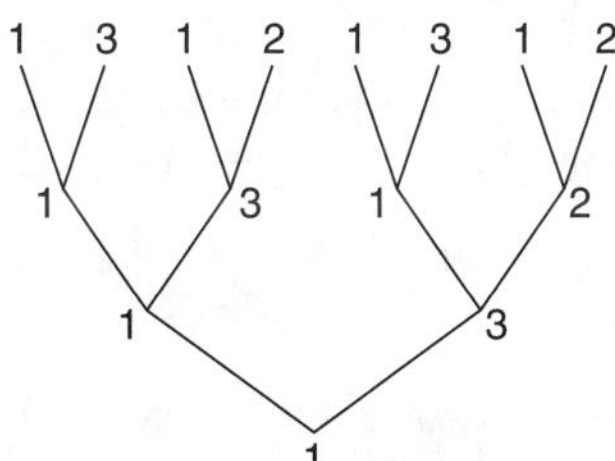

Figure 8.4   The tree code ω in Example 8.15.

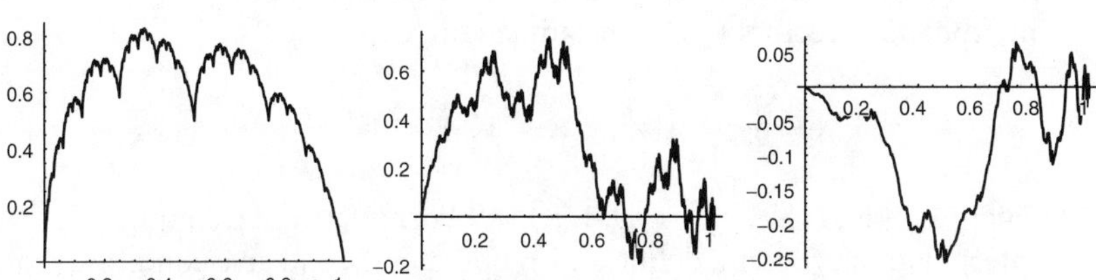

Figure 8.5   The individual attractors in Example 8.15.

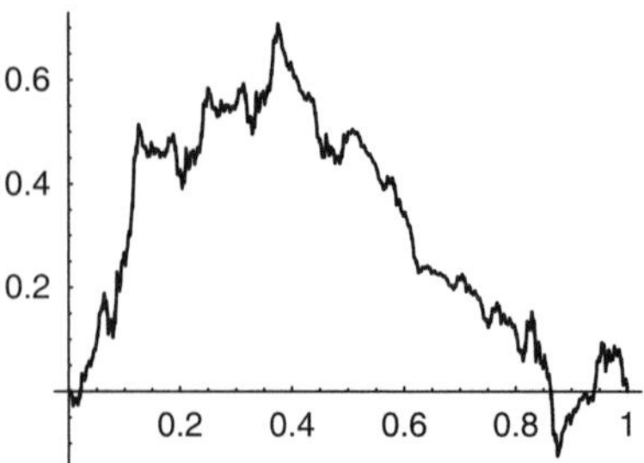

Figure 8.6 The function $\mathcal{F}(\omega)$.

fractal curves. For constructions and a discussion of these topics we refer the interested reader to [136].

Exercises

1. Show that the IFS $\mathcal{F} = (\mathbf{X}, \{f_i \mid i = 1, \ldots, N\})$, where the functions f_i are defined as in (8.3), has a unique attractor $\mathfrak{G}$, and that this attractor is the graph of a continuous function $\mathfrak{f} : I \to \mathbb{R}$ interpolating the set Y on the knots X.

2. Show that the RB operator T defined by (8.4) maps $C_{a,b}(I)$ into $C_{X \times Y}(I)$ and is contractive in the Chebychev norm. What is the constant of contractivity?

3. Prove Proposition 8.3.

4. Why is it necessary to require that in the collection of interpolating IFSs $y_0^\alpha = a$ and $y_N^\alpha = b$ for all $\alpha = 1, \ldots, A$ and fixed $a, b \in \mathbb{R}$?

5. Show that $(C_{a,b}^k(I), d_{C^k,I})$ and $(C_{X \times Y}^k(I), d_{C^k,I})$ are complete metric spaces.

6. Establish that the RB operator T introduced in Remark 8.4 maps $C_{a,b}^k(I)$ into itself, is contractive on $(C_{a,b}^k(I), d_{C^k,I})$, and its unique fixed point is an element of $C_{X \times Y}^k(I)$.

7. Generalize the construction given in Remark 8.4 to the following slightly more general Birkhoff interpolation problem. Instead of using interpolation vectors $\boldsymbol{y}_\nu$ of constant length, define

$$\boldsymbol{y}_\nu := \{(y_\nu^0, y_\nu^1, \ldots, y_\nu^{k_\nu})^\top \in \mathbb{R}^{k_\nu+1}\},$$

where $k_\nu \in \{0, 1, \ldots, k\}$, $\nu = 0, 1, \ldots, N$. Formulate the resulting interpolation problem, define an RB operator T, and give conditions so that T has a unique fixed point in an appropriately defined function space. (*Hint: To ensure continuity of the k-th derivative, you need to* impose conditions on k_0 and k_N at the endpoints $x_0 = 0$ and $x_N = 1$, respectively!)

8. Provide the details for a proof of Proposition 8.12.

9. Prove the statement made in Remark 8.7. (*Hint: Recall Theorem 5.30!*)

10. Construct the two components of the attractor $\mathfrak{G}^{(2)}$ in Example 8.8.

11. Write down the 2-variable RB operator mentioned in Example 8.11.　□

Bibliography

1 M. Abramowitz and I. Stegun, *Handbook of Mathematical Functions*, Dover Publications Inc., New York, 1972.

2 R. A. Adams, *Sobolev Spaces*, Pure and Applied Mathematics Series, Vol. 65, Academic Press, New York, 1975.

3 B. K. Alpert, "Wavelets and other bases for fast numerical linear algebra," in *Wavelets – A Tutorial in Theory and Applications*, C. K. Chui, Ed., Academic Press, 181–216 (1992).

4 M. Bajraktarević, "Sur une équation fonctionelle," *Glasnik Mat.-Fiz. Astr. Ser. II* **12**(1956), 201–205.

5 Ch. Bandt, "Self-similar sets. I. Topological Markov chains and mixed self-similar sets," *Math. Nachr.* **142**(1989), 107–123.

6 Ch. Bandt, "Self-similar sets. III. Constructions with sofic systems," *Mh. Math.* **108**(1989), 89–102.

7 M. F. Barnsley, "Fractal functions and interpolation," *Constr. Approx.* **2**(1986), 303–329.

8 M. F. Barnsley, *Fractals Everywhere*, 2nd ed., Academic Press Professional, Orlando, Florida, 1993.

9 M. F. Barnsley, *Superfractals*, Cambridge University Press, UK, 2006.

10 M. F. Barnsley, "Transformations between self-referential sets," to appear in *Mathematical Monthly*, 2009.

11 M. F. Barnsley and S. Demko, "Iterated function systems and the global construction of fractals," *Proc. R. Soc. Lond. A* **399**(1985), 243–275.

12 M. F. Barnsley and J. H. Elton, "A new class of Markov processes for image encoding," *Adv. Appl. Probab.* **20**(1) (1988), 14–32.

13 M. F. Barnsley, J. H. Elton, and D. P. Hardin, "Recurrent iterated function systems," *Constr. Approx.* **5**(1) (1989), 3–31.

14 M. F. Barnsley, J. Elton, D. P. Hardin, and P. R. Massopust, "Hidden variable fractal interpolation functions," *SIAM J. Math. Anal.* **20**(5) (1989), 1218–1242.

15 M. F. Barnsley and A. N. Harrington, "The calculus of fractal interpolation functions," *J. Approx. Th.* **57**(1989), 14–34.

16 M. F. Barnsley, J. E. Hutchinson, and Ö. Stenflo, "V-variable fractals: fractals with partial self-similarity," *arXiv preprint*, 2008.

17 M. F. Barnsley, J. E. Hutchinson, and Ö. Stenflo, "V-variable fractals and superfractals," *arXiv preprint*, 2003.

18 M. F. Barnsley, J. E. Hutchinson, and Ö. Stenflo, "A fractal valued random iteration algorithm and fractal hierarchy," *Fractals*, Vol. 13, No. 2 (2005), 111–146.

19 M. F. Barnsley, P. R. Massopust, H. Strickland and A. D. Sloan, "Fractal modeling of biological structures," *Ann. New York Acad. Sci.* **504**(1987), 179–194.

20 H. Bauer, *Measure and Integration Theory*, De Gruyter Studies in Mathematics, Berlin, Germany, 2001.

21 T. Bedford, "The box dimension of self-affine graphs and repellers," *Nonlinearity* **2**(1989), 53–71.

22 T. Bedford, "Hölder exponents and box dimension for self-affine fractal functions," *Constr. Approx.* **5**(1) (1989), 33–48.

23 T. Bedford and M. Urbański, "The box and Hausdorff dimension of self-affine sets," *Erg. Th. and Dynam. Sys.* **10**(4) (1990), 627–644.

24 J. Bergh and J. Löfström, *Interpolation Spaces*, Springer Verlag, New York, 1976.

25 A. S. Besicovitch and H. D. Ursell, "Sets of fractional dimension," *J. Lond. Math. Soc.* **12**(1937), 18–25.

26 K. Böhmer, *Spline-Funktionen*, Teubner Studienbücher Mathematik, Mannheim, 1974.

27 P. Bouboulis and L. Dalla, "Fractal interpolation surfaces derived from fractal interpolation functions," *J. Math. Anal. Appl.* **336**(2) (2007), 919–936.

28 P. Bouboulis and L. Dalla, "A general construction of fractal interpolation functions on grids of $\mathbb{R}^n$," *European J. Appl. Math.* **18**(4) (2007), 449–476.

29 P. Bouboulis and L. Dalla, "Closed fractal interpolation surfaces," *J. Math. Anal. Appl.* **327**(1) (2007), 116–126.

30 P. Bouboulis, L. Dalla, and V. Drakopoulos, "Construction of recurrent bivariate fractal interpolation surfaces and computation of their box-counting dimension," *J. Approx. Th.* **141**(2) (2006), 99–117.

31 P. Butzer, *Semi-groups of Operators and Approximation*, Springer Verlag, New York, 1967.

32 P. Butzer and R. Nessel, *Fourier Analysis and Approximation*, Birkhäuser Verlag, New York, 1971.

33 P. Butzer and K. Scherer, *Approximationsprozesse und Interpolationsmethoden*, Bibliograpisches Institut, Mannheim, Germany, 1965.

34 C. A. Cabrelli, B. Forte, U. M. Molter and E. R. Vrscay, "Iterated fuzzy set systems: A new approach to the inverse problem for fractals and other sets," *J. Math. Anal. Appl.* **171**(1992), 79–100.

35 K. Chandrasekharan, *Classical Fourier Transforms*, Springer Verlag, Berlin, Germany, 1989.

36 E. W. Cheney, *Introduction to Approximation Theory*, 2nd ed., American Mathematical Society, Providence, 2000.

37 O. Christensen, *An Introduction to Frames and Riesz Bases*, Applied and Numerical Harmonic Algebra, Birkhäuser Verlag, Boston, 2003.

38 C. K. Chui, *Multivariate Splines*, SIAM CBMS, Vol. 54, Philadelphia, 1988.

39 C. K. Chui, *An Introduction to Wavelets*, Academic Press, San Diego, 1992.

40 R. Courant and D. Hilbert, *Methods of Mathematical Physics*, Vol. I and II, Interscience Publishers, New York, 1955.

41 W. Dahmen, "B-splines in analysis, algebra, and applications," *Traveaux Mathématiques* (1999), 15–76.

42 W. Dahmen and C. A. Micchelli, "On theory and application of exponential splines," in *Topics in Multivariate Approximation*, C. K. Chui, L. L. Schumaker, and F. I. Utreras, Eds., Academic Press, New York, 1987, 37–46.

43 I. Daubechies, *Ten Lectures on Wavelets*, CBMS 61, SIAM, Philadelphia, 1992.

44 C. de Boor, *A Practical Guide to Splines*, 2nd ed., Springer Verlag, New York, 2001.

45 C. de Boor, "Splines as linear combinations of B-splines: A survey," in G. G. Lorentz, C. K. Chui, and L. L. Schumaker, Eds., *Approximation Theory II*, Academic Press, New York, 1976.

46 C. de Boor and R. DeVore, "Approximation by smooth multivariate splines," *Trans. Amer. Math. Soc.* **276**(1983), 775–788.

47 C. de Boor, K. Höllig, and S. Riemenschneider, *Box Splines*, Springer Verlag, New York, 1993.

48 M. F. Dekking, "Recurrent sets," *Adv. in Math.* **44**(1982), 78–104.

49 G. Donovan, J. S. Geronimo, D. P. Hardin and P. R. Massopust, "Construction of orthogonal wavelets using fractal interplation functions," *SIAM J. Math. Anal.* **27**(4) (1996), 1158–1192.

50 L. E. Dubins and L. J. Savage, *Inequalities for Stochastic Processes*, Dover Publications Inc., New York, 1976.

51 S. Dubuc, "Interpolation through an iterative scheme," *J. Math. Anal. Appl.* **114**(1)(1986), 185–204.

52 S. Dubuc, "Interpolation fractale," in *Fractal Geometry and Analysis*, J. Bélais and S. Dubuc, Eds., Kluwer Academic Publishers, Dordrecht, The Netherlands, 1989.

53 R. M. Dudley, *Real Analysis and Probability*, 2nd ed., Cambridge University Press, Cambridge, UK, 2002.

54 J. Duoandikoetxea, *Fourier Analysis*, GSM Vol. 29, American Mathematical Society, Providence, Rhode Island, 2001.

55 G. A. Edgar, "Kieswetter's fractal has Hausdorff dimension 3/2," *Real Analysis Exchange*, Vol. 14 (1988–89), 215–223.

56 G. A. Edgar, *Measure Theory, Topology, and Fractal Geometry*, Springer Verlag, New York, 1990.

57 K. J. Falconer, *The Geometry of Fractal Sets*, Cambridge University Press, Cambridge, UK, 1985.

58 K. J. Falconer, *Fractal Geometry – Mathematical Foundations and Applications*, John Wiley & Sons, Michester, UK, 1990.

59 K. J. Falconer, "The Hausdorff dimension of self-affine fractals," *Math. Proc. Camb. Phil. Soc.* **103**(1988), 339–350.

60 K. J. Falconer and D. T. Marsh, "The dimension of affine-invariant fractals," *J. Phys. A: Math. Gen.* **21**(1988), 121–125.

61 K. J. Falconer, "Dimensions and measures of quasi self-similar sets," *Proc. Amer. Math. Soc.* **106**(2) (1989), 543–554.

62 B. Forster, T. Blu, and M. Unser, "Complex B-splines," *Appl. Comp. Harmon. Anal.* **20**(2006), 281–292.

63 B. Forster and P. Massopust, "Some remarks about the connection between fractional divided differences, fractional B-splines, and the Hermite-Genocchi formula," *International Journal of Wavelets, Multiresolution and Information Processing*, Vol. 6, No. 2 (2008), 279–290.

64 B. Forster and P. Massopust, "Statistical encounters with complex B-splines," *Constr. Approx.*, 29(3), 325–344, 2009.

65 B. Forte and F. Mendivil, "A classical ergodic property for IFS: a simple proof," *Ergodic Theory Dynam. Systems* **18**(3) (1998), 609–611.

66 B. Forte and E. R. Vrscay, "Theory of generalized fractal transforms," in *Fractal Image Encoding and Analysis*, Y. Fisher Ed., Springer Verlag, Heidelberg, 1998.

67 B. Forte and E. R. Vrscay, "Inverse problem methods for generalized fractal transforms," in *Fractal Image Encoding and Analysis*, Y. Fisher, Ed., Springer Verlag, Heidelberg, Germany, 1998.

68 B. Forte and E. R. Vrscay, "Solving the inverse problem for measures using iterated function systems: A new approach," *Adv. Appl. Prob.*, **27**(1995), 800–820.

69 B. Forte and E. R. Vrscay, "Solving the inverse problem for function and image approximation using iterated function systems," *Dyn. Cont. Disc. Imp. Sys.* **1**(1995), 177–231.

70 J. S. Geronimo and D. P. Hardin, "An exact formula for the measure dimensions associated with a class of piecewise linear maps," *Constr. Approx.* **5**(1) (1989), 89–98.

71 J. S. Geronimo and D. P. Hardin, "Fractal interpolation surfaces and a related 2-D multiresolution analysis," *J. Math. Anal. and Appl.* **176**(2) (1993), 561–586.

72 J. S. Geronimo, D. P. Hardin, and P. R. Massopust, "Fractal functions and wavelet expansions based on several scaling functions," *J. Approx. Th.* **78**(3) (1993), 373–401.

73 J. S. Geronimo, D. P. Hardin, and P. R. Massopust, "Fractal surfaces, multiresolution analyses, and wavelet transforms," *NATO ASI Series F*, Vol. 106 (1994), 275–290.

74 J. S. Geronimo, D. P. Hardin, and P. R. Massopust, "An application of Coxeter groups to the generation of wavelet bases in $\mathbb{R}^n$," *Lecture Notes in Pure and Applied Mathematics: Fourier Analysis – Analytic and Geometric Aspects*, Vol. 157 (1994), 187–196.

75 W. J. Gilbert, "The fractal dimension of sets derived from complex bases," *Canad. Math. Bull.* **29**(4) (1986), 495–500.

76 S. Graf, "Statistically self-similar fractals," *Probab. Theory Related Fields* **74**(3) (1987), 357–392.

77 A. Haar, "Zur Theorie der orthogonalen Funktionensysteme," *Math. Ann.* **69**(1910), 331–371.

78 C. A. Hall, "Natural cubic and bicubic spline interpolation," *SIAM J. on Numerical Analysis* **10**(6) (1973), 1055–1060.

79 G. Hämmerlin and K.-H. Hoffmann, *Numerical Mathematics*, Springer Verlag, New York, 1991.

80 D. Hardin and T. Hogan, "Constructing orthogonal refinable function vectors with prescribed approximation order and smoothness," in *Wavelet Analysis and Applications, Guangzhou 1999* (2002), 139–148.

81 D. P. Hardin, B. Kessler, and P. R. Massopust, "Multiresolution analyses based on fractal functions," *J. Approx. Th.* **71**(1) (1992), 104–120.

82 D. P. Hardin and P. R. Massopust, "The capacity for a class of fractal functions," *Commun. Math. Phys.* **105**(1986), 455–460.

83 D. P. Hardin and P. R. Massopust, "Fractal interpolation functions from $\mathbb{R}^n$ to $\mathbb{R}^m$ and their projections," *Zeitschrift für Analysis u. i. Anw.* **12**(1993), 535–548.

84 D. P. Hardin and D. Roach, "Multiwavelet prefilters I: Orthogonal prefilters preserving approximation order $p \leq 2$," *IEEE Trans. Circ. and Sys. II: Anal. and Dig. Sign. Proc.* **45**(8) (1998), 1106–1112.

85 D. P. Hardin, X.-G. Xia, J. Geronimo, and B. Suter, "Design of prefilters for discrete multiwavelet transforms," *IEEE Trans. on Signal Processsing* **44**(1996), 25–35.

86 K. Höllig, "A remark on multivariate B-splines," *J. Approx. Theory* **33**(1981), 119–125.

87 J. E. Hutchinson, "Fractals and self similarity," *Indiana Univ. J. Math.* **30**(1981), 713–747.

88 E. L. Ince, *Die Integration gewöhnlicher Differentialgleichungen*, B. I. Hochschul-taschenbücher, Bibliographisches Institut, Mannheim, Germany, 1956.

89 D. Jackson, *The Theory of Approximation*, Cushing-Malloy, New York, 1958.

90 J. W. Jerome and L. L. Schumaker, "On Lg-splines," *Journal Approx. Th.* **(2)**(1969), 29–49.

91 R.-Q. Jia, "Refinable shift-invariant spaces: From splines to wavelets," *Approximation Theory VIII, Vol. 2: Wavelets and Multilevel Approximation*, Academic Press, 179–208 (1995).

92 A. Kameyama, "Self-similar sets from the topological point of view," *Japan J. Ind. Appl. Math.* **10**(1993), 85–95.

93 S. Karlin, C. Micchelli, A. Pinkus, and I. Schoenberg, *Studies in Spline Functions and Approximation Theory*, Academic Press, New York, 1976.

94 S. Karlin and Z. Ziegler, "Chebyshevian spline functions," *SIAM J. Numer. Anal.*, **3**(1966), 514–543.

95 Y. Katznelson, *An Introduction to Harmonic Analysis*, Dover Publications Inc., New York, 1976.

96 P. Kergin, "A natural interpolation of C^K functions," *J. Approx. Th.* **29**(1980), 278–293.

97 B. Kieninger, *Iterated Function Systems on Compact Hausdorff Spaces*, Ph.D. Dissertation, Shaker Verlag, Aachen, Germany, 2002.

98 K. Kiesswetter, "Ein einfaches Beispiel für eine Funktion welche überall stetig und nicht differenzierbar ist," *Math. Phys. Semesterber.* **13**(1966), 216–221.

99 J. Kigami, *Analysis on Fractals*, Cambridge University Press, Cambridge, UK, 2001.

100 K. Kono, "On self-affine functions I. and II.," *Japan J. Appl. Math.* **3**(2)(1986), 252–269, and **5**(3) (1988), 441–454.

101 P. Lancaster and K. Šalkauskas, *Transform Methods in Applied Mathematics*, John Wiley & Sons, Inc., New York, 1996.

102 E. H. Lieb and M. Loss, *Analysis*, 2nd ed., Graduate Texts in Mathematics, Vol. 14, American Mathematical Society, Providence, Rhode Island, 2001.

103 G. G. Lorentz, *Approximation of Functions*, Chelsea Publishing Company, 2nd ed., New York, 1986.

104 T. Lyche, "Knot removal for spline curves and surfaces," in *Approximation Theory VII*, E. W. Cheney, C. K. Chui, and L. L. Schumaker, Eds., 207–226, Academic Press, New York, 1992.

105 B. Mandelbrot, *The Fractal Geometry of Nature*, W. H. Freeman and Company, New York, 1977.

106 J. Marsden, "On uniform spline approximation," *J. Approx. Theory* **6**(1972), 249–253.

107 P. R. Massopust, *Space Curves Generated by Iterated Function Systems*, Ph.D. Thesis, Georgia Institute of Technology, 1986.

108 P. R. Massopust, "Dynamical systems, fractal functions, and dimension," *Topology Proc.* **12**(1987), 93–110.

109 P. R. Massopust, "Fractal Peano curves," *J. of Geometry* **34**(1989), 127–138.

110 P. R. Massopust, "Fractal surfaces," *J. Math. Anal. and Appl.* **151**(1) (1990), 275–290.

111 P. R. Massopust, "Vector-valued fractal interpolation functions and their box dimension," *Aequationes Mathematicae* **42**(1991), 1–22.

112 P. R. Massopust, "Smooth interpolating curves and surfaces generated by iterated function systems," *Zeitschrift für Analysis u. i. Anwend.* **12**(1993), 201–210.

113 P. R. Massopust, *Fractal Functions, Fractal Surfaces, and Wavelets*, Academic Press, New York, 1995.

114 P. R. Massopust, "Fractal Functions and applications," *Chaos, Solitons, and Fractals*, **8**(2) (1997), 171–190.

115 P. R. Massopust, "Fractal Functions and wavelets: Examples of multiscale theory," in *Abstract und Applied Analysis, Proceedings of the International Conference*, Hanoi, Vietnam, (N. M. Chuong, L. Nirenberg, W. Tutschke, Eds.), World Scientific Press, Singapore, 2004.

116 P. R. Massopust, "Fractal functions, splines, and Besov and Triebel-Lizorkin spaces," in *Fractals in Engineering: New trends und applications* (J. Lévy-Véhel, E. Lutton, Eds.), 21–32, Springer Verlag, London, 2005.

117 P. R. Massopust, "Multiwavelets: Some approximation-theoretic properties, sampling on the Interval, and translation invariance," in *Harmonic, Wavelet and p-Adic Analysis* (N. M. Chuong et al. Eds.), 37–57, World Scientific Publishing Co., 2007.

118 B. J. McCartin, "Theory of exponential splines," *J. Approx. Theory* **66**(1991), 1–23.

119 C. A. Micchelli, "A constructive approach to Kergin interpolation in $\mathbb{R}^k$: multivariate B-splines and Lagrange interpolation," *Rocky Mountain J. Math.* **10**(1980), 485–497.

120 J.-P. Mongeau, G. Deslauriers, and S. Dubuc, "Continuous and differentiable multidimensional iterative interpolation," *Linear Algebra and its Applications* **180**(1991), 95–120.

121 P. A. P. Moran, "Additive functions of intervals and Hausdorff measure," *Proc. Camb. Phil. Soc.* **42**(1946), 15–23.

122 M. A. Navascués, "Fractal polynomial interpolation," *Zeitschrift für Analysis u. i. Anwend.* **24**(2) (2005), 401–418.

123 M. A. Navascués and M. V. Sebastián, "Generalization of Hermite functions by fractal interpolation," *J. Approx. Th.* **131**(2004), 19–29.

124 M. A. Navascués and M. V. Sebastián, "Error bounds for affine fractal interpolation," *Math. Inequal. Appl.* **9**(2) (2006), 273–288.

125 M. A. Navascués and M. V. Sebastián, "Smooth fractal interpolation," *J. Inequal. Appl.* (2006), Art. ID 78734, 1–20.

126 P. M. Prenter, *Splines and Variational Principles*, John Wiley & Sons, New York, 1975.

127 F. Przytycki and M. Urbański, "On Hausdorff dimension of some fractal sets," *Studia Math.* **93**(1989), 155–186.

128 A. H. Read, "The solution of a functional equation," *Proc. Roy. Soc. Edinburgh Sect. A*, **63**(1951–52), 336–345.

129 T. J. Rivlin, *An Introduction to the Approximation of Functions*, Dover Publications Inc., New York, 1981.

130 D. Roach, *Multiwavelet Prefilters: Orthogonal Prefilters Preserving Approximation Order $p \leq 3$*, Ph.D. Thesis, Vanderbilt University, Nashville, Tennessee (1997).

131 H. Royden, *Real Analysis*, 3rd ed., Prentice Hall, New York, 1988.

132 W. Rudin, *Real and Complex Analysis*, 3rd ed., McGraw-Hill, New York, 1986.

133 C. Runge, "Über empirische Funktionen und die Interpolation zwischen äquidistanten Ordinaten," *Z. Math. u. Physik* **46**(1901), 224–243.

134 M. Sakai and R. A. Usmani, "On exponential splines," *J. Approx. Th.* **47**(1986), 122–131.

135 T. Sauer, *Splinekurven und -flächen in Theorie und Anwendung*, Lecture Notes, Justus-Liebig University Giessen, Germany, 2008.

136 R. Scealy, *V-variable Fractals and Interpolation*, Ph.D. Thesis, The Australian National University, Canberra, Australia, 2008.

137 I. J. Schoenberg, *Cardinal Spline Interpolation*, CBMS-NSF Regional Conference Series in Applied Mathematics, SIAM, Philadelphia, 1973.

138 I. J. Schoenberg, "On trigonometric splines," *J. Math. Mech.*, **13**(1964) 795–825.

139 I. J. Schoenberg, "Cardinal interpolation and spline functions II," *J. Approx. Th.* **6**(1972), 404–420.

140 L. L. Schumaker, *Spline Functions: Basic Theory*, Krieger Publishing Company, 1993.

141 L. L. Schumaker, "On hyperbolic splines," *J. Approx. Th.*, **38**(1983), 144–166.

142 K. Scherer, *Manuskript zu Splinefunktionen*, Lecture Notes, Institute for Applied Mathematics, University of Bonn, Germany, 2002.

143 I. W. Selesnick, "Multiwavelet bases with extra approximation properties," *IEEE Trans. on Signal Processing* **46**(11) (1998), 2898–2909.

144 I. W. Selesnick, "Balanced GHM-like multiscaling functions," *IEEE Signal Processing Letters*, **6**(5) (1999), 111–112.

145 P. Singer and P. Zajdler, "Self-affine fractal functions and wavelet series," *J. Math. Anal. Appl.* **240**(2), (1999), 518–551.

146 E. M. Stein and R. Shakarchi, *Real Analysis: Measure Theory, Integration and Hilbert Spaces*, Princeton University Press, Princeton, 2005.

147 E. Stein and G. Weiss, *Fourier Analysis on Euclidean Spaces*, Princeton University Press, Princeton, 1971.

148 J. Stoer and R. Bulirsch, *Introduction to Numerical Analysis*, 2nd ed., Springer Verlag, New York, 1993.

149 T. Takagi, "A simple example of the continuous function without derivative," *Proc. Phys. Math. Soc. Japan* **1**(1903), 176–177.

150 C. Tricot, "Two definitions of fractal dimension," *Math. Proc. Camb. Phil. Soc.* **91**(1982), 57–74.

151 H. Triebel, *Theory of Function Spaces II*, Birkhäuser Verlag, Basel, Switzerland, 1992.

152 M. Unser and T. Blu, "Cardinal exponential splines: Part I – theory and filtering algorithms," *IEEE Trans. Signal Processing* **53**(4) (2005), 1425–1438.

153 M. Unser and T. Blu, "Fractional splines and wavelets," *SIAM Review* **42**(1) (2000), 43–67.

154 M. Urbański, "Hausdorff dimension of the graphs of continuous self-affine functions," *Proc. Amer. Math. Soc.* **108**(1990), 921–930.

155 A. Voigt and J. Wloka, *Hilbert Räume und elliptische Differentialoperatoren*, B. I. Wissenschaftsverlag, Zurich, Switzerland, 1975.

156 H. von Koch, "Sur une courbe continue sans tangente obtenue par une construction géométrique élémentaire," *Arkiv für Matematik, Astronomie och Fysik* **1**(1904), 681–704.

157 H. von Koch, "Une méthode géométrique élémentaire pour l'étude de certaines questions de la théorie des courbes planes," *Acta Math.* **30**(1906), 145–174.

158 W. Walter, *Ordinary Differential Equations*, Springer Verlag, New York, 1998.

159 H. Werner and R. Schaback, *Praktische Mathematik II, Methoden der Analysis*, Springer Verlag, Berlin, Germany, 1972.

160 J. D. Young, "Numerical applications of hyperbolic spline functions," *Logistics Rev.*, **4**(1968), 17–22.

Nomenclature

(Ω, d_Ω)	metric space of code trees, page 279
$(\mathbf{X}; \mathcal{F})$	iterated function system (IFS), page 150
$(\mathbf{X}; \mathcal{F}, \mathcal{P})$	IFS with probabilities, page 171
$B[a, b]$	space of bounded functions on $[a, b]$, page 194
$B_q^s(L^p)$	Besov space, page 234
B_k	cardinal B-spline, page 58
$B_{vk,X}$	B-spline of order k, page 48
$C(K)$	space of continuous functions, page 6
$C^k(\Omega)$	space of k-times continuously differentiable functions, page 78
$C^{-1}[a, b]$	space of piecewise continuous functions, page 45
$C^{k,\alpha}(\Omega)$	Hölder space, page 233
$C_b(\Omega)$	space of bounded continuous functions, page 7
$C_{X \times Y}[a, b]$	$\{f \in C[a, b] \mid f(x_v) = y_v, \ v = 0, 1, \ldots, N\}$, page 290
$C_{\alpha,\beta}[a, b]$	$\{f \in C[a, b] \mid f(a) = \alpha \text{ and } f(b) = \beta\}$, page 192
D	differential operator, page 45

E_a	cardinal exponential spline, page 103
E_n	approximation error, page 31
$F_q^s(L^p)$	Triebel–Lizorkin space, page 234
G	Green's function, page 101
$H^{k,p}(\Omega)$	Sobolev space, page 80
I	unit interval $[0, 1]$, page 4
K	nonempty compact set, page 4
$L^p(\Omega)$	Lebesgue space, page 7
L_k	fundamental cardinal spline, page 65
$M_n(f)$	moment of a function, page 200
$S(\boldsymbol{X}_n)$	space of box splines, page 130
$S(\mathcal{F})$	finitely generated shift-invariant space, page 116
$S^k(\boldsymbol{X})$	tensor product spline space, page 133
$S^k(X_n)$	space of polynomial splines, page 45
$S^k(\mathbb{Z})$	cardinal spline space, page 57
$S^{k,\alpha}(\mathbb{Z})$	cardinal spline space of polynomial decay, page 65
$S_{\mathcal{L}}^m(\boldsymbol{\xi})$	space of $\mathcal{L}$-splines, page 112
$W^{k,p}(\Omega)$	Sobolev spaces, page 82
$[x_0, \ldots, x_v](\bullet)$	divided difference operator, page 16
Δ_t^k	k-th order difference operator, page 38
$\mathrm{Lip}(f)$	Lipschitz constant of f, page 32
$\mathrm{Lip}^\alpha(\boldsymbol{X})$	space of Hölder continuous functions, page 32
Ω	nonempty open connected set, page 4
$\Omega_{\mathcal{F}}$	tops code space, page 163
$\Pi_{\boldsymbol{\xi}}^k$	space of piecewise polynomial functions, page 68
Π_s^m	linear space of polynomials over $\mathbb{R}^s$ of degree less than $\boldsymbol{m}$, page 125
Π^k	linear space of polynomials of order k, page 13
Π_s^n	linear space of polynomials over $\mathbb{R}^s$ of total degree n, page 125
$\Pi_{\boldsymbol{\xi},r}^{2r}$	space of Hermite splines, page 74

Σ^n	n-simplex, page 42
Σ_N	code space of an IFS, page 160
$\mathrm{AC}^k(M)$	space of absolutely continuous functions of order k, page 82
$*$	convolution, page 34
$\mathbf{K}(\boldsymbol{X};f)$	Kergin interpolant, page 137
$\boldsymbol{T}^{(V)}$	V-variable superRB operator, page 297
$\boldsymbol{X}_n$	direction set, page 127
$\mathcal{F}^{(1)}$	1-variable IFS, page 269
$\mathcal{F}^{(V)}$	V-variable IFS, page 276
$\mathcal{F}_{\sqcup}$	underlying IFS, page 270
$\mathcal{F}_{X\times Y}$	interpolating IFS, page 291
$\mathcal{F}_{X\times\{Y_\alpha\}}$	interpolating superIFS, page 293
β_α	exponential B-spline, page 106
mult	vector-valued multiplicity, page 132
$\mathcal{A}$	σ-algebra, page 167
$\mathcal{B}(\mathbf{X})$	Borel σ-algebra, page 168
$\mathcal{C}_{\mathcal{F}}$	address structure, page 165
$\mathcal{O}$	Landau symbol, page 31
$\mathcal{Q}_\sigma$	restricted address structure, page 271
$\mathcal{H}(\mathbf{X})$	hyperspace of nonempty compact subsets, page 150
χ_A	characteristic function, page 5
$\mathrm{conv}\,X$	convex hull of set X, page 12
δ_a	Dirac delta distribution, page 96
$\delta_{\alpha\beta}$	Kronecker Delta, page 15
diam	diameter of a set, page 22
$\dim_H$	Hausdorff–Besicovitch dimension, page 145
$\dim$	topological dimension, page 143
$\dim_B$	box dimension, page 148
$\dot{\bigcup}$	disjoint union, page 157
ℓ^p	sequence space, page 6

$\mathfrak{A}^{(V)}$	V-variable superfractal, page 276
$\mathfrak{A}^{\sharp(V)}$	V-variable superfractal set of measures, page 277
$\mathfrak{A}_{\sqcup}$	attractor of underlying IFS, page 270
$\mathfrak{F}(\Omega)$	set of standard random fractals, page 285
$\mathfrak{G}^{(V)}$	V-variable superfractal (of graphs) of fractal interpolation functions, page 294
$\mathfrak{K}_{\infty}(\omega)$	standard random fractal set, page 282
$\mathfrak{M}(\Omega)$	set of standard random fractal measures, page 285
$\mathfrak{S}^{k}[a,b]$	space of polynomial fractal functions of order k, page 197
$\mathfrak{T}_{\mathcal{FG}}$	fractal transformation, page 164
$\mathfrak{m}_{\infty}(\omega)$	random fractal measure, page 284
$\widehat{f}$	Fourier transform of f, page 60
$\mathrm{jump}_{\xi}f$	jump of a function at ξ, page 55
$\|\ \|_{C^{k},[a,b]}$	C^{k}-norm, page 212
$\|\ \|_{L^{p}}$	L^{p}-norm, page 7
$\|\ \|_{k,p}$	Sobolev norm, page 80
$\|\ \|_{\infty,K}$	Chebyshev norm, page 7
$\lfloor\ \rfloor$	floor function , page 233
$|\ |_{k}$	Sobolev semi-norm , page 81
$\mathbf{N}(X;f)$	Newton Interpolant of f, page 17
$\mathcal{N}^{2k}(\Xi)$	space of natural splines, page 87
$\mathcal{S}^{k}(X)$	space of interpolation splines, page 77
$\mathcal{D}$	dyadic dilation operator, page 114
$\mathcal{L}$	linear differential operator, page 99
$\mathcal{T}$	unit translation operator, page 114
∇^{k}	backward difference operator of order k, page 58
∇_{α}	exponential difference operator, page 107
$\omega(f,h)$	modulus of continuity, page 37
$\omega\lfloor n$	finite code subtree of height n, page 280
$\omega\rfloor\sigma$	code subtree of ω with base node σ, page 283

$\omega_k(f, h)$	modulus of smoothness, page 39
$\overline{\mathbb{R}}$	two-point completion of $\mathbb{R}$, page 31
π_Ξ	interpolation projector, page 77
$\mathscr{D}(\mathbb{R})$	space of test functions, page 96
$\mathscr{F}$	Fourier transform, page 60
$\mathscr{P}(S)$	power set of S, page 4
$\mathscr{S}(\Omega, \mathbb{R}^n)$	space of Schwartz functions, page 63
$\mathscr{W}$	discrete wavelet transform, page 114
$\mathsf{P}(\mathsf{X})$	metric space of Borel measures on X, page 172
T	tree, page 278
$\square$	unit square, page 167
$\tau_{\mathcal{F}}$	tops function, page 163
τ_σ	restricted tops function, page 271
τ_y	translation operator, page 113
$\mathbf{IFS}_{X \times Y}$	class of interpolating IFSs, page 291
$\mathbf{RB}_{X \times Y}$	class of RB operators, page 291
ϱ	code tree measure, page 281
d_F	Fréchet metric, page 161
$d_{\mathcal{H}}$	Hausdorff metric, page 150
$f^{(k)}$	k-order distributional derivative, page 97
x_+^m	truncated power function, page 46

Index